TRANSPORTATION 2003: MAKING TRACKS FOR TOMORROW'S TRANSPORTATION

presented at
THE 2003 ASME INTERNATIONAL MECHANICAL ENGINEERING CONGRESS
NOVEMBER 15–21, 2003
WASHINGTON, D.C.

organized by
S. K. (JOHN) PUNWANI
FEDERAL RAILROAD ADMINISTRATION

THE AMERICAN SOCIETY OF MECHANICAL ENGINEERS
Three Park Avenue / New York, N.Y. 10016

ISBN No. 0-7918-3730-0

FOREWORD

The national transportation system includes many constituent forms of transportation, all of which are served by members of ASME and the staff of ASME. There are 3,900,000 highway miles with 205 million commercial and private vehicles generating 4 trillion passenger miles, 920 billion ton-miles of freight, 41,798 fatalities, 3.4 million injuries and consuming 18,000 trillion BTUs. Other transportation modes listed above have similar statistics. Mechanical engineers have opportunities to impact these statistics in a positive way by learning and using newer technologies and implementing them. This centennial year of the "age of flight" has brought forth reminders of the pioneering sprit of the Wright brothers and their early exploits. Today we have continuing challenges to make transportation more user-friendly and environmentally friendly. Smart vehicles and intelligent transportation are an integral part of our future goals.

In the interest of industry-focused dissemination of knowledge, ASME leaders Jack Whitehead and others have continued to make it possible to have a "Track" at Congress 2003 focused on transportation needs for research, development, and other forms of knowledge. The track includes sessions sponsored and arranged by the Transportation Track as well as some sessions developed under the Transportation Track with the help of other ASME Technical Divisions. The sessions include:

- Papers, presentations, panel discussions, educational mini-seminars, symposia. These are intended to be of interest to the practicing mechanical engineer who works in transportation-related industries.
- Highways — trucks, buses, automobiles
- Mass transit
- Railroads passenger, freight, commuter, tourist
- Aviation — commercial and passenger
- Water — barges, etc
- Pipelines

Topics are generally intended to cover:

- Intelligent transportation systems, including vehicle and infrastructure
- Next generation global air transportation
- Enhanced goods and freight movement at domestic and international gateways
- Terminal security at airports, stations, ports, etc.
- Next generation vehicles to improve safety and productivity — aircraft, barges, locomotives, freight, passenger and transit cars, automobiles, buses, and special purpose vehicles
- Communications and control equipment to improve safety of the subject transportation modes
- Tools for modeling design, construction, and operation
- Energy conservation equipment and efficient fuel systems
- Advanced materials and design methods, especially to improve crashworthiness
- Advanced manufacturing, and inspection methods
- Vehicle tracking and integrity monitoring
- Human factors in transportation

Topics related to enabling technologies, which can impact the transportation industries are: MEMS devices, GPS, remote sensing, and nanosystems, to name a few.

The 2003 Transportation Track placed continuing emphasis on crashworthiness and occupant protection, and noise and vibration control with a symposium addressing several aspects related to noise and vibration.

Major contributors to the Transportation Track include: Mardi Hastings, Mehdi Ahmadian, Gopal Samavedam, Kash Kasturi, Monique Stewart, Alan Browne, Hikmat Mahmood, and M. Ridha Baccouche. The Transportation Track acknowledges their vision and efforts in making this year's Transportation Track possible.

S. K. (John) Punwani
Transportation Track Manager

CONTENTS

TRANSPORTATION CRASHWORTHINESS APPLICATIONS FOR LS-DYNA

TRANSPORTATION NOISE AND VIBRATION

VEHICLE CRASHWORTHINESS

Proceedings of IMECE'03
2003 ASME International Mechanical Engineering Congress
Washington, D.C., November 15–21, 2003

IMECE2003-55100

ASSESSMENT OF FAILURE MODELING WITH THINNING STRAIN CRITERION IN LS-DYNA

Ching-Shan Cheng
General Motors R&D Center
30500 Mound Rd., M/C 480-106-256
Warren, MI 48090-9055

ABSTRACT

Element deletion based on various failure criteria has been implemented in commercial finite element analysis packages, e.g. LS-DYNA. However, due to the localized and nonlinear nature of the material failure, especially for the ductile materials, a good failure criterion needs to be robust for different mesh definitions and loading conditions. In the present work, the material model with enhanced failure criterion in LS-DYNA, material type 123 (MAT123), is investigated. The equations for determining the parameter of the thinning strain failure criterion are derived based on the assumptions that Yeh et al. [7] proposed. Simulations of different tests using MAT123 with the thinning strain at failure calculated show better correlation with the experimental results than the other material models examined.

INTRODUCTION

While the components in general are designed not to fail before the end of their life cycle, failure prediction will help to reduce the physical component tests, the time needed for the design and development cycle and eventually achieve the optimized design while meeting the requirements. Unfortunately, due to the highly localized nature of crack formation, it has been a challenge to the engineering community to predict the crack initiation and propagation. Although there are numerical methods proposed by researchers, e.g. adaptive meshing for crack growth [1], cohesive interface element (CIE) [2,3,4], meshfree method [5] and extended finite element method (XFEM) [6], they are either computational expensive or still under development.

Element deletion is an existing approach to model the crack initiation and growth. The crack growth is approximated with a number of linked deleted elements. The fundamental problem with the element deletion method is that conservation of mass and energy is not satisfied. Also, because elements are removed from the model, the contact conditions between the newly created crack surfaces following the crack growth are not the same as reality. It is a phenomenological model and will not be able to provide accurate stress strain information around the crack. However, it could provide an engineering approximation for the failure modeling if the appropriate element deletion criterion is utilized.

Yeh et al. [7] derived the equations for the lateral strain in the thickness direction in a standard tensile test. They concluded that the total engineering thinning strain ($e_{t,eng}$) is a function of the engineering stress and strain at the maximum load, s_m and e_m respectively, and the engineering stress at the fracture load, s_f.

$$e_{t,eng} = f(e_m, s_m, s_f) \quad (1)$$

Furthermore, the critical fracture strain (*CFS*) was derived as follows.

$$CFS_{Alcoa} = -\ln(1 - e_{t,eng}) \quad (2)$$

Element deletion criterion based on *CFS* is then applied to the integration points in the tension side of an element whereas the failure (effective plastic) strain criterion is used in the compressive side. Since the failure criterion needed for element deletion in finite element method is known to be mesh size dependent, Yeh et al. utilized a constant element size in the analyses to avoid the mesh-size effects. The examples reported include various loading conditions, namely a hat-shaped crush test, aluminum extrusion axial crush tests and a three-point bending test. They demonstrated successfully the failure modeling capability of the material model with *CFS* and/or the thinning strain as the failure criterion.

The modified piecewise linear plasticity model (MAT123) is a material model in LS-DYNA [8] similar to the piecewise linear plasticity model (MAT24) but with several additional failure criteria. Among those, there is a failure criterion that uses the thinning strain and allows users to define the number of failed integration points in the thickness direction when deleting the element. In this paper, the simulation results with MAT123 are compared to those with other material models (e.g. MAT24) and the experimental results. Five different tests are evaluated. The materials examined are extruded and cast aluminum and co-extruded multi-layer high-density polyethylene (HDPE). For HDPE, a damage model is also analyzed for comparison.

THINNING STRAIN AT FAILURE

The engineering assumptions proposed by Yeh et al. [7] are employed here to calculate the thinning strain at failure, one of the enhanced failure criteria in MAT123. First, since the deformation localization (necking) occurs after the maximum load (Pmax) in the force-deflection curve is reached, the strains in the thickness and width directions are assumed to be the same before the maximum load. Secondly, similar to the assumption for the elastic-perfectly plastic material, the true stress after Pmax is assumed to be constant for simplification. Thus, the true stress at failure is the same as the true stress when the maximum load occurs. Finally, in order to determine the strain in the thickness direction from a simple tensile test, the strain in the width direction is assumed to be constant after Pmax. In other words, the strain in the thickness direction is assumed to be the only contributor to the localized deformation.

The values for the thinning strain failure criterion in MAT123 are listed in Table 1 for different materials. Since the aluminum alloys are not very strain rate sensitive in the strain rate range of interest, the values listed in Table 1 are taken from the quasi-static uniaxial tensile tests of corresponding materials. However, it has to be noted that MAT123 only uses a constant value for the thinning strain criterion but the co-extruded HDPE is known to be strain rate dependent [9]. Therefore, the strain rate level of the punch test (described in the following section) is estimated to be around 16.67 1/s and the values are taken from the tensile test at that strain rate. The thinning strain of co-extruded multi-layer HDPE at 166.67 1/s is calculated as well. Since the failure stress of some engineering stress-strain curves cannot be clearly identified, approximation for the failure stress is determined by taking the stress at the onset of the localized deformation (i.e. when the stress just starts to drop dramatically).

Material	Al6061-T6 (Extruded)	Al 6063-T6 (Extruded)	Cast Aluminum	Co-extruded HDPE (16.67 1/s)	Co-extruded HDPE (166.67 1/s)
e_t	0.11	0.12	0.05	0.84	0.63
CFS_C	0.175	0.171	0.104	1.967	1.099

Table 1. The calculated thinning strain at failure and the CFS_c values of the materials investigated.

SIMULATIONS WITH MAT123

Several tests with a wide variety of loading conditions are conducted and modeled using LS-DYNA in order to validate the failure modeling with the thinning strain failure criterion of material type 123, which is currently only available for shell elements (Belytschko-Tsay shell elements are used in the present work). In addition to the thinning strain, MAT123 provides the option to select the number of through thickness integration points that must fail before deleting the element. Basically, with thinning strain criterion, the integration point, whose z-direction strain reaches the specified thinning strain, is considered failed. The element is deleted when the number of the failed integration points equals to the selected value.

The tests described in the following sections are axial crush tests of aluminum extrusions, the cast aluminum cradle crush test and the multi-layer high-density polyethylene (HDPE) punch test. The mesh sizes used in the different simulations are selected to be between 3 mm and 5 mm. The piecewise linear effective stress-effective plastic strain curves from the quasi-static tensile tests of the materials are used as the constitutive laws in the corresponding material models. Since the co-extruded HDPE material is found to be strain-rate dependent (Cheng and Strom [9]), an additional curve is defined to describe the rate dependent behavior. Simulations are also done with other material models for comparison.

1. Aluminum extrusion axial crush test:

The single-cell extrusions of aluminum 6063-T6, with crush initiators are placed between two rigid plates. The external dimension of the cross-sectional profiles of the extrusions is the same, 135 mm by 75 mm. More detailed information regarding the test setup and the extrusions was reported by Mayer et al. [10].

The finite element model of the aluminum extrusion with the crush initiator is crushed between two analytical rigid walls of infinite size. Figure 1 depicts the model setup including the crush initiators used in the single-cell extrusion. Except for the elements representing the crush initiators, the element sizes are 3.125 mm by 3.155 mm for the single-cell extrusion.

Analyses with the same finite element models but different material laws, MAT123 and MAT24 in LS-DYNA, are performed for comparison. The basic mechanical properties of the two material models, e.g. elastic constants and the post-yield behavior, are the same except the failure criteria are different. The simulations with MAT123 can model the crack initiating at the corners of the extrusion and that at the folds of the aluminum walls. Figure 2 illustrates the deformation of Al 6063 single-cell extrusions as well as the locations of the deleted elements (in blue). When the simulation results are compared against those with MAT24 using the effective failure strain of 0.40, which is more than twice the experimental failure strain, it can be easily seen that the failure modes are completely different as shown in Figure 3. The load time history curves are shown in Figure 4. The model with MAT24 and failure strain of 0.40 does not have the crush mode with nice folding observed in the tests, while the load time history (curve MAT24_F40 in Figure 4) is close to the test result. It is interesting to note that, when an even larger effective failure strain or no failure criterion is used in the MAT24 material model, curves MAT24_F70 and MAT24 respectively in Figure 4, it results in the better deformation mode but the predictive capability of the crack initiation location deteriorates. It has to be noted that there is no guideline or equation to identify the appropriate effective failure strain for MAT24 that will result in good correlation.

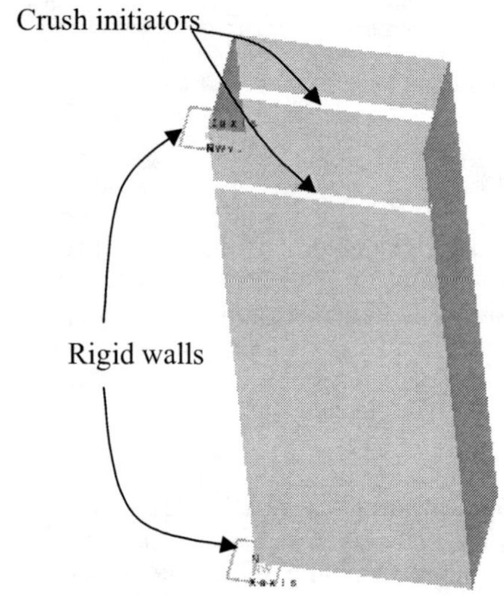

Figure 1. The model setup and the crush initiators for the single-cell extrusion axial crush test.

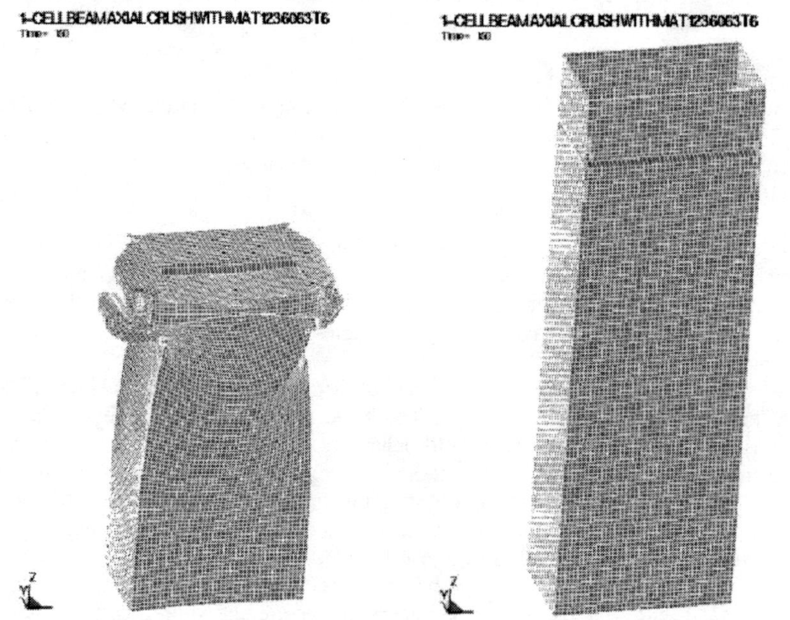

Figure 2. The deformation modes and the deleted elements (elements in blue) of Al 6063 single-cell extrusion axial crush with 150 mm crush distance.

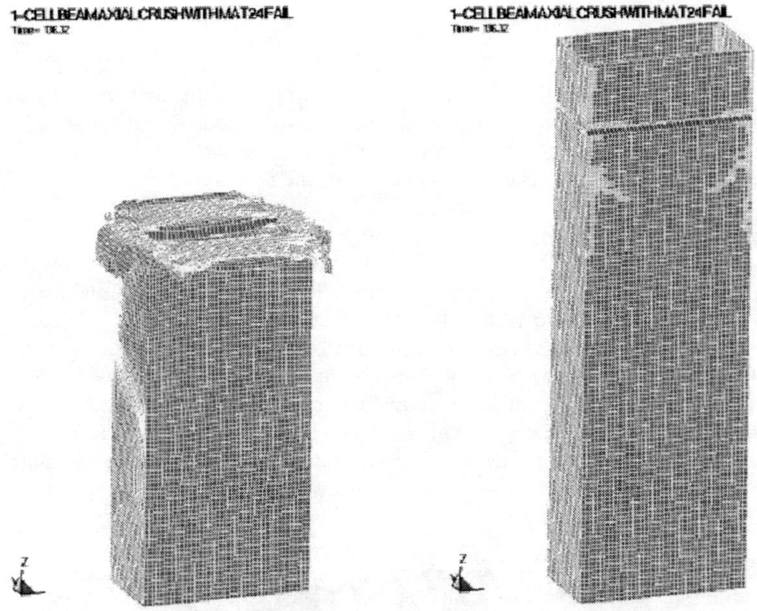

Figure 3. The single-cell Al 6063-T6 extrusion axial crush test modeled using MAT24 with effective failure strain of 0.40.

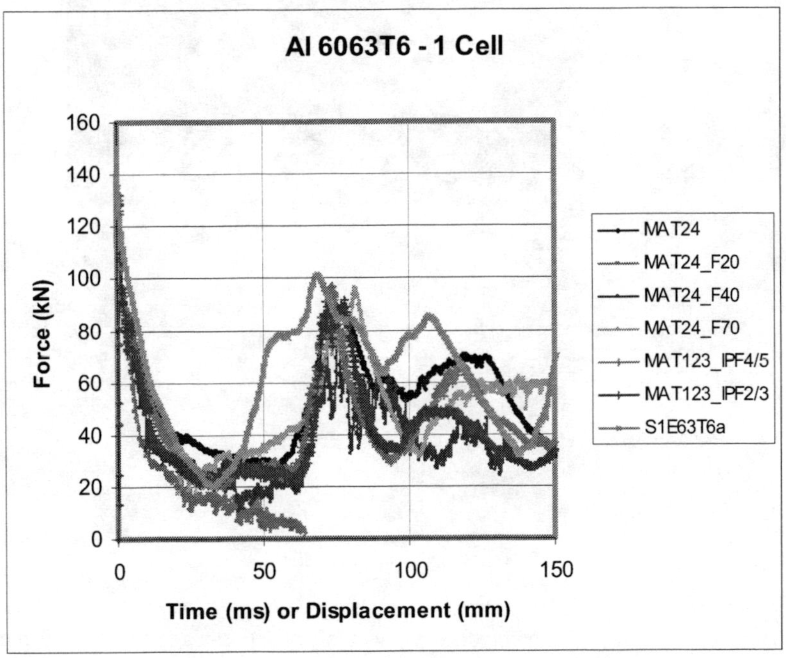

Figure 4. The force-deflection histograms of the Al 6063 single-cell extrusion axial crush test and simulations with different material models.

2. Cast aluminum cradle crush test:

A test is done on a cast aluminum cradle to purposely generate fracture to check the thinning strain failure criterion. This is not a regularly performed test, nor does it represent actual loadings seen during use. The cradle is mounted to the platform of the servo-hydraulic material tester at four points and the 3-inch diameter cylindrical ram is moving downward at a velocity of 25 mm per minute as shown in Figure 5 (a). The load and displacement histogram of the ram is recorded.

In the finite element analysis, the four mounting points are rigidly fixed in the space. Since the bolt connecting the ram to the load washer was slightly bent in the experiments, the bolt is modeled with solid elements using mild steel material properties and the cylindrical ram is modeled with shell elements and assumed to be rigid. Due to the design, the cradle is originally modeled with solid elements (Figure 5 (b)). Since MAT123 is currently only available for shell elements, a portion of the cradle is converted to shell elements. The element sizes are on the order of 5 mm by 5 mm. Because the reinforcement ribs of the cradle do not have uniform thickness, they are modeled with several strips of shell elements with appropriate thickness. As shown in Figure 5 (c), some of the reinforcement ribs are presented by shell elements using different colors for different shell thickness.

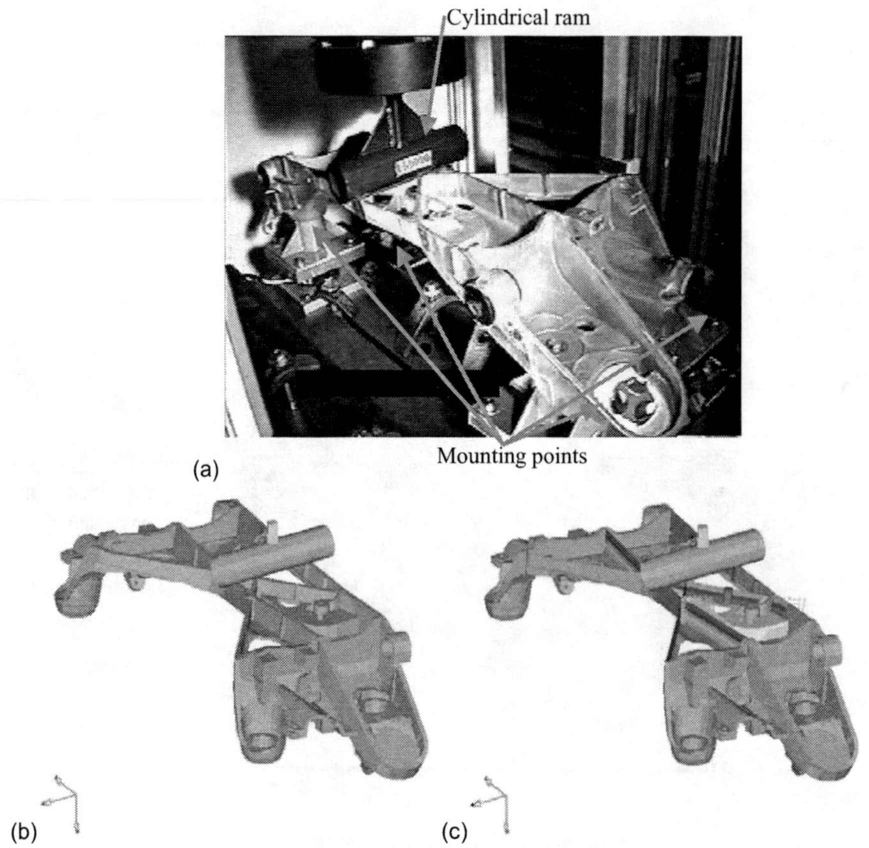

Figure 5. The test setup, (a), and the models, (b) original solid model and (c) shell-solid model, for the cast aluminum cradle crush test.

The same uniaxial tensile test result is used as the constitutive relation in MAT123 and MAT24. The thinning strain at failure listed in Table 1 is used as the failure criterion for MAT123. The element will be deleted when the strain at two of the three through-thickness integration points reach this value. The effective failure strain of MAT24 is 0.10, which is about the same as the experimental result from one of the uniaxial tensile tests. Since the model with MAT24 experiences much smaller time steps due to the distorted elements adjacent to some of the deleted element and the inertia effects are not significant, mass scaling is used to have a constant time step size of 5E-4 millisecond in one of the analyses.

Figure 6 shows the force-displacement curves of the crush test and the simulation results. Only one test result is shown while the repeatability is verified by the two tests conducted. The simulation results with MAT123 and MAT24 seem to correlate well with the test result, after the test result is shifted to match the elastic deformation to reflect the compliance of the test fixtures. When comparing curves MAT24F10 and MAT24F10_ms, it is clear that the mass scaling does not degrade the accuracy of the result in this case. On the other hand, it helped to eliminate the oscillation due to the severely distorted mesh after some elements were deleted.

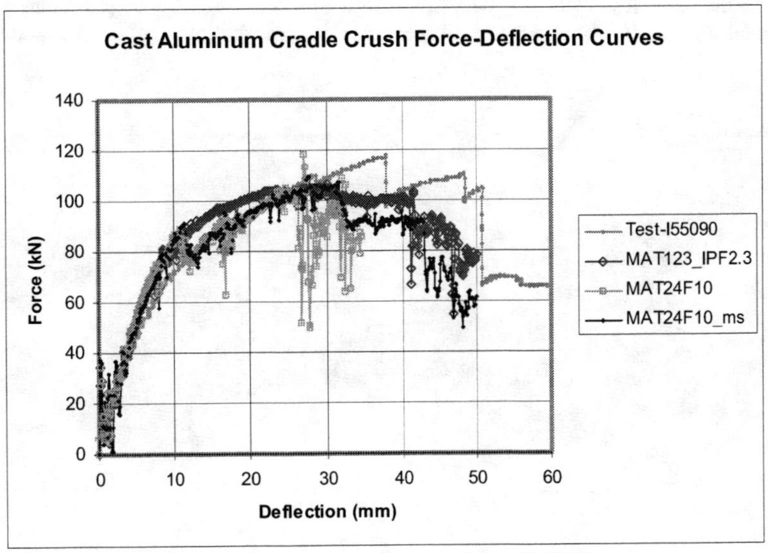

Figure 6. The force-deflection histograms of the cast aluminum cradle crush test and simulations with MAT123 and MAT24.

When it comes to failure modeling, MAT123 performs better than MAT24 based on the simulation results. The cradle crushed 75 mm is shown in Figure 7. The cradle is flipped from its tested condition to show the four cracks labeled in sequence of occurrence. Figures 8 and 9 illustrate the cracks developed in the sequential order in the analyses with MAT123 and MAT24 respectively. Due to the sampling rate of the result (d3plot) file, only the deformed cradle at the times following the crack initiation are shown. With MAT123, the first crack initiates when the crush distance is between 28 mm and 32 mm, as shown in Figure 8 (a), and the rest of the cracks are developed before a crush distance of 44 mm. The locations of the cracks are very close to the test results shown in Figure 7. It is interesting to note that the first significant drop of the experimental load-deflection histogram, indicating the first crack development, occurs when the crush distance is about 28 mm.

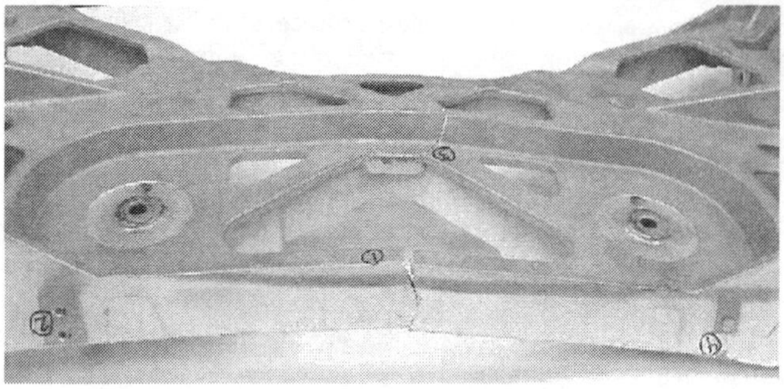

Figure 7. The cracked cast aluminum cradle with crush distance of 75 mm.

While the force deflection curve from the simulation with MAT24 seems to fit the test result well, the cracks modeled are not the same as the ones seen in the test. The first crack develops when the crush distance is between 12 mm and 16 mm as shown in Figure 9 (a). The additional cracks in Figure 9 (b) actually develop between 32 mm and 44 mm. The cracks identified with the smaller red circles are not found in the test. It is clear that MAT24 with the effective failure criterion can not be used to accurately model the cracks because it will not able to differentiate the tensile and compressive conditions.

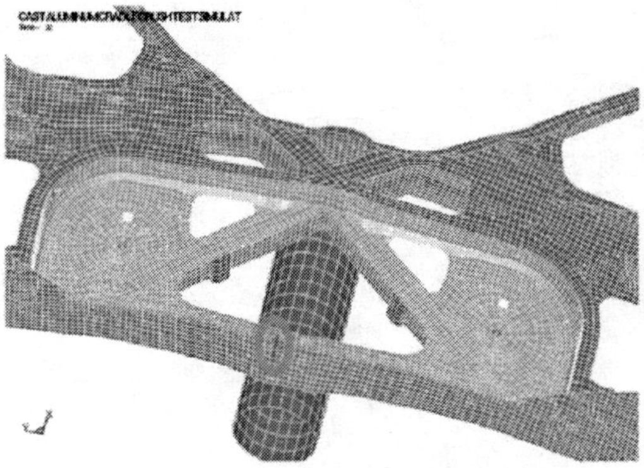

(a)

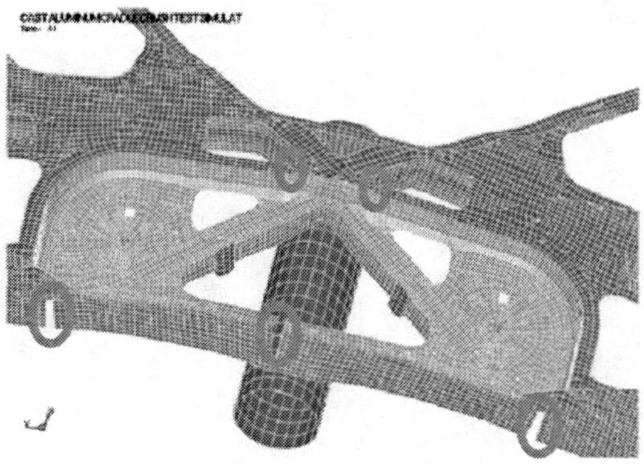

(b)

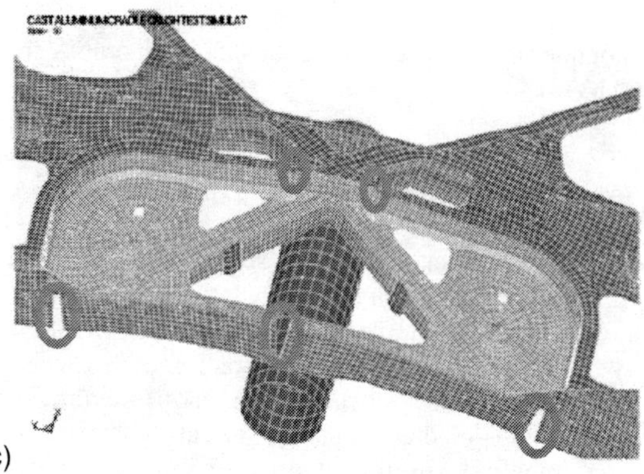

(c)

Figure 8. The cracks developed in the analysis with MAT123. Crash distances are (a) 32 mm, (b) 44 mm and (c) 50 mm.

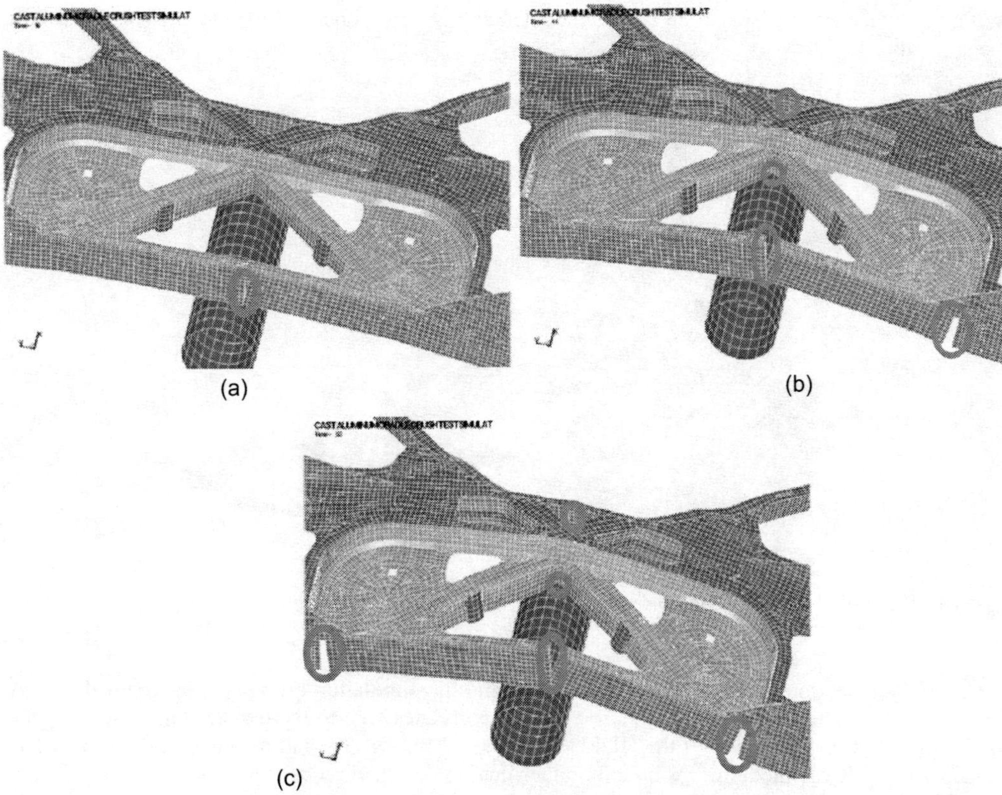

Figure 9. The cracks developed in the analysis with MAT24. Crash distances are (a) 16 mm, (b) 44 mm and (c) 50 mm.

3. Co-extruded multi-layer HDPE punch tests:

The dynamic punch tests are conducted on a servo-hydraulic machine. The samples of multi-layer HDPE material are cut in 3.75 inches by 3.75 inches square from the locations that are found to have the most uniform thickness. The specimens are then inserted into the punch test fixture as shown in Figure 10. Four screw bolts are used to hold the specimen in between the top and bottom rings. During the test, the specimen is moved into a fixed 0.5-inch diameter titanium punch located at the center of the specimen at a velocity of 1 m/s.

As described in the previous section, shell elements and MAT123 with the thinning strain value for HDPE listed in Table 1 are used for the HDPE specimen in the punch test simulation. The strain rate dependence is accounted for by taking the relevant values from a dynamic tensile test at 16.67 1/s. The other fundamental material properties, including elastic properties, plasticity and the strain rate dependence, used in MAT123 are the same as what was used for the MAT24 developed by Cheng and Strom [9]. The punch is modeled with shell elements and rigid material. In the simulation, a portion of the top and bottom specimen support rings is modeled with two sets of fully constrained rigid shell elements as shown in Figure 10. While there is no force applied to the support rings in the simulation, the friction is modeled to mimic the friction caused by the four finger-tight bolts described previously. With this approach of allowing the sliding between the specimen and the support rings to occur, the initial slope of the force deflection curve is found to correlate to the test result well. The response of the multi-layer HDPE specimen is stiffer if the supported portion of the specimen is just rigidly constrained. In addition to the friction between the specimen and the support rings, the friction between the punch and the specimen is also modeled and is very important to how the specimen fails in the simulation.

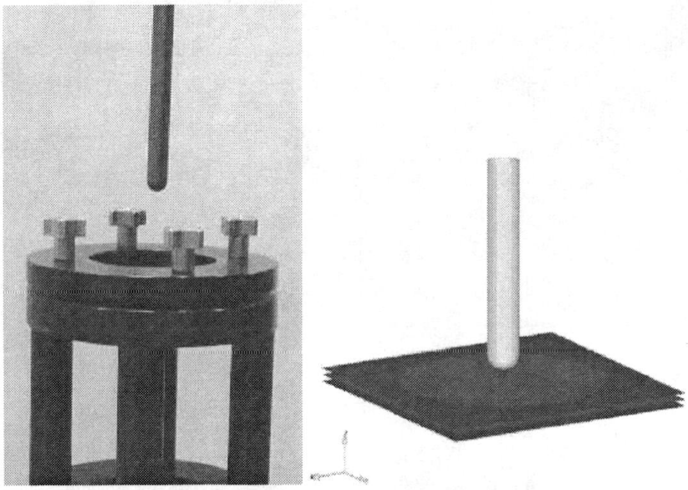

Figure 10. The model setup for the co-extruded multi-layer HDPE punch test.

The force-deflection curves from the punch test and the simulation are shown in Figure 11. It is clear that the simulation with MAT123 (curve M123) correlates well to the experiment up to the peak load, i.e. the onset of the tearing of the HDPE specimen. Although the failure starts at the same time, the load capacity of the model drops quickly rather than the gradual reduction seen in the experiment. Again, it indicates that MAT123 with element deletion method is currently not able to accurately model the sequential material response following the initial failure. It is probably due to the unrealistic contact condition after element being deleted in this case.

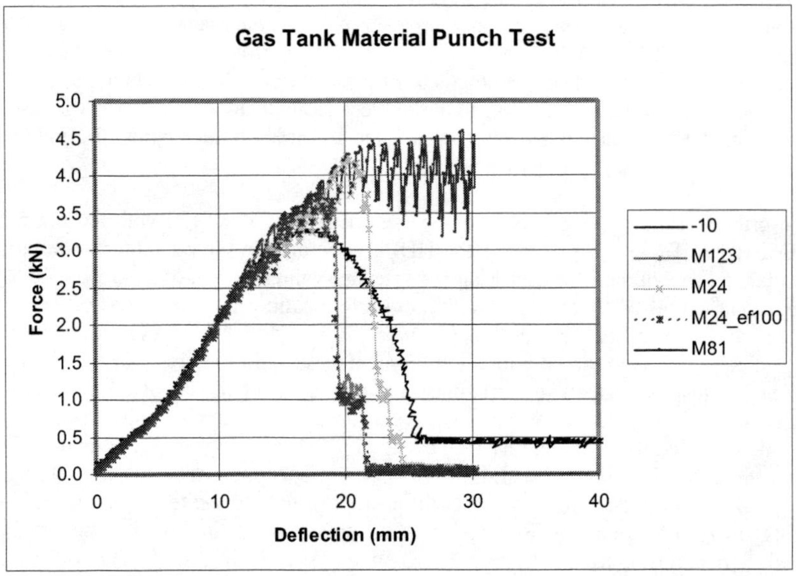

Figure 11. The test and simulation results of the co-extruded multi-layer HDPE punch test.

The MAT24 and MAT81 models discussed by Cheng and Strom [9] are also used in the simulations for comparison. The only difference between the three material models is the criteria for element deletion. MAT24 has an effective failure strain of 1.25 and MAT81 uses a damage curve for the damage evolving in the HDPE material. Without further tuning of the failure criteria, both MAT24 and MAT81, curves M24 and M81 in Figure 11 respectively, over-predict the peak load. Additional simulations are conducted using MAT24 and different effective failure strain values to investigate how the model behaves with a tuned failure criterion. As shown in Figure 12, the result with MAT123 is very close to that with MAT24 and effective failure strain of 1.00 (curve M24_efl00).

The failed punch test specimen is shown in Figure 12. Due to the ductile nature of the co-extruded multi-layer HDPE material, there is severe localized plastic deformation in the area around the punch point. As mentioned previously, the friction between the punch and the specimen affects how the specimen fails in the simulation. If the friction coefficient is too small (e.g. 0.2), i.e. the material is easy to slide over the punch tip, the most significant localized deformation (largest thinning strain) will occur in the area next to the tip of the punch. As shown in Figure 12, a piece of the HDPE material is held by the punch tip. The deformation starts to localize around the punch tip and the material finally fails. Figure 13 illustrates the model of the failed specimen when the static and dynamic friction coefficients are 0.8 and 0.4 respectively.

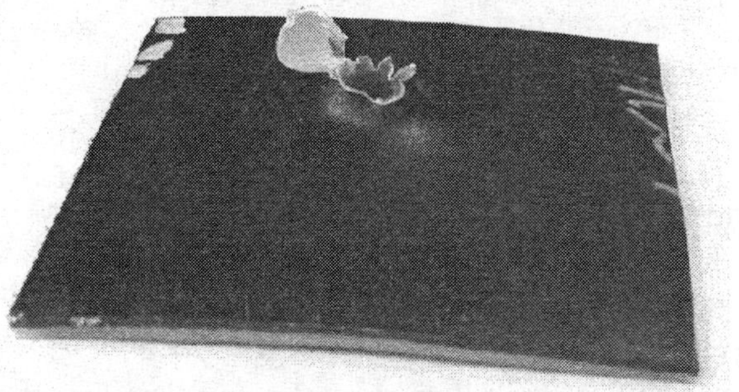

Figure 12. The deformed co-extruded multi-layer HDPE punch test specimen.

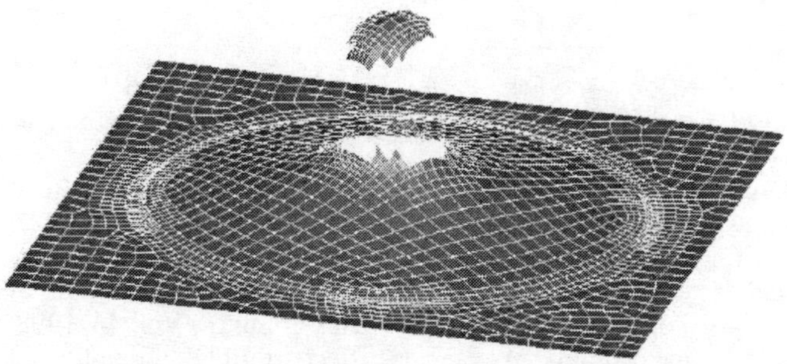

Figure 13. The model of the failed co-extruded multi-layer HDPE punch test specimen.

CONCLUSIONS

It is recognized that the element deletion method is not the perfect representation of a crack. This is especially true when the contact between the crack surfaces is imminent following the crack initiation. However, it is still a good approximation for the cases investigated as long as the right material properties, an appropriate failure criterion and correct boundary conditions are used.

The thinning strain failure criterion based on the three engineering assumptions is one of many available failure criteria for element deletion in LS-DYNA. For the cases investigated, the thinning strain failure criterion is able to model the cracks (at least the initiation) when calculated in the same way for different materials. It seems to be less sensitive to the range of the mesh size (3 mm to 5 mm) used. When the model is set up correctly, no additional tuning is needed for MAT123. On the other hand, the effective failure strain for MAT24 and the damage curve for MAT81 need to be tuned for a different mesh size. With MAT24, elements will fail whenever the effective strain reaches the specified value. Because of the definition of the effective strain, this failure criterion cannot distinguish between the tensile and compressive state of stress. Therefore, elements will be deleted even it is under compression, while it is not true for the cases modeled in the present work.

Furthermore, when MAT123 is used, no significant reduction of the time step size is observed. For the same model, when MAT24 with the effective failure strain criterion is used, the time step size often drops to a much smaller value. For example, in the cast aluminum cradle crush simulation, it drops to as low as 6E-6 ms because of the severely distorted mesh whereas that of the same model with MAT123 is around 1E-4 ms.

We recognize that, even with a simple tensile test, the material properties for calculating the thinning strain at failure may still vary. It is especially possible to observe a large variation with the engineering stress at failure. Nevertheless, according to the simulation results presented, the thinning strain is a more robust failure criterion than the effective strain at failure and the damage model investigated. It is worth being emphasized again that the failure modeling capability of MAT123 still depends on the failed-to-total through-thickness integration point ratio for element deletion and, most importantly, the accurate setup of the model (for example, friction). An enhancement on the strain rate dependence of the thinning strain at failure is also required.

ACKNOWLEDGMENTS

The author would like to thank Bob Mayer for providing the test results of the aluminum extrusions. He is also grateful for the information related to the cast aluminum cradle provided by Brad Stricklin and Soo Tan. Nancy Johnson's valuable comments on this work are also much appreciated.

REFERENCES

1. Bittencourt, T. N., Wawrzynek, P. A., Ingraffea, A. R. and Sousa, J. L., Quasi-automatic Simulation of Crack Propagation for 2D LEFM Problems, Engineering Fracture Mechanics, Vol. 55, No. 2, pp. 321-334, 1996.
2. Camacho, G. T. and Ortiz, M., Computational Modelling of Impact Damage in Brittle Materials, International Journal of Solids Structures, Vol. 33, No. 20-22, pp. 2899-2938, 1996.
3. Espinosa, H. D., Zavattieri, P. D. and Dwivedi, S. K., A Finite Deformation Continuum/Discrete Model for the Description of Fragmentation and Damage in Brittle Materials, Journal of the Mechanics and Physics of Solids, Vol. 46, 1998.
4. Needleman, A. and Rosakis, A. J., The Effect of Bond Strength and Loading Rate on the Conditions Governing the Attainment of Intersonic Crack Growth along Interfaces, Journal of the Mechanics and Physics of Solids, Vol. 47, 1999.
5. Belytschko, T., Organ, D. and Gerlach, C., Element-free Galerkin Methods for Dynamic Fracture in Concrete, Computer Methods in Applied Mechanics and Engineering, Vol. 187, pp. 385-399, 2000.

6. Dolbow, J., Möes, N. and Belytschko, T., Modeling Fracture in Mindlin-Reissner Plates with the Extended Finite Element Method, International Journal of Solids and Structures, Vol. 37, pp. 7161-7183, 2000.
7. 7. Yeh, J. R., Summe, T. L. and Seksaria, D. C., The Development of an Aluminum Failure Model for Crashworthiness Design, Crashworthiness, Occupant Protection and Biomechanics in Transportation Systems, American Society of Mechanical Engineers, AMD-Vol. 237/BED-Vol. 45, pp. 97-105, 1999.
8. LS-DYNA User's Manual, Version 960, March, 2001.
9. Cheng, C.S. and Strom, K. A., Dynamic Material Characterization for Multi-layer High Density Polyethylene Material, to appear in Society of Experimental Mechanics 2003 Annual Conference.
10. Mayer, R. R., Chen, W. and Sachdev, A. K., Crashworthiness Performance of Mass-Efficient Extruded Structures, Proceedings of 2002 ASME International Congress & Exposition, Nov. 2002.

Proceedings of IMECE'03
2003 ASME International Mechanical Engineering Congress
Washington, D.C., November 15–21, 2003

IMECE2003-55104

SIMULATION AND BOGIE TESTING OF A NEW CABLE BARRIER TERMINAL

John D. Reid, Nicholas R. Hiser and Tony J. Paulsen
Mechanical Engineering Department
University of Nebraska-Lincoln

ABSTRACT

Roadside barriers of various designs are extensively used for the purpose of shielding obstacles along the road from impact with errant vehicles. One commonly applied roadside barrier system is a cable system, consisting of three steel cables supported by weak steel posts. Due to an increase in the use of cable systems, a tangent to the roadway, as opposed to flared away from the roadway, crashworthy end terminal has been designed by the Midwest Roadside Safety Facility. The design goal was to effectively disengage the pre-tensioned cables at the end anchor point when impacted on the end by a vehicle, allowing the vehicle to pass through the system virtually unobstructed. It is the objective of this study to help evaluate the new design through bogie testing and nonlinear finite element analysis using LS-DYNA. Based on bogie test results and detailed analysis of the simulation results, it was concluded that the new cable end terminal was ready for full-scale crash testing.

INTRODUCTION

Roadside barriers of various designs are extensively used for the purpose of shielding obstacles along the road from impact with errant vehicles. One commonly applied roadside barrier system is the 3-strand cable system. This system consists of three steel cables supported by weak steel posts. The steel cables are pre-tensioned with a specified initial load, and are anchored at both ends of the system. When an errant vehicle obliquely impacts the 3-strand cable system, sufficient tension is developed within the cables to redirect the vehicle, effectively shielding the roadside hazard and increasing the safety of the vehicle operator. The weak steel posts supporting the cables offer very little resistance to the impacting vehicle. Due to the likelihood of a longitudinal impact with the end of the 3-strand cable system, a crashworthy end terminal has been designed by the Midwest Roadside Safety Facility (MwRSF). The design goal was to effectively disengage the pre-tensioned cables at the end anchor point upon impact with a vehicle, allowing the vehicle to pass through the system virtually unobstructed. To assist in the development of the 3-strand cable system end terminal, the CTB-4 bogie crash test was conducted. It is the objective of this study to simulate the cable terminal bogie test CTB-4 using LS-DYNA, a nonlinear finite element analysis code (1), and validate the simulation results with that of the physical test.

3-STRAND CABLE END TERMINAL DESIGN

To accomplish the task of disengaging the cable end fittings upon impact, the standard cable anchor bracket was modified to accommodate a cable release lever. The cable release lever is a mechanism that was designed to transmit the longitudinal force of the impacting vehicle to an upward force that lifts the cable end fittings from notches in the cable anchor bracket. Even though the system uses weak steel posts to support the cable, the end post is attached to a slip-base to further reduce its resistance to an impacting vehicle. Once the cable end fittings are dislodged from the cable anchor bracket, the tension in the cables is released, and the vehicle is allowed to pass through the system with minimal obstruction. Refer to Figure 1 for an illustration of the end terminal design.

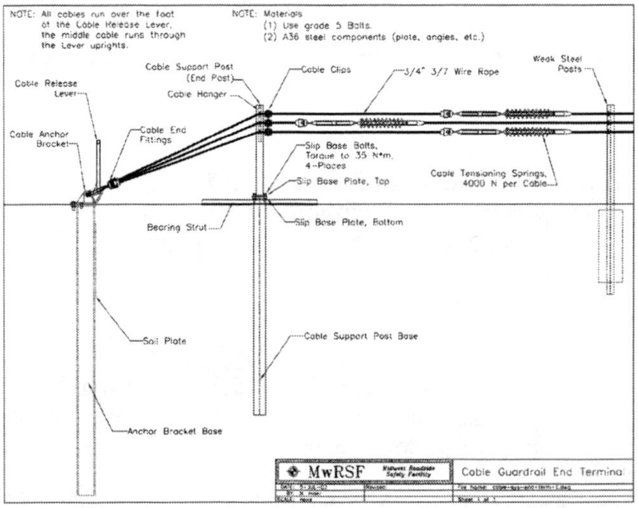

Figure 1. 3-Strand Cable End Terminal Design

BOGIE CRASH TESTING (CTB-4)

In order to test the initial end terminal design concept, the CTB-4 bogie crash test was conducted. A 3-strand cable system was built with a prototype crashworthy end terminal as previously described, and was impacted longitudinally with a 909 kg bogie vehicle traveling at 20.1 m/s (45 mph). The bogie vehicle bumper, which contacts the cable release lever upon impact, is constructed of high-density polyethylene (HDPE) to cushion peak impact forces. The contact point between the bogie bumper and the cable release lever was at 400 mm above the ground. The CTB-4 crash test was conducted without any data acquisition equipment other than the use of high-speed film, and still photographs prior to and following the impact. Refer to Figure 2 for photos of the 3-strand cable system, prototype end terminal, and bogie vehicle used in bogie test CTB-4.

SYSTEM MODELING IN HYPERMESH AND LS-DYNA

To further assist in the development of the cable system end terminal, impact simulation using LS-DYNA was performed. In order to simulate the CTB-4 crash test in LS-DYNA, the cable system end terminal had to first be modeled in HyperMesh and LS-DYNA, during which many approximations and simplifications were made. The following sections detail the modeling of the cable system.

Figure 2. CTB-4, Prototype End Terminal and Bogie Vehicle

Cable Anchor Bracket and Release Lever

The cable anchor bracket was modeled using deformable shell elements to represent the grade 250 steel plate from which the actual bracket was constructed. All of the nodes on the base-plate of the anchor bracket model were constrained in all 6 degrees-of-freedom to simulate a rigid anchor bracket base. If a more accurate method to constrain the anchor bracket is desired, it may be beneficial to constrain only those nodes near the bolts that attach it to the anchor bracket base. In LS-DYNA, fillet welds were added to the anchor bracket at the top two nodes of the vertical gussets where weld failure was observed in the tests.

The cable release lever was also modeled entirely with deformable shell elements to represent the grade 250 rolled steel tube and steel plate from which the actual lever was constructed. The lever was placed upright on the anchor bracket gussets without any constraints.

Cable

A cable model was utilized in this study. Both solid and discrete beam elements were used to model the cable. A circular cross-section was cut into eight pie-shaped elements, and was then extruded to create solid, 6-node elements. Discrete beam elements were placed at the center of the solid elements, and each beam element spanned the length of one solid element. The beam elements were used to provide the majority of the tensile load developed in the cable, and the solid elements were used to provide mass distribution, minor bending stiffness, and improved contact with other components.

Discrete spring elements, with a spring rate of 80 N/mm and an initial offset of 50.8 mm, were attached to the end of each cable downstream of the end post to simulate the spring compensators, which tension the cables. Spring compensators provide a specified preload of 4064 N per cable. The cables were extended 3 feet past the end post in order to obtain acceptable global system behavior while limiting computational time.

Cable End Fittings

The cable end fittings were modeled with deformable solid elements and a pie-shaped cross-section consistent with the cable. Additional solid elements were added to represent the volume and mass of the threaded rod, nuts, washers, and shackle which make up the cable end fitting. Using LS-DYNA, a rigid body spherical joint with 20° rotation was placed between the cable and the cable end fitting to simulate the actual degrees of freedom allowed due to the presence of the shackle. In order to incorporate a spherical rigid body joint between the cable and cable end fittings, rigid elements were required at the ends of both cable and end fittings. Figure 3 illustrates the cable end fitting modeling approximation.

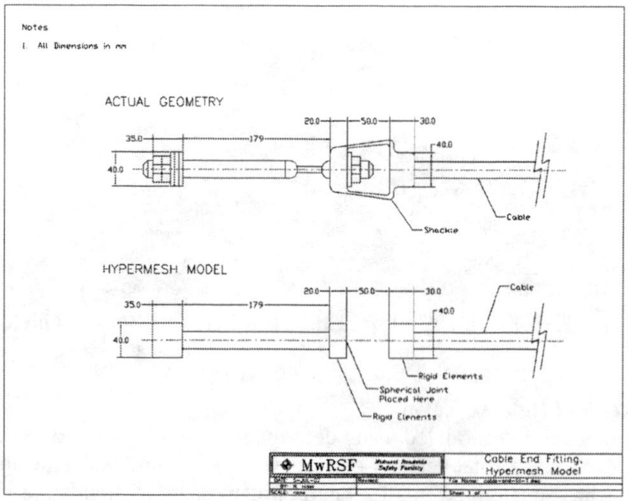

Figure 3. Cable End Fitting Modeling Approximation

End Post

The end post, which sits on a slip base, was modeled using solid and shell elements. Both the top and bottom slip base plate, as well as the I-beam post, were modeled with deformable shell elements. Initially shell elements were used to model the cable hanger, but the final model used solid elements, which provided better contact behavior with the cable. The solid cable hanger elements were also made rigid because no deformations were observed after the bogie test. The bottom slip-base plate was constrained in all 6-dof to simulate a rigid cable support post base with bearing strut, as is shown in Figure 1.

Slip Base

In the actual slip base joint, a normal force between top and bottom slip base plates is induced and controlled by a specified torque applied to four bolts that clamp the plates together. The frictional force between the two surfaces is a product of the frictional coefficient and the normal force induced by the bolt preload. Finally, activation of the slip base joint is dependent on an impacting force overcoming the frictional force present between the two slip base plates.

To model the slip base behavior in LS-DYNA, the bolt tension was simulated using four nonlinear discrete springs placed in the location of the bolts, in a similar manner as used by G.W. Paulsen (2). The springs were given a non-linear load curve with a load of 5.33 kN for a spring displacement between 0 and 10 mm, decreasing linearly to 0 kN between 10 and 20 mm displacement. The spring load was defined to coincide with the bolt preload due to torque, and was given a failure displacement of 20 mm, which is the slip distance of the actual slip base. Steel-on-steel friction was defined between the slip base plates. Upon simulation initialization, the springs create a normal clamping force equivalent to the bolt clamping force between the slip base plates, generating a frictional force that constrains the post until impacted by the bogie vehicle. During the course of the impact, the springs deflect past 20 mm there-by eliminating the spring load, activating the simulated slip base joint, releasing the post, and yielding acceptable global behavior between the post and the bogie.

Bogie

A bogie model from previous research projects at the MwRSF was used in this study, but required a bumper modification. The bumper used in CTB-4 was constructed of HDPE, and it was mounted at a height of 400 mm to match that of a small car bumper, such as a Geo Metro. The majority of the bumper was modeled with rigid, solid elements primarily for mass and geometric correlation. Deformable, solid elements with the *MAT_ELASTIC material definition were placed along the leading edge of the bumper. The presence of relatively soft, deformable elements on the leading edge of the bumper helped to reduce the peak contact forces that typically occur upon impact. A photograph of the bogie used in CTB-4 is shown in Figure 2.

STEP-BY-STEP SIMULATION PROCESS

The 3-strand cable system was simulated incrementally, with various parts of the system simulated separately, and then later combined to recreate the complete system. An impact between the bogie, cable release lever, anchor bracket, and cable end fittings was simulated first. The purpose of this was to investigate the contact between the bumper and the cable release lever, as well as the effectiveness of the cable release lever in lifting the end fittings out of the anchor bracket. *CONTACT_AUTOMATIC_SINGLE_SURFACE was used as the global contact definition for all parts. There were some initial contact problems, as seen in Figure 4, which resulted in unusual deformations in the cable release lever and the cable end fittings. This was attributed to the large difference in stiffness between the HDPE bumper (E=0.930 GPa) and the steel end terminal components (E=200.0 GPa). Pinball segment-based contact was added to the contact definition along with the edge-to-edge penetration check. These changes increased the penalty stiffness of the impacting surfaces, and reduced the effects of the edge-to-edge contact that occurs between the lower part of the bumper and the gussets of the cable release lever. However, it is noted that contact problems were apparent during a later simulation, so the bumper stiffness was increased from E=0.930 GPa to E=1.860 Gpa, which helped minimize the contact instability.

The next step in the modeling process was to add the cables and the end post to the system model. The cable end fittings were connected to the end of the cables near the anchor bracket using spherical joints as previously described. Spot welds were used to fix the post to the slip base for the initial simulation. The simulation was run for 10 ms. This was sufficient time for the cable release lever to lift the end fittings out of the slots in the anchor bracket, and for the cable to respond to the release of tension. A cross section through the cables was used to verify that the cables were being tensioned properly to 4 kN by the discrete springs.

A contact problem was discovered between the cable and the post cable hanger. When tension was applied to the cables, the cable penetrated into the cable hanger, as illustrated in Figure 5. This penetration caused the cable to snag on the cable hanger, constraining it from sliding within the cable hanger groove. The cable hanger mesh was refined in the location where it made contact with the cable. The refined mesh fixed the visible penetration problem, but a longer simulation showed that the cables were still snagging on the cable hanger. Rigid, solid elements were then used to model the cable hanger. When the solid elements were created in HyperMesh, interference was apparent between the cables and the cable hanger. It is likely that this interference resulted in initial penetrations, which caused the snagging between the cable and the cable hanger. The cable routing was adjusted slightly to avoid interference with the solid elements of the cable hanger. These changes eliminated the cable snagging problem.

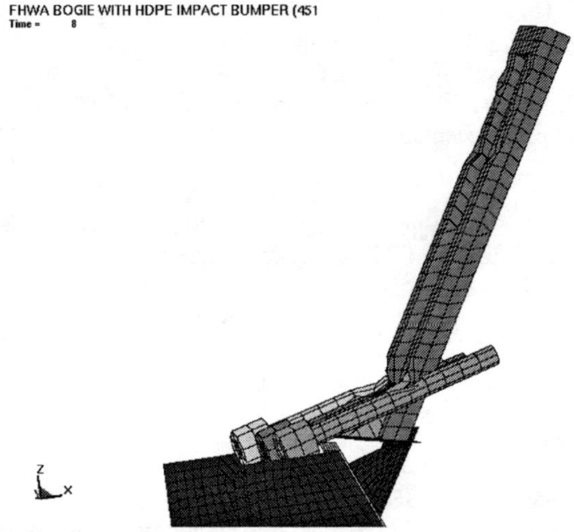

Figure 4. Deformations Prior to Soft Contact Definitions

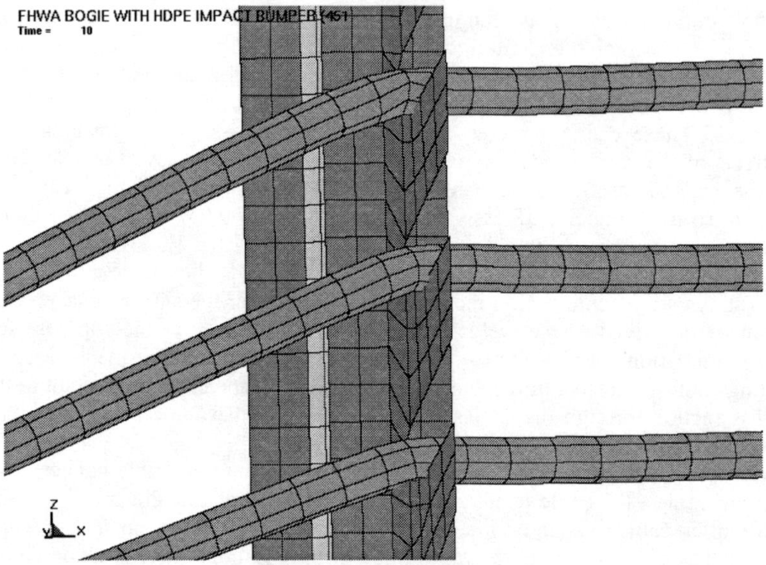

Figure 5. Cable Penetrating and Snagging on Cable Hanger

Next, spot welds constraining the slip base joint were eliminated and replaced with the slip-base model described previously. Upon subsequent simulations, it was determined that the slip-base model using discrete springs and friction produced desired system behavior.

A global friction was then added to the model in *CONTACT_AUTOMATIC_SINGLE_SURFACE. The static friction coefficient was set to 0.74 and dynamic friction coefficient was set to 0.57 per published values of mild steel-on-steel friction (3). An exponential decay rate of 1.0 was used for the transition between static and dynamic friction. This friction coefficient was apparently too high for some of the contact locations, because overall system behavior was poor. The post was twisting severely due to cable tension, and the plastic deformation seen in the cable release lever and anchor bracket were too severe. However, adding friction improved rotation of the cable release lever, and improved slip base joint behavior.

Friction was then removed from the global contact definition and inserted into *PART_CONTACT cards in order to define friction within specific parts. It was determined that friction would be assigned to specific parts based on how they interact with the other parts in the simulation. The anchor bracket was given friction coefficients of fs = 0.74 and fd = 0.57, which adds friction between the anchor bracket, cable release lever, and the cable end fittings. The same friction coefficients were added to the post base plate in order to include frictional effects within the slip base joint. In all cases, an exponential decay rate of 1.0 was used for the transition between static and dynamic friction. Initially friction was defined on the cable hanger, but was later removed entirely because the frictional forces between the cable and the cable hanger were causing excessive initial twisting of the post. Without friction applied to the cable hanger, it was determined that the cable snagging on the sharp corner of the cable hanger resulted in acceptable deflections of the post.

After friction was added to the necessary parts of the model, it was determined that the interaction between the cable release lever and anchor bracket wasn't behaving as desired. The cable release lever was not rotating as observed in the film of CTB-4, and the fillet welds were failing prematurely using failure strain values published by Shigley and Mischke (4). An increase in the fillet weld strain failure criteria from 0.20 to 0.30 produced an increase in the rotation of the cable release lever, but excessive deformation occurred in the anchor bracket near the fillet welds. A third node was added to the fillet welds on each vertical gusset and the failure strain was reduced to simulate brittle weld failure. After several iterations, a failure strain of 0.06 was found to produce acceptable global system behavior. The welds failed in a brittle manner as desired, and the increased strength of the welded joint caused the cable release lever to rotate in a manner more consistent with CTB-4.

SIMULATION VALIDATION

Ideally, accelerometer data would be used to compare energy loss in the actual bogie to the simulated bogie, however, since this data was not recorded, the next best validation technique is visual comparison. Time comparison between digital video of CTB-4 and the simulation is the main validation used for this model. Plastic deformation of the cable end fittings, end post, and anchor bracket was a second method used to validate the simulation of CTB-4.

Certain physical characteristics of the bogie test were chosen and used for validation of the simulation. Rotation of the cable release lever, the time of fillet weld failure, and the time of contact with the post can all be closely inspected using high-speed film. In a more subjective manner, the global impact behavior of the cable end fittings, cables, and post were used as secondary validation techniques. Figure 6 shows time comparisons of the actual and simulated end terminal test CTB-4.

When viewing the time comparisons, certain details were inspected for simulation validation. The lower part of the bumper contacts the cable release lever at 8 ms in both the simulation and the video. The fillet welds fail at 9 ms into the simulation, but they didn't fail in the video until 12 ms. However, the video was recorded at 500 frames per second, so it is possible that the weld failed within the range of 10 to 12 ms. At 18 ms, the cable release lever has almost lost contact with the anchor bracket, and it was almost parallel with the ground, which is evident in both the simulation and the video. The simulation accurately shows that the bogie made contact with the post at approximately 70 ms.

Permanent post, cable end fitting, and anchor bracket deformations were also compared for simulation validation, as shown in Figure 7. It is apparent from the visual comparison that the deformations observed in the damage photos coincides reasonably well with the simulation results.

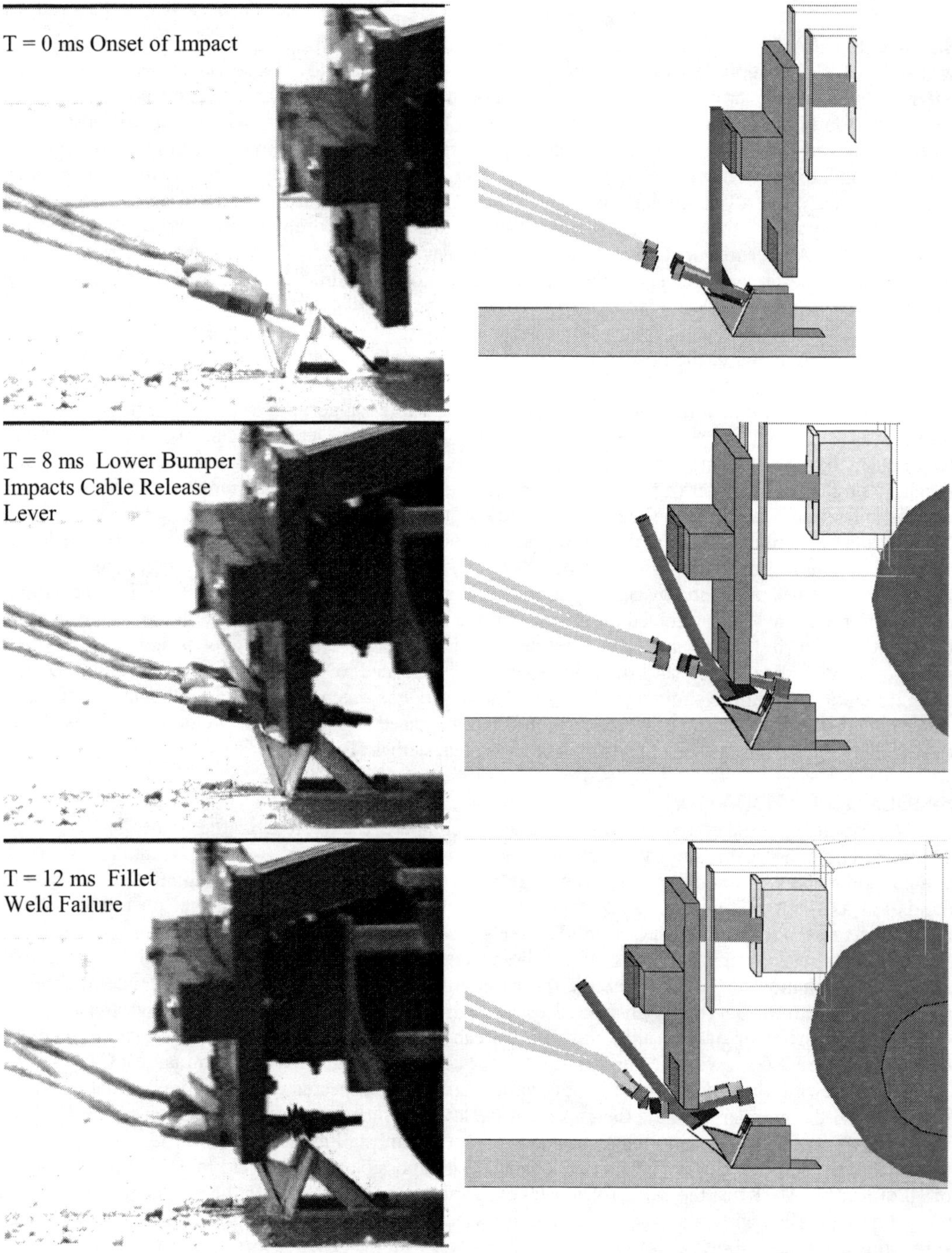

Figure 6. Time Comparison, High-Speed Film vs. LS-DYNA Simulation

T = 18 ms Cable Release Lever Loses Contact with Anchor Bracket

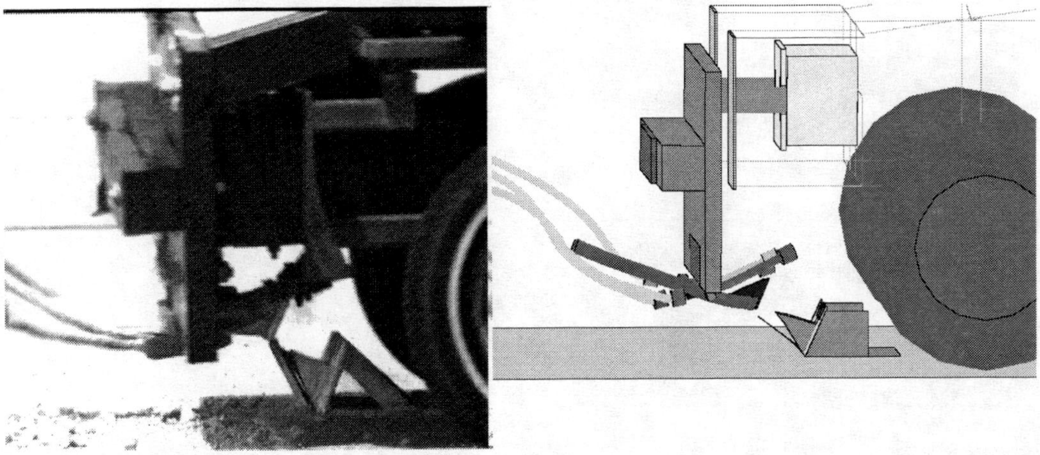

T = 70 ms Post impact

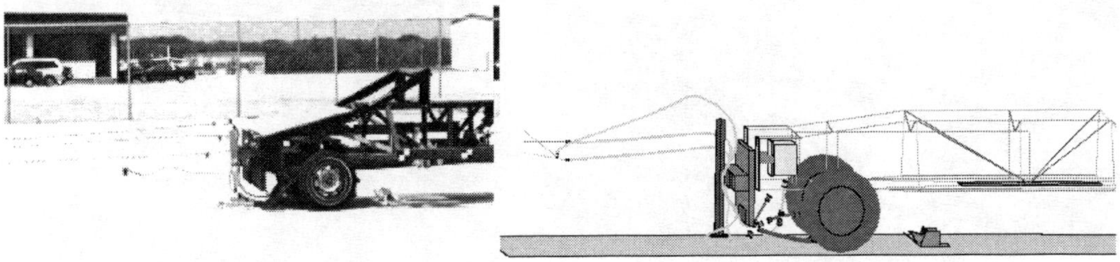

T = 80 ms Post Releases from Slip Base

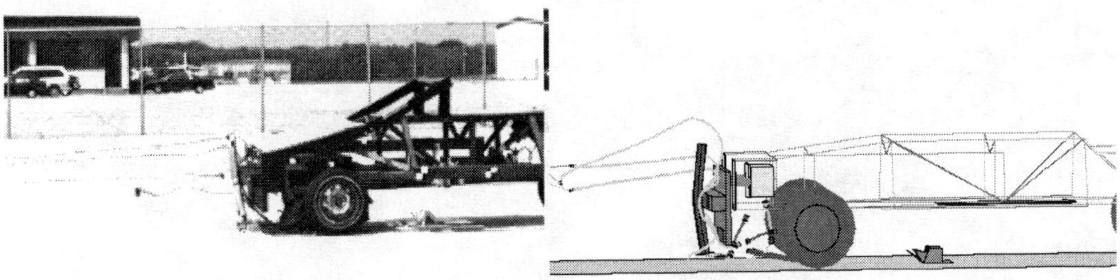

Figure 6 (Continued). Time Comparison, High-Speed Film vs. LS-DYNA Simulation

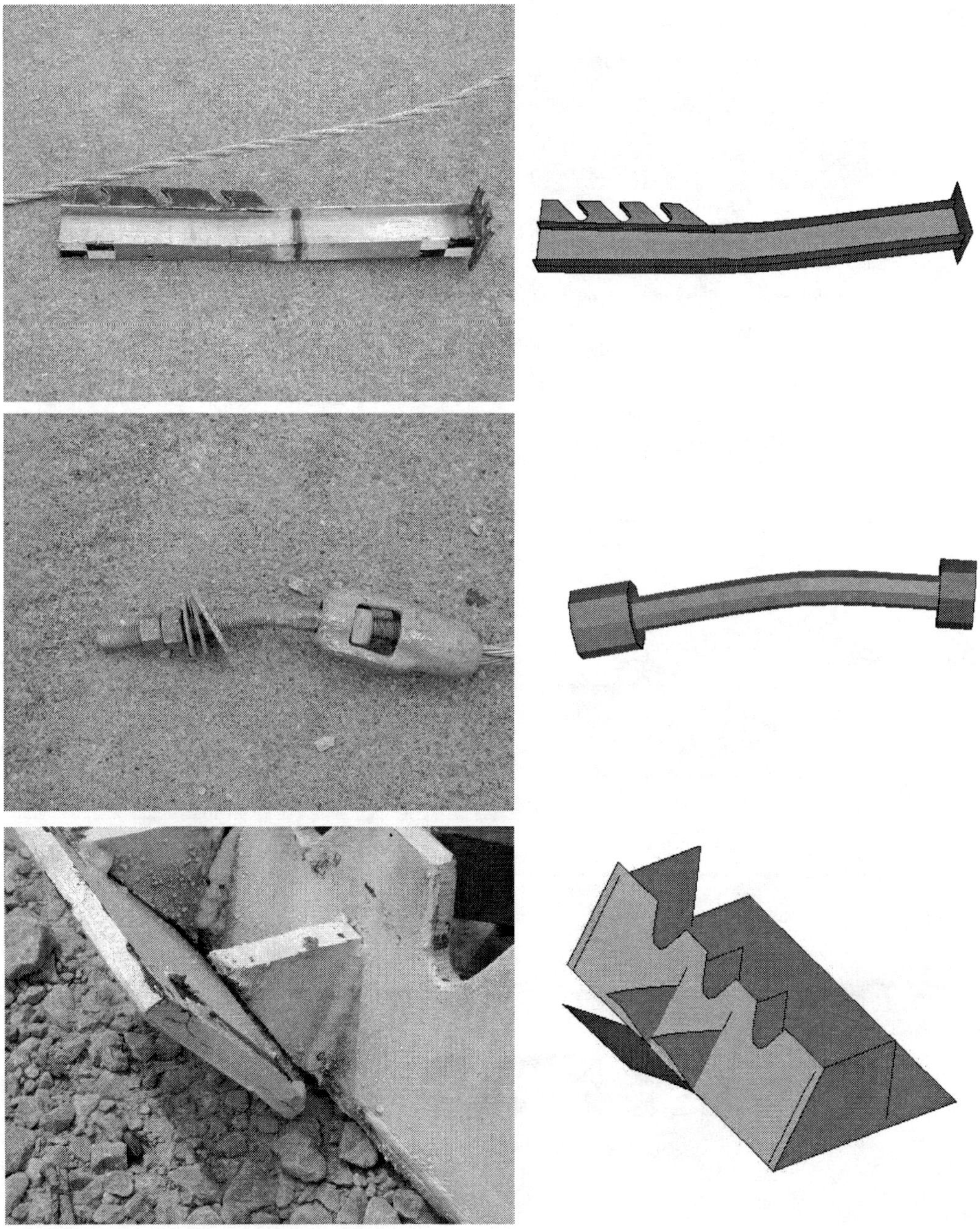

Figure 7. Deformation Comparison - Post, Cable End Fitting, and Anchor Bracket

FURTHER ANALYSIS

Since the simulation has been validated through visual means, it is useful to obtain other data from the simulation that was not, or could not be obtained from the bogie test. Contact forces and accelerometer data can be easily obtained using LS-POST. Figure 8 is a force-displacement plot of the contact forces on the bogie bumper, and Figure 9 is a time plot of the velocity of the bogie.

Major peaks in contact forces, as shown in Figure 8, coincide with noticeable drops in velocity. The reductions shown in the velocity curve coincide with the impacts with the end terminal components. The first drop in velocity represents initial impact with the cable release lever. Velocity reductions between 8 ms and 18 ms indicate the lower part of the bumper interacting with the cable release lever and failure of the anchor bracket fillet welds. The final velocity drop at 70 ms signifies bogie impact with the end post. Initially the bogie is at 20.1 m/s and the velocity drops to just below 19.0 m/s, indicating a relatively minor decrease in velocity.

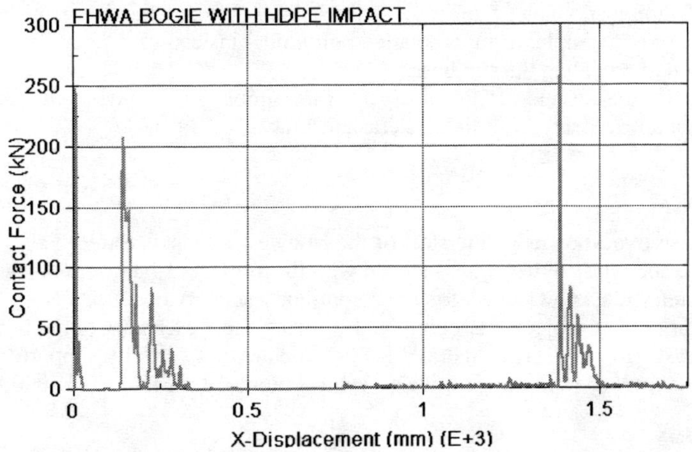

Figure 8. Bogie Bumper Contact Forces, Force vs. Displacement

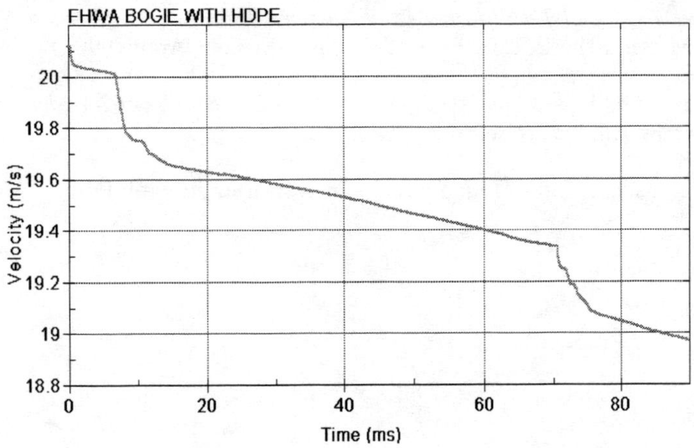

Figure 9. Bogie Velocity vs Time

CONCLUSIONS AND RECOMMENDATIONS

LS-DYNA was used to simulate a new 3-strand cable terminal system. A 900-kg bogie vehicle with a bumper height of 400 mm impacted the system end-on at an initial velocity of 20.1 m/s (45 mph). Visual comparison of simulation results with high-speed video from the bogie test was used to validate the simulation. The cable end fittings were properly released from the anchor bracket with the cable release lever, thus eliminating the tension in the cables, and preventing the vehicle from ramping up the tensioned cables. The simulated slip base joint constrained the post during the simulation until activation by bogie impact, failing in manner consistent with observed behavior from the test film. Deformations in the simulated cable end fittings, end post, and anchor bracket were similar to the actual deformations seen in the damage photos. With this validation, it is reasonable to gather other results from the simulation that could not be produced in the actual crash test. The peak contact force was about 300 kN, and the velocity of the bogie after impact with thc end post was 19.0 m/s.

Further work could be done in the modeling of the slip base on the end post. While the slip base model in this simulation works fairly well to replicate overall system behavior, a more realistic model could be developed. For example, using a detailed bolt model instead of discrete springs might prove to be a better method of modeling the slip base.

Based on bogie test results and detailed analysis of the simulation results, it was concluded that the new cable end terminal was ready for full-scale crash testing.

ACKNOWLEDGMENTS

The authors wish to acknowledge the staff of the Midwest Roadside Safety Facility for providing the design guidance and physical testing associated with this project. Livermore Software Technology Corporation is gratefully acknowledged for their continued support of LS-DYNA. The simulation work performed during this project was completed utilizing the Research Computing Facility of the University of Nebraska-Lincoln. This material is based upon work partially supported by the Federal Highway Administration under the cooperative agreement number DTFH61-00-X-00084.

REFERENCES

1. Hallquist, J.O., *LS-DYNA Keyword User's Manual*, Livermore Software Technology Corporation, Livermore, CA, 2001.

2. Paulsen, G.W., *Design and Testing of a Dual Support Breakaway Sign using Non-linear Finite Element Analysis*, Master's Thesis, University of Nebraska-Lincoln, April 1996.

3. E.A. Avallone, E.A. and Baumeister III, T., *Marks' Standard Handbook for Mechanical Engineers*, Tenth Edition, McGraw-Hill Inc., New York, 1996.

4. Shigley, J.E. and Mischke, C.R., *Mechanical Engineering Design*, Fifth Edition, McGraw-Hill Inc., New York, 1989.

Proceedings of IMECE'03
2003 ASME International Mechanical Engineering Congress
Washington, D.C., November 15–21, 2003

IMECE2003-55106

Investigation of Structural and Material Effects on Crashworthiness of Advanced High Strength Steel Columns

Faycal Ben-Yahia and James A. Nemes
Department of Mechanical Engineering
McGill University
Montreal, QC H3A 2K6
Canada

Farid Hassani
Stelco, Inc.
P.O. Box 2030,
Hamilton, ON L8N 3T1
Canada

ABSTRACT

An experimental and numerical study was performed to evaluate the crashworthiness of several advanced high strength steels. The behavior of two Dual Phase (DP) steels and an HSLA steel are compared by examining the crush response of longeron column specimens, experimentally and computationally. The closed section columns, fabricated by spot welding formed channel sections, in both single hat and double hat configurations were exposed to 182 kg and 454 kg axial impacts at different velocities. Final column height and impact force history were recorded and compared with results of finite element simulation of the columns. Good agreement was found between experiments and computations.

INTRODUCTION

The evolution of environmental and safety requirements as well as international economic pressures introduces new challenges to the transportation industry. As a result, automotive vehicles are required to be cleaner and safer. This continuing effort of reducing mass and cost of vehicles on one hand and increasing their safety and crashworthiness efficiency on the other has led to several structural and material investigations. Isotropic, anisotropic, and combined composite aluminum materials, as well as sandwich structures have been the objects of recent numerical and experimental research projects [1-3]. In addition, the merits of single-cell, multi-cell, and foam filled, thin-walled, energy absorbing structures have been highlighted by other researchers [4-6]. Nevertheless, the development of new high strength steel materials, such as dual phase (DP) steels have made them attractive for vehicle manufacturers. DP steels exhibit high strength, high strain hardening rates, and high total elongation. The combination of these properties provide excellent crashworthiness characteristics [7,8]. Hence, proper implementation of dual phase steel in vehicle design will result in lighter and more fuel-efficient vehicles.

In this paper we focus on the development of a robust finite element simulation of the column crush experiment that permits investigation of a wide range of material and structural variables that can be validated by a reduced set of experiments. Experiments, in this work, are performed using three high strength steels in single hat and double hat geometries.

MATERIALS AND EXPERIMENTAL METHOD

The chemistry of the three types of 1.7 mm galvannealed steel sheets used for fabrication of single and double hat column specimens is shown in Table 1, along with the yield strength, tensile strength, total elongation, and n-value. Stelmax50 is a Stelco conventional precipitation hardened 350 MPa (50 ksi) minimum yield strength HSLA steel, which is commonly used for manufacturing automobile front rails. The StelPhaseII600 is a newly developed 600 MPa tensile strength DP steel. Two variants of the StelPhaseII600, having yield strengths of 300 MPa and 400 MPa were considered. The 400 MPa yield strength steel was attained by stretching or tension levelling in the longitudinal direction after the coil was produced, resulting in the increased yield, with a slight reduction in the n-value and total elongation.

The geometry of the single hat and double hat columns is shown in Fig. 1. The column length was 300 mm and the longitudinal axis of the columns is parallel to the rolling direction of the sheets. The half-columns were made by bending the blanks in the transverse direction, without stretching the material. The columns were then manufactured by spot-welding either the flat back piece or the two halves. The weld nugget size was 5.4 mm with a weld spacing of 35 mm. A drop weight machine with a maximum drop height of 4.26 m was used for crushing the columns. Impactor weights of 454 kg and 182 kg were used. Experiments using the 182 kg mass were instrumented using a 222.4 kN (50000 lb) impact quartz force transducer [force capacity of 333.6 kN (75000 lb)] to measure force-time histories.

Table 1 Chemistry and static mechanical properties of steels

Steels			Chemistry (wt%)				Mechanical properties			
Label	Name	Type	C	Mn	Si	Other	YS (MPa)	TS (MPa)	El (%)	n[1]
I	Stelmax50	HSLA	0.07	0.90	0.13	Nb	360	430	35	0.20
II	StelPhaseII600	DP	0.06	1.90	0.08	Mo	300	620	25	0.15
III	StelPhaseII600	DP	0.06	1.90	0.08	Mo	400	620	23	0.13

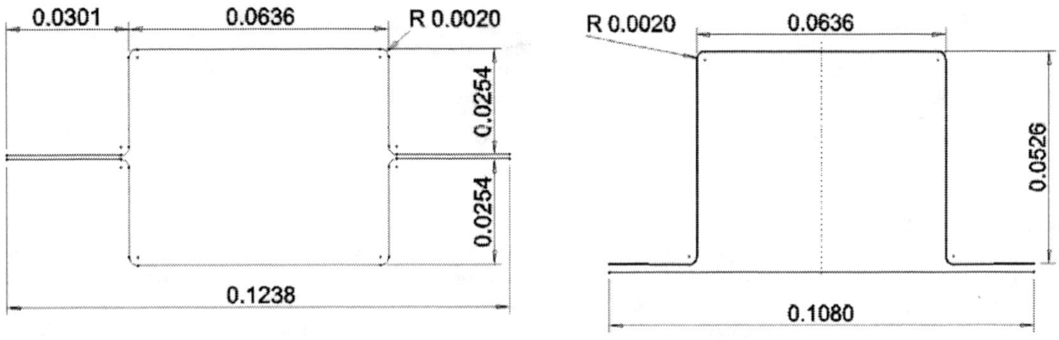

a) Double Hat b) Single Hat

Figure 1 Geometry of columns (dimensions in m)

NUMERICAL MODEL

The commercial finite element code ABAQUS is used in the numerical computations made for this study. Both standard and explicit algorithms were used. The first ten structural eigenmodes are initially determined using a static buckling perturbation step. The nodal displacements of the ten modes are stored and scaled by a selected factor to perturb the mesh before explicitly simulating the crush. A linear combination of the symmetric nodal displacement modes is created to account for imperfections. The largest scale factor used for the first mode, which is most critical for the post buckling behavior, is approximately 2% of the shell thickness. The other scale factors are chosen smaller unless the following modes are close to a smaller eigenmode, in which case an imperfection sensitivity study should take place to examine the post buckling behavior of the structure. In our case, the first eigenmode is sufficiently different from the following ones, and a small variation of the scale factors of the higher modes is found to have a negligible effect on post buckling results.

Elements and Meshing

The single hat column finite element model, shown in Fig. 2 is made of 3960 4-node S4RSW doubly curved shell hourglass control elements. These elements consider warping and have the advantage of being well-suited for many impact dynamics problems, including structures undergoing large-scale buckling behavior, and involving small-strains but large rotations and severe bending. The moving rigid plate (impactor) and the fixed plate are meshed with 196 R3D4 rigid discrete elements each. The mass of the impactor is attached to the moving plate through its reference node using a mass element. The same element types are used to mesh the double hat column with 4560 deformable shell elements and 392 rigid discrete elements.

One of the difficulties in obtaining the correct eigenmodes is the difference in how spot welds and contact are treated in ABAQUS/Standard and ABAQUS/Explicit. The parts that need to be in contact in the buckling step should stand initially without clearance and a contact should be defined between them. Thus, parts in contact are completely joined during the computational step as if they have been tied. As a result, the column parts were tied which makes the frame somewhat more rigid than it should be and does not give the exact buckling shape of the spot-welded structure. Nevertheless, this method was used to initiate the dynamic buckling in as accurate way as possible. The buckling modes for the double hat and single hat geometries are shown in Figs. 3 and 4, respectively.

[1] n-value is found from the slope of the log stress vs. log strain curve between a strain of 10% and the smaller of 20% or the uniform strain.

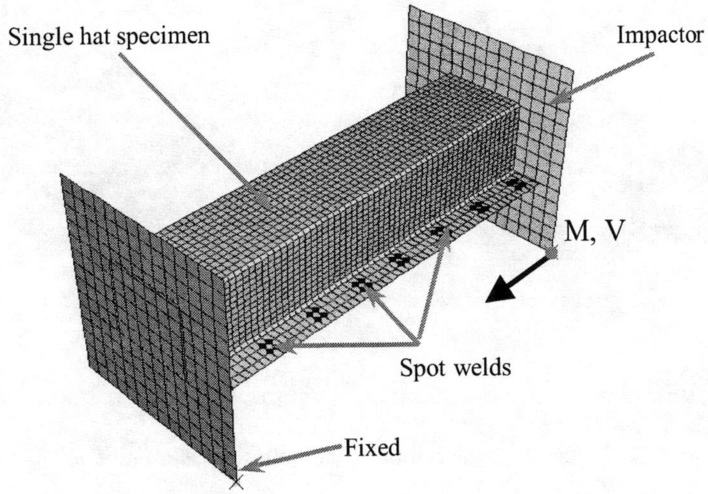

Figure 2 Single hat numerical model

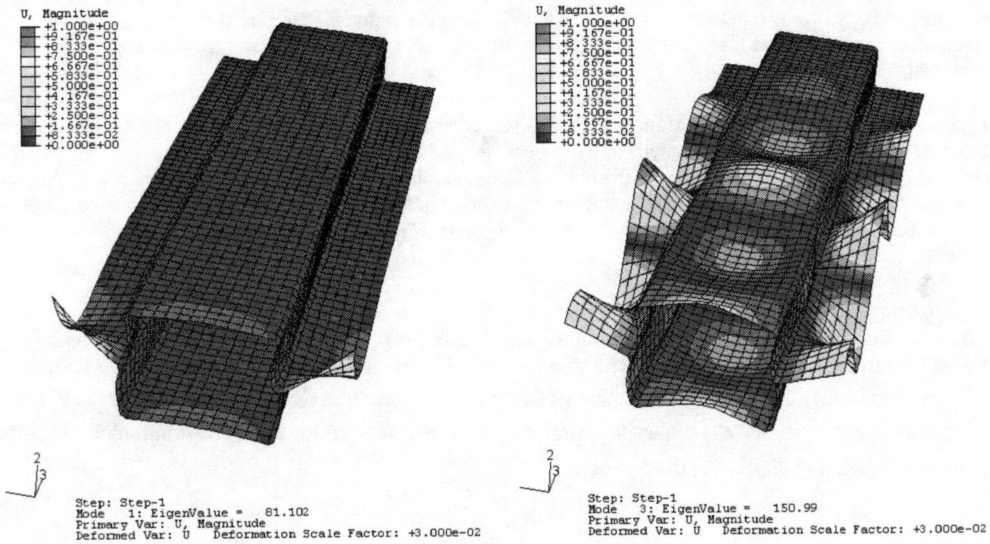

Figure 3 Double hat first symmetric buckling modes

Spot Welds

Two ways of modeling spot welds are available in ABAQUS/Explicit. The mesh dependent method consists of specifying the nodes to be tied in the spot-weld region. The mesh independent procedure has been introduced in version 6.3 and is used by specifying the location and the properties of the spot weld in a simpler way. However, unlike the first method, the latter doesn't allow for spot weld failure. Since spot weld failure was not observed in the crush experiments, the mesh independent approach is used in our case. The forces acting on the spot welds can be plotted and compared to appropriate failure criteria.

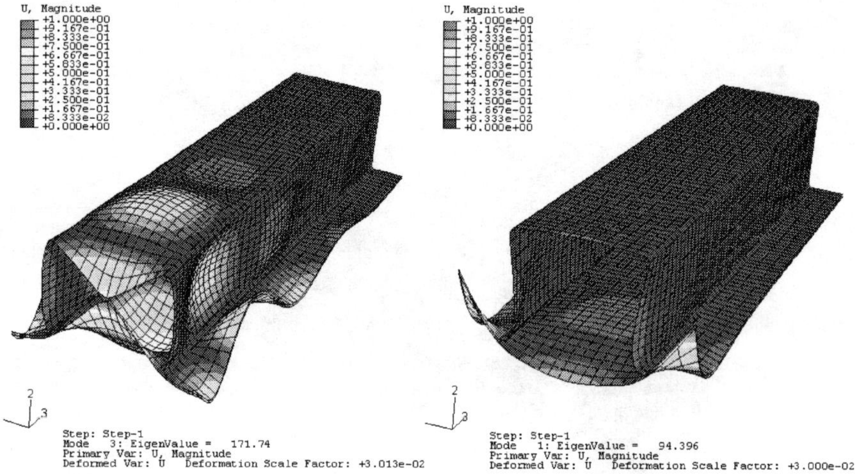

Figure 4 Single hat first symmetric buckling modes

Contact

A new general contact algorithm has been used in this study. This contact has the advantage of being very easy to use and providing excellent energy balance in the model. It enforces contact constraints using a penalty contact method, which searches for node-into-face and edge-into-edge penetrations in the current configuration. ABAQUS/Explicit then chooses the penalty stiffness that relates the contact force to the penetration distance automatically so that the effect on time increment is minimal yet the penetration is not significant. This penalty stiffness has been kept by default in all our computations.

The contact pair algorithm, based on the master-slave approach, has been tested and compared to the general contact behavior. The external work generated by the contact pair algorithm is found to be very high compared to the internal one, which needs a modification of the penalty factor, and hence a reduction in the computational time step. In addition, the force history issued from the contact pair algorithm is found to be very noisy and to generate high peaks. In contrast, the general contact algorithm produces a fairly smooth force history, although sometimes underestimates the mean force.

Material Constitutive Model

Knowing that the different steels studied in this paper are rate-dependent solids and assuming a Von Mises yield criteria with isotropic hardening, the evolution of the yield surface is described in terms of a single scalar variable, which is taken to be the equivalent stress. In this work, we consider the equivalent stress, σ, to be given as a function of the equivalent plastic strain, ε_p, equivalent strain rate, $\dot{\varepsilon}_p$, and temperature, T, from adiabatic deformation by the form proposed by Johnson and Cook [9] as

$$\sigma = (A + B\varepsilon_p{}^{n'})(1 + C\ln\frac{\dot{\varepsilon}_p}{\dot{\varepsilon}_o})\left(1 - \theta^m\right) \quad , \quad \theta = \frac{T - T_0}{T_m - T_0} \tag{1}$$

where A, B, C, n', and m are material parameters. T_m and T_0 are the melt and room temperatures, respectively, and $\dot{\varepsilon}_0$ is taken as 1.0 s^{-1}. Parameters for the three steels are given in Table 2. Parameters A, B and n' were obtained from static tensile tests and C from literature data on similar materials.

Table 2 Parameters describing Johnson-Cook model for steel materials used

Label	ρ (kg/m^3)	E (GPa)	ν	A (MPa)	B (MPa)	C	n'	m
I	*7800*	*207*	*0.3*	*344*	*200*	*0.025*	*0.33*	*1.0*
II	*7800*	*207*	*0.3*	*270*	*875*	*0.080*	*0.36*	*1.0*
III	*7800*	*207*	*0.3*	*300*	*630*	*0.030*	*0.30*	*1.0*

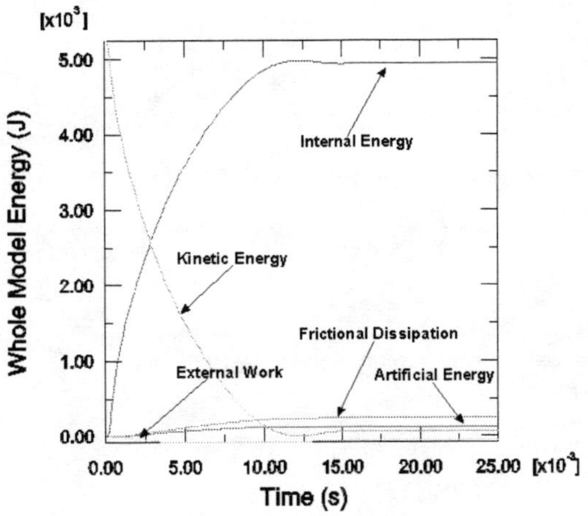

Figure 5 Energy balance

RESULTS

Model Validation

The validation of the model is based on three main criteria, namely, energy balance, extent and mode of deformation, and force history. The energy balance, which is given in Fig. 5, shows that the kinetic energy is almost totally transformed to internal energy, including plastic work. The external work remains less than 5% of the internal energy, which indicates good contact simulation. Moreover, the artificial work is also shown to be very low compared to the internal energy indicating low hourglass energy in the model and hence showing an acceptable mesh refinement. The crushed shape of HSLA and DP (type II) steel columns subjected to a mass of 182 kg impacting at a velocity of 7.6 m/s is shown in Fig. 6 compared to computed contour plots of equivalent stress. As seen in the figure, the crushed shape from the simulation closely agrees with that seen in the post-test observations. Figure 7 shows the computed force history compared to that measured during tests conducted with a 182 kg mass at 7.6 m/s impact velocity for the 3 steels considered. Where repeated tests have been performed, both are shown to give an indication of the variability of the force history under nominally equivalent conditions. As seen in the figure, the computed results compare well with the measured forces, particularly in light of the scatter in the experimental results. In general the difference seen is a slight underprediction of the peak load and overprediction in the computed crush force after the peak load. The results presented in Figs. 5-7, illustrate the robustness of the developed simulation method, which allows for further investigation of material and structural parameters.

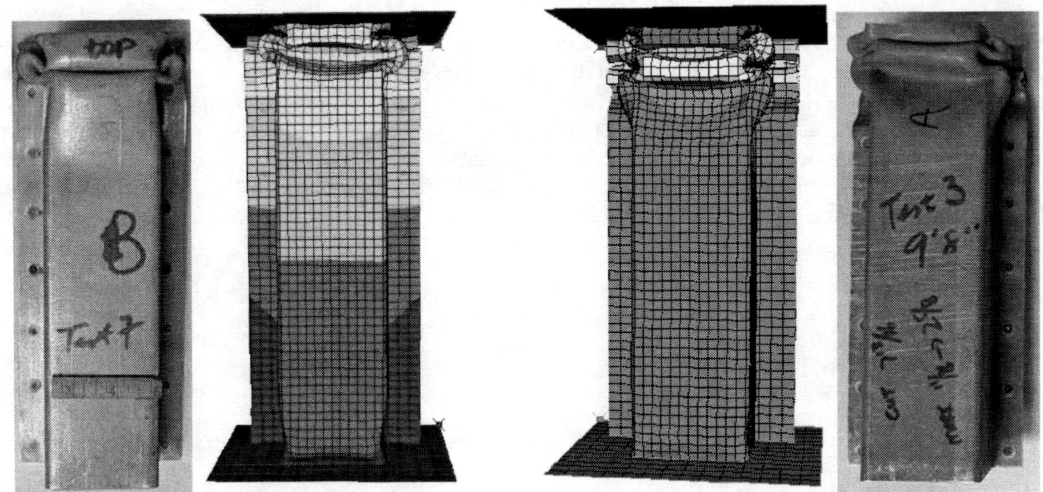

a) DP steel (II) b) HSLA steel (I)

Figure 6 DP steel (II) & HSLA steel (I) crushed by a mass of 182 kg at 7.6 m/s

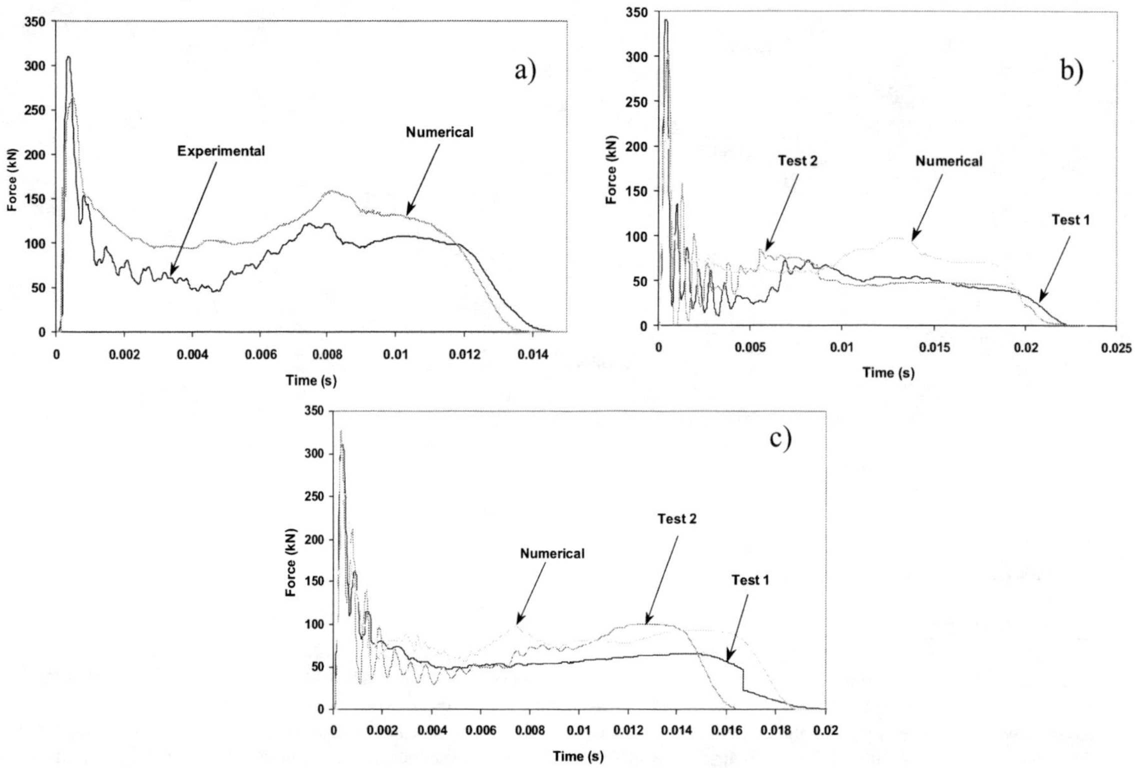

Figure 7 Experimental and numerical force histories for single hat columns struck by 182kg at 7.6m/s for a) DP 600-400, b) HSLA and c) DP 600-300

Mesh Perturbation and Geometric Imperfections

As described previously, two steps are needed to perform the crush simulation. A buckling step is first made to find the first 10 eigenmodes of the structure. The nodal displacements are stored and used to perturb the mesh of the model to allow smooth post buckling behaviour. The nodal displacements calculated in the buckling step are usually scaled by a maximum factor of 2% of the shell thickness. However, this value sometimes needs to be increased to represent the effect of the real initial imperfection of the specimens and not only to start a smooth numerical postbuckling behavior. This scale factor has been seen to have some effect on the initial force peak, but negligible effect on overall energy absorption. Other imperfections may not adequately be accounted for by the eigenmode procedure. Dimensional variations that are not manifested in the first 10 eigenmodes may have a significant effect on the crush response. Deviation in the dimensions of the formed channel, for example, which may result from factors such as springback are not accounted for by the eigenmode procedure. In order to estimate the effect of this dimensional tolerance a parametric study of the dimension at the base of the channel has been conducted by varying this dimension from 0.0600 m to 0.0680 m as shown in Fig. 8. The significant effect on computed force history is shown in Fig. 9.

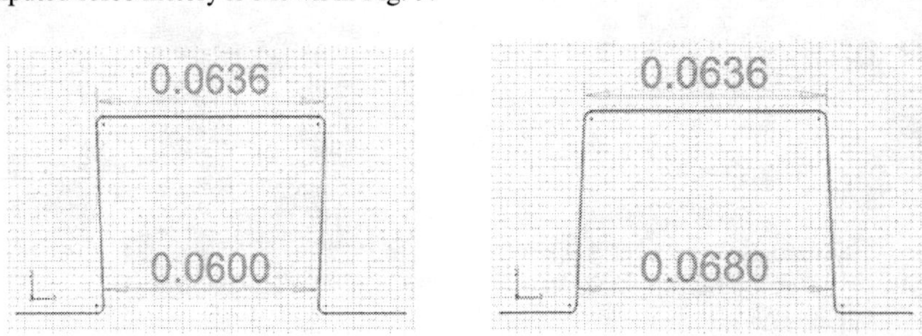

Figure 8 Initial geometrical imperfection (m)

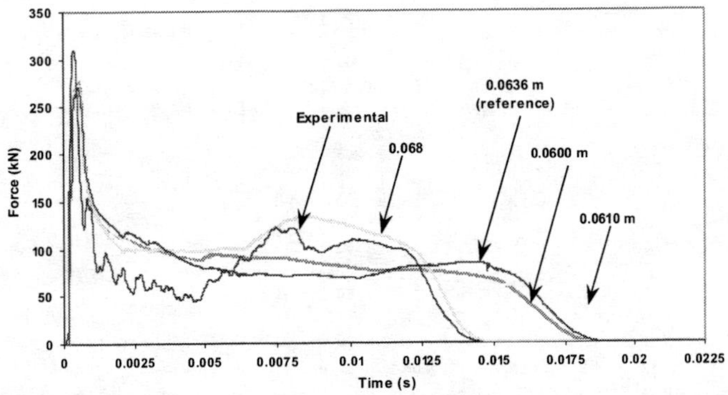

Figure 9 The effect of initial imperfections on the force history (single hat columns)

Structural and Material Effects

The effect that the type and grade of steel has on energy absorption can be seen in the calculated force histories shown in Fig. 10 for the single hat geometry subjected to a range of impact mass and velocity. In all cases the crush time is shortest (shortest crush length) for the type III steel, followed by the type II and then type I, indicating the greater energy absorption of the DP steels. The data also indicate that the difference between the energy absorption of HSLA and DP becomes larger as the impact velocity and crash energy are increased, and is more pronounced as the yield strength of dual phase steel is increased. Crush force is similar for the three steels, although the peak is somewhat higher in the DP steels. It should also be noted that the peak force and crush force are quite similar even for significantly different input energies, with differences predominately observed in crush duration.

A similar comparison is shown in Fig. 11 for the double hat geometry. The same trends are observed as for the single hat for the higher velocity impacts, however, at the lower velocity, the crush behavior of the type III steel is quite different. Peak force is considerably lower and crush duration considerably longer. Examination of the response shows this change in behavior arises from a change in response mode from local crushing to global buckling, as illustrated in Fig. 12. Global buckling of some double hat columns was also observed experimentally.

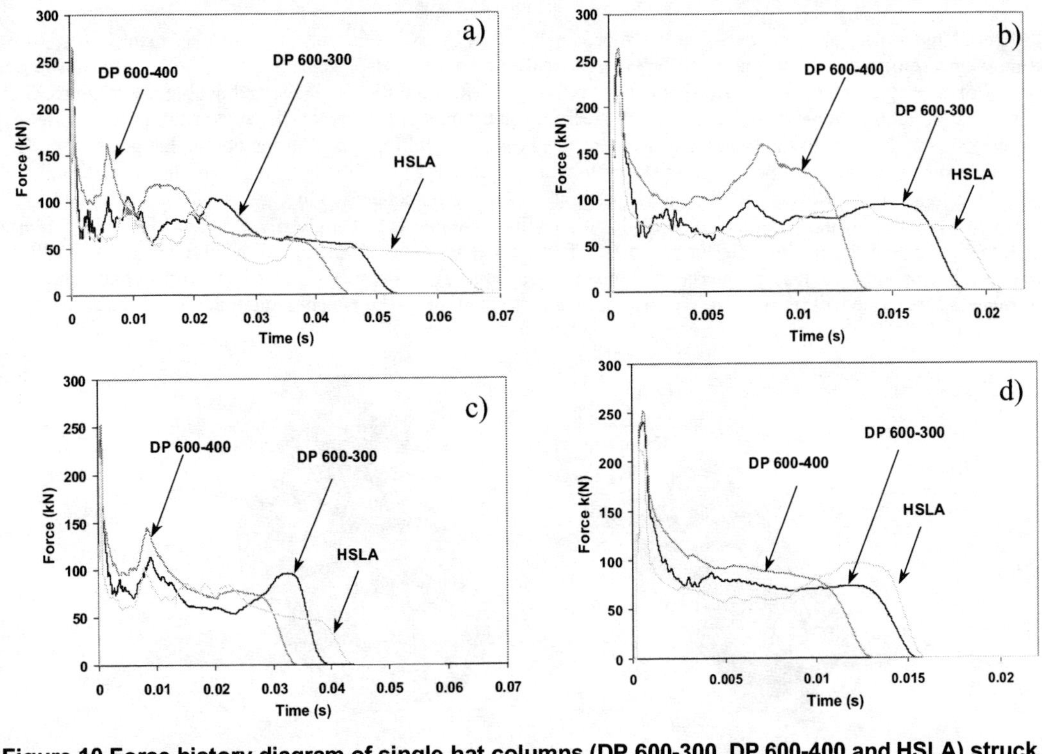

Figure 10 Force history diagram of single hat columns (DP 600-300, DP 600-400 and HSLA) struck by a) 454kg at 7.6m/s, b) 182kg at 7.6m/s, c) 454kg at 5.8m/s and d) 182kg at 5.8m/s

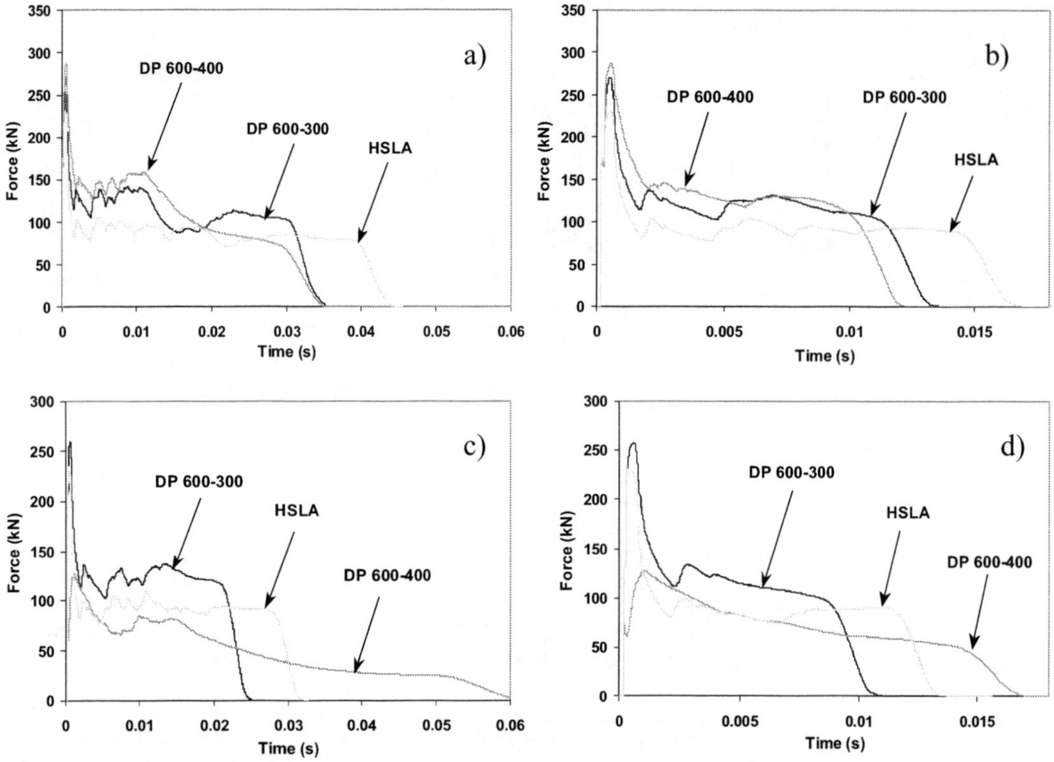

Figure 11 Force history diagram of double hat columns (DP 600-300, DP 600-400 and HSLA) struck by a) 454kg at 7.6m/s, b) 182kg at 7.6m/s, c) 454kg at 5.8m/s and d) 182kg at 5.8m/s

Material and structural comparisons can be more readily seen by examining the crush displacement for the same input mass and velocity, with the energy absorbed in a smaller amount of displacement usually being most desirable. The effect of steel type on crush displacement is shown in Fig. 13 for both the single hat and double hat columns. For the single hat columns, the advantages of the DP steel are quite apparent, although the previously noted change in response mode can be seen in the larger displacement for the higher yield DP steel in the double hat geometry. The material parameters, column geometries and deformation rates that will change the buckling mode from local to global is under investigation. However, with the limited current data it can be concluded that the global buckling mode will be activated when the stiffness of the column for local buckling is higher than the global stiffness. Material with high yield strength and high strain hardening characteristics, formed to a very stiff geometry will be prone to global buckling mode in crash situations. Reducing material thickness (down gauging) will reduce geometrical stiffness and hence this undesirable failure mode will be avoided. This solution is in harmony with the weight reduction goal of the automotive industry.

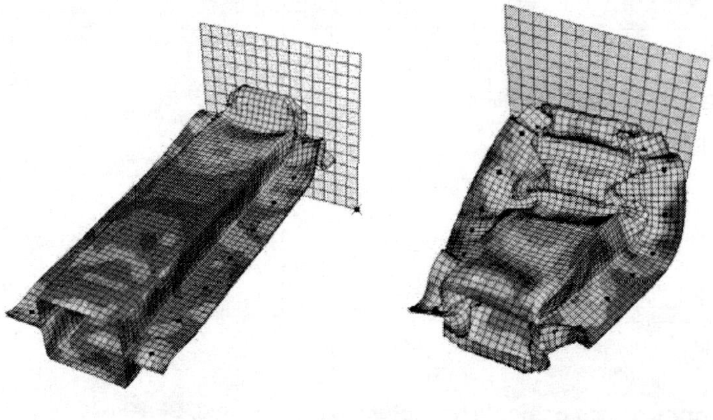

Figure 12 Comparison of local crushing vs. global buckling in double hat column

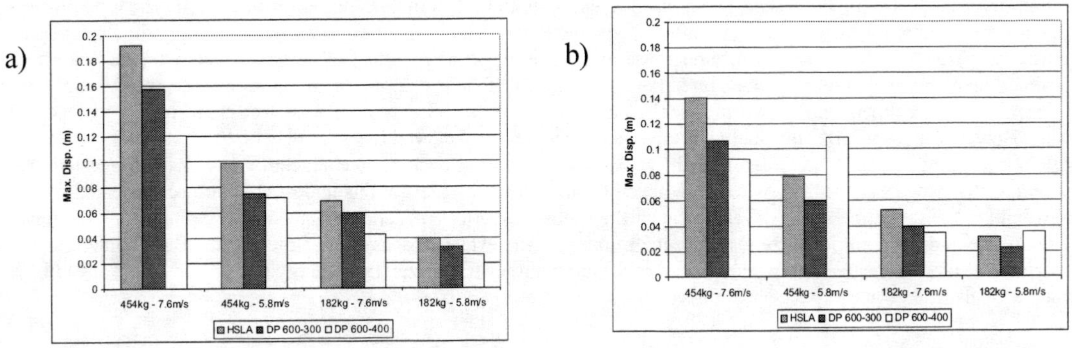

Figure 13 Comparison of maximum displacements for a) single hat columns and b) double hat columns

The crush displacement of single hat and double hat geometries are compared for the three different steels in Fig. 14 for different impact conditions. Other than cases in which global buckling occurred, the crush displacement for the double hat is significantly less, and therefore more efficient in terms of energy absorption, although it should be noted that the mass of the double hat specimens is approximately 8% higher.

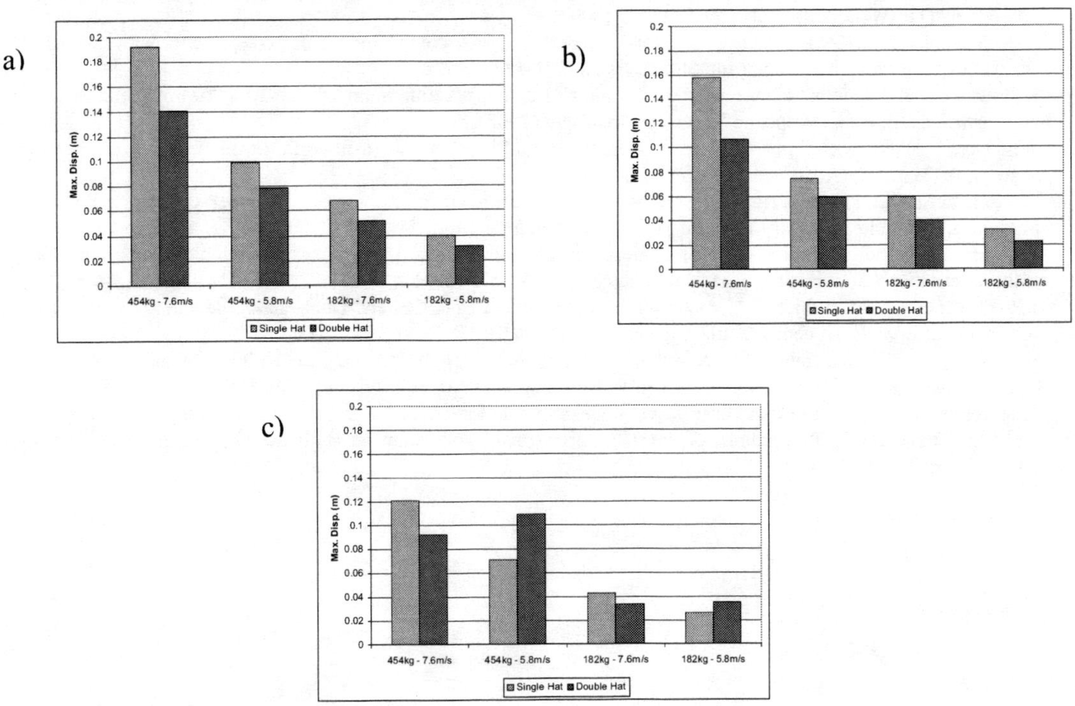

Figure 14 Comparison of maximum displacements of single and double hat columns for a) HSLA, b) DP 600-300 and c) DP 600-400

CONCLUSIONS

A robust numerical model has been developed for simulating the response of longeron column specimens subjected to axial impacts. The computed force history response and deformation behaviour agree well with experiments performed on HSLA and DP steels. Both numerical and experimental results illustrate the greater energy absorbing capacity of DP steels. Furthermore, the energy absorption of DP steels becomes greater as their yield strength increased. The model has also been used to examine the effect of variation of critical geometrical parameters that can arise from springback, for example. It is seen that relatively small changes in certain dimensions can significantly affect the force time history response and crush length. The effect of structural geometry has also been examined by comparing single hat and double hat specimens, which have approximately the same cross-sectional area. The double hat specimens are shown to absorb the same amount of energy over a shorter crush length under most cases. However in some cases the double hat specimens have been observed experimentally and numerically to exhibit a global buckling mode, which significantly reduces their energy absorbing characteristics. It is suggested that reducing the geometrical stiffness by reducing the sheet thickness can prevent this undesirable failure mode and reduce the weight at the same time. The development of the robust numerical model allows for investigation of a wide range of geometric and material parameters, and hence minimizes the extent of actual testing required. Both numerical and experimental work is being continued to better understand and characterize crash energy absorption of advanced high strength steels.

ACKNOWLEDGEMENT

This work has been carried out under the Auto21 Networks of Center of Excellence. Financial support of Auto21 is appreciated. The authors would also like to thank Stelco for providing the steels and permission to publish this work. FH would like to thank Stelco Cold Roll and Coated R&D Manger and Technology Manager, Messrs Peter Badgley and Dennis McCutcheon, respectively, for their supports and valuable comments.

REFERENCES

1. H.-S. Kim, New extruded multi-cell aluminum profile for maximum crash energy absorption and weight efficiency, Thin-Walled Structures 40, 311-327 (2002).
2. A.G. Mamalis et al., Axial collapse of hybrid square sandwich composite tubular components with corrugated core: numerical modeling, Composite Structures 58, 571-582 (2002).
3. J. Bouchet et al., Dynamic axial crushing of combined composite aluminum tube: the role of both reinforcement and surface treatments, Composite Structures 56, 87-96 (2002).
4. W. Chen, T. Wierzbicki, Relative merits of single-cell, multi-cell and foam-filled thin-walled structures in energy absorption, Thin-Walled Structures 39, 287-306 (2001).
5. M. Seitzberger et al., Experimental studies on the quasi-static axial crushing of steel columns filled with aluminum foam, International Journal of Solids and Structures 37, 4125-4147 (2000).
6. A.G. Hanssen et al., Optimum design for energy absorption of square aluminum columns with aluminum foam filler, International Journal of Mechanical Sciences 43, 153-176 (2001).
7. J.R. Fekete and A. M. Stibich, A Comparison of the Response of HSLA and Dual Phase Sheet in Dynamic Crush, International Body Engineering Conference & Exposition (IBEC), Detroit MI, 2001.
8. K. Miura et al, High Strain Rate Deformation Behaviour of High Strength Steel Sheet for Automobile, 29th International Symposium on Automotive Technology and Automation; Florence, Italy, 1996.
9. G.R. Johnson and W.H. Cook, A constitutive model and data for metals subjected to large strains, high strain rates and high temperatures, Proceedings of the 7th International Symposium on Ballistics, The Hague, Netherlands, 541-547, 1983.

Proceedings of IMECE'03
2003 ASME International Mechanical Engineering Congress
Washington, D.C., November 15–21, 2003

IMECE2003-55107

DEVELOPMENT OF A STOCHASTIC APPROACH FOR VEHICLE FE MODELS

David Riha
Southwest Research Institute

Joseph Hassan, Marlon Forrest and Ke Ding
Chrysler Group, DaimlerChrysler Corporation

ABSTRACT

This paper describes the development of a mathematical model capable of providing realistic simulations of vehicle crashes by accounting for uncertainty in the model input parameters. The approach taken was to couple advanced and efficient probabilistic and reliability analysis methods with well-established, high fidelity finite element and occupant modeling software.

Southwest Research Institute has developed probabilistic analysis software called NESSUS. This code was used as the framework for a stochastic crashworthiness FE model. The LS-DYNA finite element model of vehicle frontal offset impact and the

MADYMO model of a 50th percentile male Hybrid III dummy were integrated with NESSUS to comprise the crashworthiness characteristics.

The system reliability of the vehicle is computed by defining ten acceptance criteria performance functions; four occupant injury criteria and six compartment intrusion criteria.

The reliability for each acceptance criteria was computed using NESSUS to identify the dominant acceptance criteria of the original design. The femur axial load acceptance criteria event has the lowest reliability (46%) followed by the HIC event (58%) and the door aperture closure event (73%). One approach to improve the reliability is to change vehicle parameters to improve the reliability for the dominant criteria. However, a parameter change such as vehicle strength/stiffness may have a beneficial effect on certain acceptance criteria but be detrimental to others. A system reliability analysis was used to include the contribution of all acceptance criteria to correctly quantify the vehicle reliability and identify important parameters.

A redesign analysis was performed using the computed probabilistic sensitivity factors. These sensitivities were used to identify the most effective changes in model parameters to improve the reliability. A redesign using 11 design modifications was performed that increased the original reliability from 23% to 86%. Several of the design changes include increasing the rail material yield strength and reducing its variation, reducing the variation of the bumper and rail installation tolerances, and increasing the rail weld stiffness and reducing its variation.

The results show that major reliability improvements for occupant injury and compartment intrusion can be realized by certain specific modifications to the model input parameters. A traditional (deterministic) method of analysis would not have suggested these modifications.

INTRODUCTION

Numerical methods are widely used for the analysis of the impact behavior. The most commonly used comuputational methods for vehicle imapct analysis is the rigid body formulation implemented in codes such as MADYMO and PAM-SAFE and the finite element method implemented in codes such as LSDYNA, PAM-CRASH and RADIOSS. Using these codes, designs of almost any arbitrary complexity and physical behaviour can be investigated for their impact responses. This way, large number of variations and design modififcations can be compared effectively. From the results of computational simulations, conclusions can be drawn to develop understanding of the impact response. The determination of such changes is usually driven by the experience of the engineer.

A typical CAE simulation for crash and safety analysis requires the use of a wide range of imput parameter classes such as dimensions, material properties, geometrical constraints, boundary approximations and numerical assumptions. Geometry is obtained from a "typical" test setup, which is considered to be an accurate representation of the test being simulated. However, a single impact test is insufficient to capture the relevent physics of the charecteristic response of a vehicle and occupants subjected to impact conditions.

Metalic material properies used in crash analysis suffer from uncertainty in characteristics such as yield, strength, strain hardening and others. Typical scutter values is between 5 to 10 %. Uncertainity is even higher for properties such as strain rate, foam material definitions and weld charachtaristics.

With regard to reliability, the design procedure usually entails comparing analysis results to those of a design guide, or for the case of safety and crash, some NHTSA requirement. The problem with this approach is that it does not take into account (in a quantifiable manner) the fact that there is inevitably some element of uncertainty in the basic design parameters, such as material properties, tolerances, and loadings. In this work, the influence of parametrs such as uncertaininty in weld quality (stiffness, failure stregth), uncertainity in various matrial properties (yield, ultimate streght, strain hardening), uncertainity in local thickness of stamped parts and finally imprefections due to actual assembly processs, were used as input variables.

STOCHASTIC ANALYSIS OF VEHICLE IMPACT SIMULATION

Uncertainties and non-deterministic behavior exist in all physical processes which are known as the core applications of occupant response simulation using numerical formulations such as the case in LSDYNA [6] and MADYMO [7]. Car crash and occupant safety are applications of impact-type problems that are highly non-deterministic, as such repeated collision tests of the same vehicle type will always lead to different results.

The idea behind probabilistic structural analysis is to use the information about the probability of random variables, along with the structural behavior, in order to quantify the scatter in the structural response. The result from the analyses is a distribution in the form of a Probability Distribution Functions (PDF) and Cumulative Distribution Function (CDF), of the structural response. Thus, the analysis gives a more complete picture of the actual simulation. With this method, the probability of achieving a certain level of structural response can be computed. While the conventional Probabilistic Structural analysis (PSA) considers the uncertain quantities as *random variables*, in Stochastic Structural analysis (SSA) they are modeled as *random fields,* as such, they are considered as sequences of random variables spread over the volume of the structure object with a defined correlation structure. Therefore, the consideration of randomness in this regard is more elaborate since it is assumed that the parameters of the model have a random spatial variation [4].

The concept of a stochastic crash simulation can be explained due to the fact that the problem comprises a set of stochastic structural parameters, stochastic external forces and boundary conditions and, finally, the stochastic output variables such as displacements, accelerations, and internal energies. The stochastic crash problem may be stated, in general terms as follows: *Given the Probability Density Functions (PDFs) of the stochastic structural parameters, external forces, boundary and initial conditions, determine the corresponding PDFs of the output variables.*

In generic mathematical terms, crash is a dynamic phenomenon that may be formally described by a set of nonlinear first order vector differential equations in the form;

$$x = f(x,F,p)$$
$$y = g(x,p)$$

where $x \in R^N$ is the state vector of displacements and velocities, $F \in R^p$ represents the stochastic external forcing terms, $p \in R^n$ is a vector of stochastic structural parameters and $y \in R^q$ the measurement vector (e.g. accelerations, strains, etc.). A classical problem in stochastic mechanics is the computation of the Probability Distribution Functions (PDFs) of the output variables given the (PDFs) of the input variables. This can be accomplished by knowing the joint density function of all the random variables implied in a specific problem. The most widely used technique to estimate the required joint density function is the Nataf model [5], which uses a multidimensional Gaussian distribution with correlation coefficients modified according to the non linear transformation linking the given marginal and the Gaussian densities. The corresponding samples of the correlated variables can then be generated by means of this approximate distribution.

The objective of this work was to develop a mathematical model capable of providing realistic simulations of vehicle crashes by accounting for uncertainty in the model input parameters. The stochastic crashworthiness model consists of ten acceptance criteria. The performance models for each acceptance criteria are defined using stochastic based software called NESSUS [1]. NESSUS is an advanced probabilistic analysis code developed by Southwest Research Institute. The LS-DYNA explicit dynamic finite element software is used to compute the time-dependent structural response of the vehicle due to frontal impact loads. Response quantities such as acceleration and displacement are used for the compartment intrusion performance measures and as input to the MADYMO occupant response program to compute occupant injury measures. The LS-DYNA finite element model and the MADYMO model of a 50th percentile male Hybrid III dummy are integrated with NESSUS to comprise the crashworthiness model. This integrated model provides an automatic and flexible uncertainty analysis procedure to allow the modeler the ability to simulate uncertainties in any LS-DYNA and MADYMO input variables. The modeler can easily select any relevant model responses to define a wide range of acceptance criteria measures. The problem setup, analysis and interpretation of results are all handled through the NESSUS graphical user interface.

STOCHASTIC MODEL

For the analysis reported herein, ten acceptance criteria are used to evaluate the crashworthiness of the vehicle-to-vehicle frontal offset impact model. These criteria include four occupant injury and six compartment intrusion measures as listed in table [1].

Uncertainty inputs to the model consist of 20 random variables. These random variables include parameters that define key energy absorbing components of the vehicles such as material properties for bumpers and rails, test environment uncertainties such as impact velocity and angle, and manufacturing variations in the form of rail and bumper installation parameters. Each of these random variables is characterized by a statistical distribution defined from manufacturing data, literature and/or expert opinion.

A probabilistic system reliability analysis is required to evaluate correctly the vehicle performance, especially for computing the probabilistic sensitivity factors at the system level for redesign analysis. For example, certain parameters such as stiffness/strength parameters can improve the reliability for compartment intrusion performance measures, but they may be detrimental to the crash pulse attenuation seen by the vehicle's occupant. The probabilistic system model correctly accounts for events with common variables (correlated events) and thus correctly identifies the important variables on the system level.

The probabilistic fault tree capability was used to develop the system reliability model. System reliability for this analysis is the probability of meeting the combination of all ten acceptance criterion. The model consists of the ten acceptance criteria defined by the RSM performance functions. The probabilistic fault tree analysis was also used to compute probabilistic sensitivity factors. These sensitivities are the derivatives of the probability of failure with respect to the random variable parameters (mean and standard deviation). The magnitude and direction of the sensitivities indicate the variables that contribute most to the reliability. For the original design it was determined that the yield stress of the small vehicle rail material contributes most to the reliability of the design.

To facilitate the probabilistic analysis and a reliability improvement study, a response surface model (RSM) is developed for each acceptance criteria performance model. The RSM approach aids in reducing the number of required LS-DYNA and MADYMO analyses and provides a fast running function that can be used for design trade off studies.

Component Probabilistic and Reliability Analysis

One of the key outputs of this approach is the cumulative distribution function (CDF) of system performance (e.g. stress or strain). The CDF is the probability that the performance value is less than or equal to a specified value. Traditional reliability analysis involves computing the probability that the stress (S) will exceed the strength (R), or $P[g<0]$ where

$$g = R - S$$

R and S may be complex models involving other random variables such as $R(X_i)$ and $S(Y_i)$. The g-function is formulated such that $g=g(X_i,Y_i)$ and thus correctly accounts for possible correlation between the stress and strength parts of the performance measure (i.e., common random variables in R and S). This approach provides a general formulation of the g-function that allows different analysis codes and analytical functions to be linked together in a hierarchical fashion. For example, a stress or strain from a finite element analysis can be used with a fatigue life equation or S-N curve to define the performance of a structure.

System Reliability Analysis

Most engineering structures have multiple conditions that are evaluated for acceptable performance. These events are defined by different acceptable conditions of a component or multiple component/conditions of a system. The nonperformance of one or a combination of events can lead to nonperformance of the system. System reliability considers acceptable performance of multiple components of a system and/or acceptable conditions of a component. System reliability in NESSUS is currently addressed using a probabilistic fault tree analysis (PFTA) method [4].

System reliability is defined through a fault tree by defining the bottom events and their combination with "AND" and "OR" gates. Each bottom event considers a single performance event (component reliability) and can be defined by a finite element model and performance function or as an analytical equation. An example of a fault tree for a three-event system is shown in Figure (1).

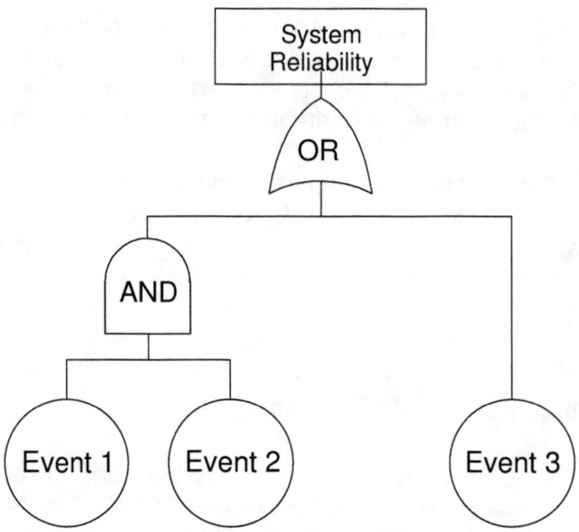

Figure 1. Example of a Fault Tree for System Reliability Analysis in NESSUS

Response Surface Methods

The response surface approach for probabilistic analysis can be used to replace a computationally intensive function evaluation with a fast-running polynomial expression. Several components are required for performing probabilistic analysis using a response surface as shown in Figure (2).

- Experimental design
- Model fitting/ANOVA
- Probabilistic Analysis (usually simulation)

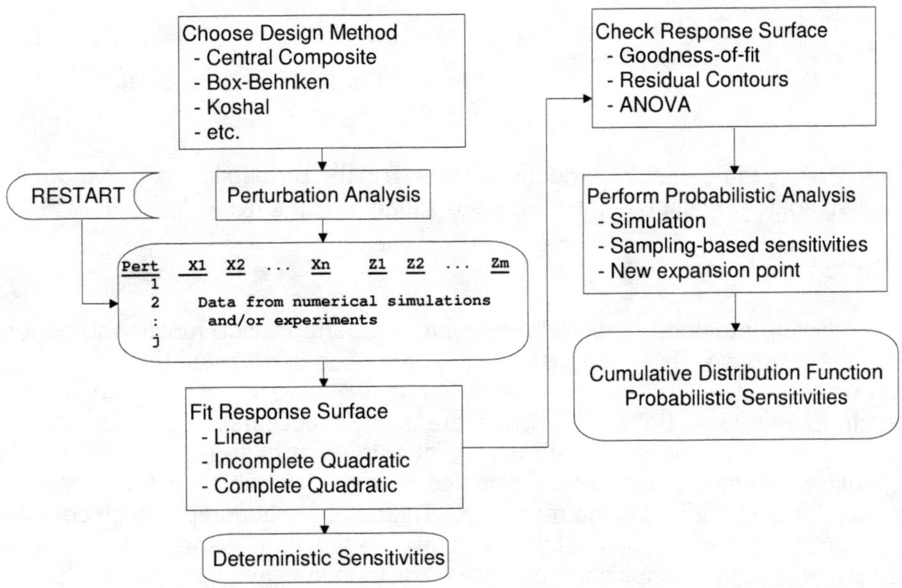

Figure 2. Probabilistic Analysis using a Response Surface Model

Response surface models (RSM) have some unique advantages for performing probabilistic analysis:

- Easy transfer of information between modelers
- Best suited for estimating mean and standard deviation
- Can incorporate test and numerical data
- Quickly generate accurate g-functions for system analysis
- Reduced number of function evaluations in some cases
- Fast running function that can be used for design tradeoff studies

While the RSM is a powerful tool for the probabilistic analyst, there are some disadvantages to its use:

- May not be accurate for small probabilities
- Difficult to assess error without rerunning the original function in the design region
- Typically not as efficient as AMV+

Probabilistic Sensitivity Factors

As part of the probabilistic analysis, the probabilistic sensitivity measures are computed in the form of the derivatives of the probability of failure with respect to the random variable parameters:

$$\frac{\partial p}{\partial \mu_i} \text{ and } \frac{\partial p}{\partial \sigma_i}$$

Where i is the particular random variable, p is the probability, and μ and σ are the mean and standard deviation of the ith random variable respectively.

Monte Carlo Simulation

Simulation methods repeatedly evaluate the performance function to generate a sample of the response. The response samples are used to estimate the mean, standard deviation or other statistics. For simulation methods, there is no requirement that the response function be well behaved or that there is any practical restriction on the number of random variables as there are with most probable point based methods. The Monte Carlo simulation method uses random samples from the probability distributions of each random variable and computes the response. The reliability is computed by counting the number of acceptable evaluations of the performance function. Other information can be obtained such as the empirical cumulative distribution function (CDF), probability distribution of the response, and moments of the response. The method is exact as the number of simulations approaches infinity. The major disadvantage to Monte Carlo simulation is that the analysis can be time consuming for complex deterministic models. Even for fast running models the analysis can require a large number of samples for large or small reliabilities as indicated using a sampling error equation

$$\%error = 100 \cdot \Phi^{-1}[1-\alpha/2]\sqrt{\frac{1-p}{k \cdot p}}$$

In this equation, $(1-\alpha)$ is the confidence level, p is the probability, and k is the required sample size. The equation states that there is a $(1-\alpha)$ chance that the %error in the estimated probability will be less than that given in the equation. This equation is valid for probabilities less than 0.5. If the probability is greater than 0.5 then 1- p is used to evaluate the error.

System Reliability Analysis for Independent Events

Defining the event as a desirable outcome, then the intersection of all events produces the system reliability as:

$$\text{Reliability} = P(E_1 \cap E_2 \cap E_3 ... \cap E_n)$$

where E_i is the occurance of the ith event. In many system reliability analyses, the correlation between events is ignored and the events are assumed to be independent If the events are assumed independent, then the system reliability is the product of the independent event reliabilities such as:

$$P(E_1 \cap E_2 \cap E_3 ... \cap E_n) = \prod_{i=1}^{n} P(E_i)$$

Deterministic Model

The crashworthiness model used in this analysis simulates vehicle-to-vehicle frontal offset impact. The vehicles include a small car and larger SUV style vehicle as shown in figure (3). The LS-DYNA finite element model used in this analysis was built by the National Crash Analysis Center (NCAC). It consists of over 250,000 nodes and 240,000 elements and was analyzed using LS-DYN version 960 on SGI and HP parallel platforms. The analysis time was approximately 30 CPU hours using 8 processors.

The LS-DYNA model is used to compute the structure responses of the vehicle such as accelerations and displacements. These response quantities are used for the compartment intrusion performance measures and as input to MADYMO to compute occupant injury measures. The LS-DYNA finite element model and the MADYMO model with a 50th percentile male Hybrid III dummy are integrated with NESSUS to comprise the crashworthiness model. This integration framework is shown in figure (4).

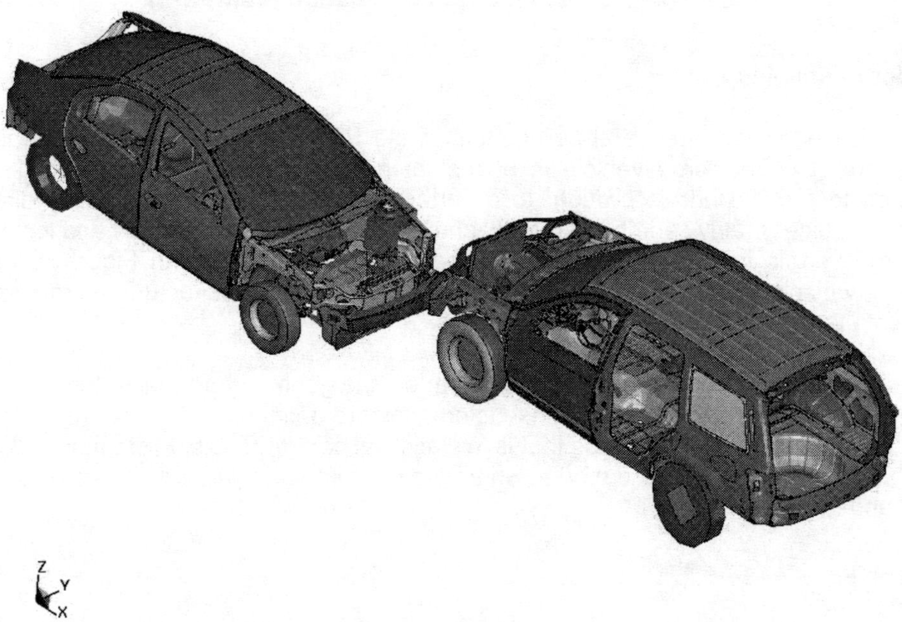

Figure 3. Vehicle-to-Vehicle Frontal Offset Crash Simulation Model.

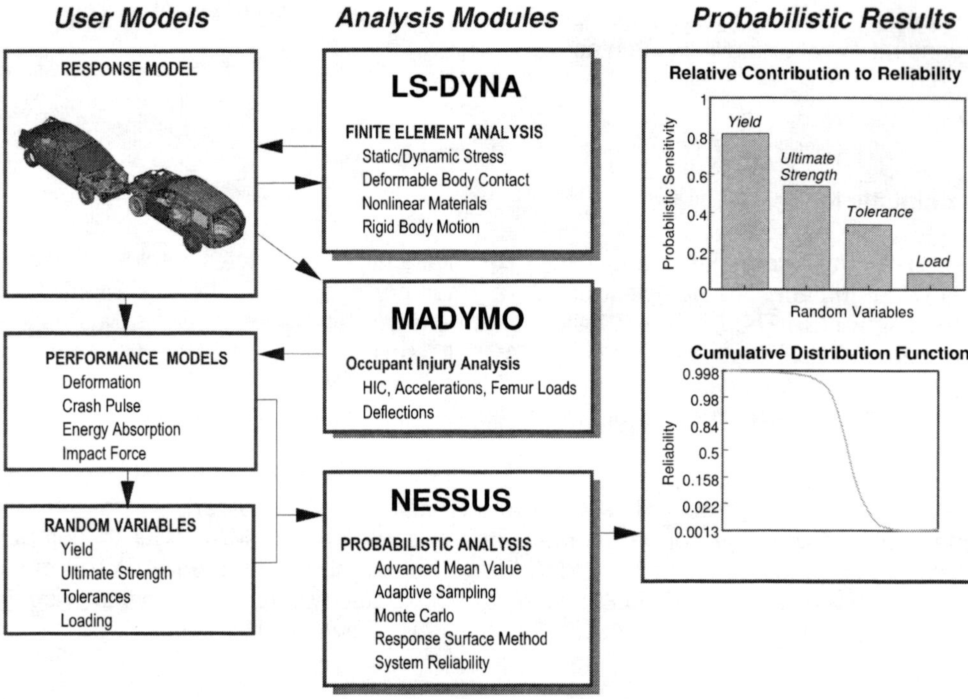

Figure 4. Stochastic Crash Simulation Framework

Random Variables

Uncertainties of several parameters in the FE model are defined for controlled test conditions. The large vehicle is at rest at a specified location. The small car is attached to a guide and is brought to a certain impact velocity. The velocity direction along the guide is fairly certain but the attachment of the vehicle to the rail and location of the large vehicle introduces uncertainty. The test setup is shown in Figure (5). Other random variables include uncertainties that arise in the manufacturing process and inherent uncertainty of material characteristics.

The distributions for parameters that affect the geometry are based on design/manufacturing tolerances. These tolerances are defined to ±3 standard deviations from the mean value for a process that is in statistical control. Table 1 lists the random variable definitions used in the model, where the random variables are separated for the small and larger vehicles.

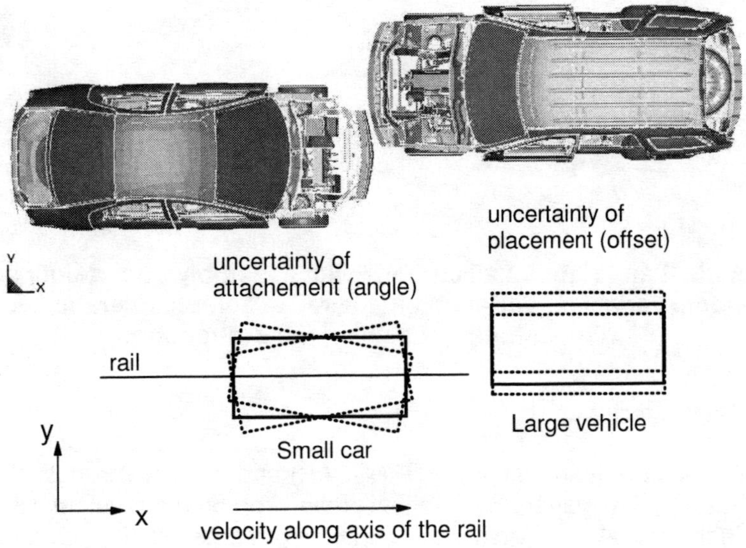

Figure 5. Model Setup

SMALL VEHICLE RANDOM VARIABLES

Rail Characteristics

Vehicle rails are critical components to the energy absorbing characteristics of the vehicle. Uncertainties in the rail material properties such as yield strength, elastic modulus, metal thickness, and weld connections are considered to be the main parameters driving the energy absorption. The yield strength of high strength steel for rail material (YIELDCAR) is modeled as a random variable as well as the elastic modulus (EMODCAR). Rail thickness (RAILTCAR) is also modeled as a random variable. Variations in sheet metal parts occur from the manufacturing and forming processes. The left and right rails are generally manufactured using the same lot of material and at the same time. Therefore the left and right rail thickness variables are modeled as a single random variable (fully correlated). Uncertainty also exists in the attachment of the rails during manufacturing. The right (RTOLCARR) and left rail assembly tolerances, (RTOLCARL) are modeled as random variables. The uncertainty is modeled as shown in Figure (6) where each rail may vary in the Z-direction of the model. Left and right small car rails with assembly uncertainty are modeled as two random variables. Variations in the rail assembly tolerance are modeled by shifting the rail in the Z-direction The uncertainty of each rail placement is independent and based on design tolerances.

Figure 6. Left and right small car rails with assembly uncertainty are modeled as two random variables. Variations in the rail assembly tolerance are modeled by shifting the rail in the Z-direction.

Bumper Characteristics

The bumper foam density (FDENSCAR) and ultimate strength (FOAMUCAR) are modeled as random variables. The uncertainty of bumper height (BUMBER) is also included in the model. Uncertainty in bumper height arises during the attachment process in manufacturing, it is modeled by moving the bumper nodal coordinates in the Z direction The bumper is depicted in Figure (7a).

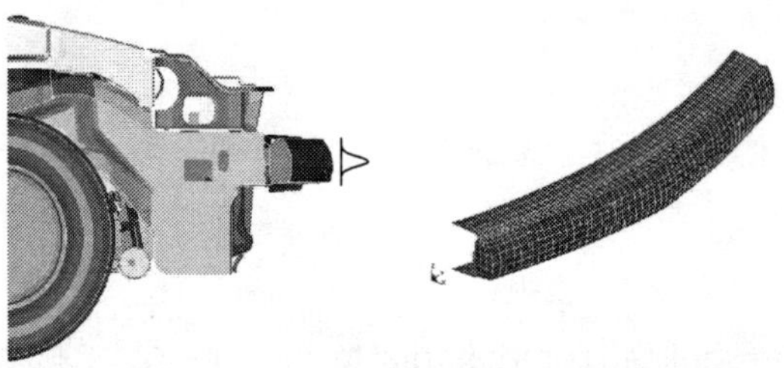

Figure 7a. Bumper height assembly tolerance

Vehicle Height

Another uncertainty is the vertical height of the vehicle which arises from tire inflation (TIRE). This uncertainty is modeled by changing the Z coordinate of the small vehicle with respect to the larger vehicle is shown in Figure (7b). This uncertainty is based on design tolerances and test procedure requirements (±0.5 inches).

Table 1. Random Variables

	Description	Name	Mean	Standard Deviation	Units	Dist.
Small Vehicle						
1	HSS yield strength	YIELDCAR	275	15.1250	N/mm^2	Lognormal
2	Elastic modulus	EMODCAR	210000	10500.0000	N/mm^2	Normal
3	Foam density	FDENSCAR	8.00E-10	1.2000E-10	Ton/mm^3	Normal
4	Foam ultimate strength	FOAMUCAR	25	3.7500	N/mm^2	Normal
5	Right rail assembly tolerance	RTOLCARR	0[1]	4.23333	mm	Normal
6	Left rail assembly tolerance	RTOLCARL	0[1]	4.23333	mm	Normal
7	Impact velocity	IMPVEL	15643	156.43	mm/sec	Normal
8	Impact angle	IMPANGL	0	3.3333	deg	Normal
9	Bumper height	BUMPER	0[1]	4.23333	mm	Normal
10	Tire height	TIRE	0[1]	4.23333	mm	Normal
11	Front rail weld (effective plastic strain)	FRWSTIFF	1.33E-03	1.33E-04	-	Lognormal
12	Front left rail weld (effective plastic strain)	LRWSTIFF	1.33E-03	1.33E-04	-	Lognormal
13	Front right rail weld (effective plastic strain)	RRWSTIFF	1.33E-03	1.33E-04	-	Lognormal
14	Rail thickness	RAILTCAR	1.9	0.0342	mm	Normal
Large Vehicle						
15	Foam density	FDENSTRK	4.95E-11	7.4184E-12	Ton/mm^3	Normal
16	Foam ultimate strength	FOAMUTRK	210	31.5000	N/mm^2	Normal
17	Right rail assembly tolerance	RTOLTRKR	0[1]	4.23333	mm	Normal
18	Left rail assembly tolerance	RTOLTRKL	0[1]	4.23333	mm	Normal
19	Rail thickness	RAILTTRK	1.73	0.03114	mm	Normal
20	Overlap	OVERLAP	0[1]	0	mm	Normal

[1]mean value is relative to the nominal installation position

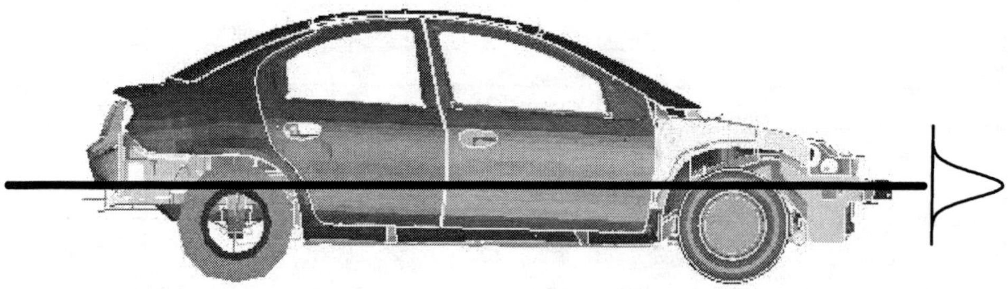

Figure 7b. Tire height uncertainty is modeled by moving the small car nodal coordinates in the Z direction (distribution is not to scale)

The rails are attached to the vehicle using spot welds. The number, size and quality of the welds affect the vehicles energy absorbing characteristics. The weld stiffness for the front rail (FRWSTIFF), front left rail (LRWSTIFF), and front right rail (RRWSTIFF) were considered as independent random variables. The same distribution (lognormal) and coefficient of variation as the weld diameters were used for the weld stiffness.

Parameter Uncertainties

Several parameters cannot be controlled completely in a crash event and can be expected to lead to uncertainty in the crashworthiness performance. The impact velocity (IMPVEL) is modeled as random variable and the uncertainty is defined by the test requirements (±3%). The impact angle (IMPANGLE) is another uncertain parameter in the test and also defined by the test requirements (±10º). The impact angle variation is simulated about the small car center of gravity as shown in Figure (8).

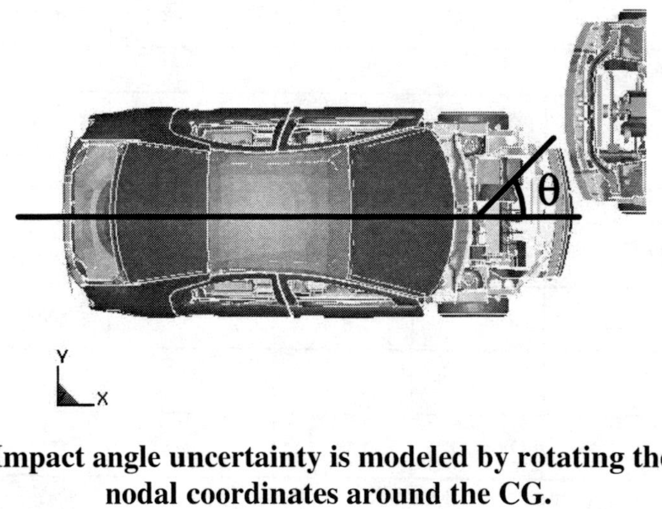

Figure 8. Impact angle uncertainty is modeled by rotating the small car nodal coordinates around the CG.

Large Vehicle Random Variables

All acceptance criteria for this model are for the small vehicle. However, uncertainties of several critical energy absorbing parameters in the large vehicle were also considered to determine their contribution to the event. These variables include the bumper foam density (FDENSTRK) and ultimate strength (FOAMUTRK), the left and right rail assembly tolerances (RTOLTRKL, RTOLTRKR), and the rail thickness. The distribution definitions for these parameters were obtained from supplier's data.

Acceptance Criteria

Ten acceptance criteria were used to evaluate the crashworthiness of the vehicle-to-vehicle frontal offset impact test. All acceptance criteria are evaluated for the small vehicle. The criteria include four occupant injury and six compartment intrusion measures as listed in table 2. Injury measures are based on a 50th percentile male Hybrid III dummy in the driver seat modeled with MADYMO.

Table 2. Acceptance Criteria for Crashworthiness Model

	Acceptance Criteria	**Assessment Values**
Occupant Injury		
1	Head Injury Criteria (HIC36)	1000 (36 msec clip)[1]
2	Chest acceleration (g's)	60g[1,2]
3	Chest deflection	64 (mm)[2]
4	Femur axial load	10000 (N)[1]
Passenger Compartment Intrusion		
5	Footrest intrusion	150 mm
6	Average of left/center/right toe pan deflection	150 mm
7	Brake pedal location	150 mm
8	Average of left/right instrument panel (IP) deflection	50 mm
9	Door aperture closure (measured at beltline on struck side)	50 mm
10	Engine displacement relative to left B-pillar	200 mm

[1]FMVSS 208
[2]"Frontal Offset Crashworthiness Evaluation: Guidelines for Rating Structural Performance," Insurance Institute for Highway Safety, April 2002.

The performance of each acceptance criteria is defined by LS-DYNA or MADYMO computed quantities. Information describing each of these performance measures is listed in Table 3.

Table 3. LS-DYNA and MADYMO Model Entities that Define Acceptance Criteria

Acceptance Criteria	MADYMO Response	LS-DYNA Response
HIC	HIC	Negative of X acceleration (crash pulse)
Chest Acceleration	Chest g's	Negative of X acceleration (crash pulse)
Chest Deflection	Chest Deflection	Negative of X acceleration (crash pulse)
Femur Axial Load	Right Femur Compression	Negative of X acceleration (crash pulse)
Footrest Intrusion	N/A	X-displacement B-pillar X-displacement footrest
Toe Pan Deflection	N/A	X-displacement B-pillar X-displacement right toe pan X-displacement center toe pan X-displacement left toe pan
Brake Location	N/A	X-displacement B-pillar X-displacement Brake location
Instrument Panel Deflection	N/A	X-displacement B-pillar X-displacement left instrument panel X-displacement right instrument panel
Door Aperture Closure	N/A	X-displacement B-pillar X-displacement forward door frame X-displacement rear door frame
Engine Displacement	N/A	X-displacement B-pillar X-displacement Engine

Probabilistic Crashworthiness Model

Probabilistic analysis algorithms require response solutions for specific combinations of the random variable values. These perturbed responses created by changing the variables in the response models are used for random sampling or may be used for finite difference approximations of gradients used in fast probability integration methods.

The overall probabilistic crashworthiness model is defined by NESSUS, LS-DYNA and MADYMO as in Figure (2). The combination of these models allow an automatic procedure to relate changes in the random variables to the LS-DYNA and MADYMO input and easily select model responses such as displacements used to define the acceptance criteria measures. The model allows the variation of any of the defined random variables.

A large portion of the development of this model involved defining how a change of each random variable is mapped to the LS-DYNA input. The mapping was defined and verified for each of the twenty variables. As an example, when the rail tolerance is perturbed, the resulting change in the LS-DYNA finite element model is automatically made by NESSUS based on the mapping definition (modifying the Z-coordinates of the nodes for the rail), the analysis rerun, and finally the response of interest extracted from the LS-DYNA results file. This procedure is fully automated in the developed model for all defined random variables.

Analysis Procedure

Solutions to nonlinear dynamic analyses are sensitive to changes in the geometry and material property parameters. In some cases these changes can be severe enough to cause numerical problems that terminate the analysis and, in others, the solution time step can become so small that the analysis time becomes prohibitive. The first attempt of computing the reliability for the footrest intrusion acceptance criteria using fast probability integration methods resulted in incomplete LS-DYNA solutions due to model sensitivity to these combined random variables. The cause of error of these cases was not investigated in detail. One suspected cause is the alignment of contact surfaces when geometry variables are changed. Remedying these cases would require portions of the finite element model to be regenerated resulting in a slightly different model and substantial development time. It was deemed more expedient to use an approach that would lend itself to less rigid requirements on the variable perturbations for this analysis.

A response surface approach was used to create functions defining each acceptance criteria measure because of the difficulty in automatically computing perturbed solutions required for fast probability integration methods. The response surface approach was also used since it would reduce the number of required LS-DYNA analyses and facilitates the redesign analysis. The response surface approach will not be as accurate as fast probability integration methods if the derived function does not represent the response over the design region. For this demonstration, the response surface approach was used to represent the general behavior of the acceptance criteria measures and also allowed the procedure to be demonstrated within a reasonable amount of computational effort. The probabilistic results computed using response surfaces could be checked and refined as needed.

Response Surface Model Generation

Response surface models (RSM) were created for each of the ten acceptance criteria measures using a linear and quadratic term for each random variable (mixed terms were not obtained due to the computational expenses required to compute them). The approach was to obtain LS-DYNA and MADYMO solutions for each random variable perturbed ±2 standard deviations from the mean value. For several random variables only one perturbed solution was obtained. For these cases, only the linear term is included in the regression model.

Large vehicle rail thickness (RAILTTRK) was perturbed ±11.111 standard deviations. This change resulted from an improved definition of the standard deviation for this variable after the LS-DYNA analyses were complete. The perturbed responses for all acceptance criteria measures indicated that the change in the response was small and fairly linear and thus sufficient for the response surface models. Table 4 lists the perturbations used to create the response surface models for each acceptance criteria.

Table 4. Perturbations used to Create Response Surface Models.

#	rtolcarr	impvel	bumper	tire	railttrk	impan	yieldcar	emodcar	fdenscar	foamuc	foamtrk	rtoltrkr	rtoltrkl	frwstiff	rrwstif	railtcar
1	2															2
2	2															2
3		2														
4		-2														
5			-2													
6			-2													
7				2												
8				2												
9					11.111											
10					-11.111											
11						2										
12						-2										
13							0.331									
14							-2									
15								1.714								
16								-2								
17									-2							
18									-2							
19										2						
20										-2						
21											2					
22											-2					
23												2				
24												-2				
25													2			
26													-2			
27														2		
28														-2		
29															2	
30															-2	
31	2															2
32	2															2

Finally, it was noted that several runs had been completed where the rail weld stiffness was not included in the LS-DYNA model. It was found that including the weld stiffness did not affect the mean value solution of the acceptance measures by a significant amount. Therefore, rather than rerunning this cases at a considerable computational expense, the regression models were corrected using a first order approximation (a ratio of each acceptance measure computed with and without the weld stiffness). For probabilistic analysis, the important aspect of the response surface is accurate derivatives in the response surface function. This first order correction provides reasonable accuracy of the derivatives for the response surface models.

A stochastic model was defined that included each of the acceptance criteria responses. Each response surface model was generated by selecting the appropriate response. This model works directly with the LS-DYNA and MADYMO analysis programs.

System Models

A system reliability analysis is critical to correct evaluation of the vehicle performance especially for identifying the probabilistic sensitivity factors at the system level for redesign analysis. It is expected that certain parameters such as stiffness/strength can improve reliability for compartment intrusion performance measures but may be detrimental to the crash pulse attenuation as measured at the vehicle occupant. The system model correctly accounts for events with common variables (correlated events) and thus correctly identifies the important variables on the system level. System reliability for the model was computed using a probabilistic fault tree analysis (PFTA) method shown in figure (9). The performance of each event is modeled using a response surface model as described in Section. 0. Monte Carlo simulation is used to compute the system reliability since each performance function is defined by a fast running response surface model. The system reliability is also computed using the individual event reliabilities based on the independent event assumption described earlier.

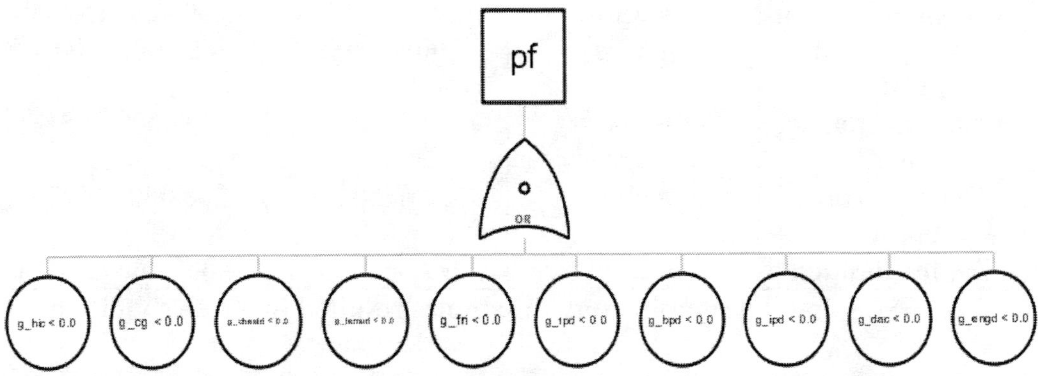

Figure 9. Probabilistic Fault Tree of the Mathematical Model

Baseline Results:

The system reliability was computed using the Monte Carlo simulation method in NESSUS with 100,000 samples. The computed system reliability for the original design is 23% (1.1% error, 95% confidence). The system reliability assuming independent events is 18%, which is a conservative estimate in this case. These results are summarized in Table 5.

Table 5. System Reliability of the Original Design

System Reliability Approach	Reliability
NESSUS PFTA	23% (1.1% error)
Independent Events	18%

A Monte Carlo analysis was performed for each event and the results are shown in table (6).The error in the probability for a 95% confidence interval is also included in the table. The femur axial load acceptance criteria event has the lowest reliability followed by the HIC event and the door aperture closure event. All other acceptance criteria have relatively high reliability.

Table 6. Event Reliability for Original Design

Acceptance Criteria		Reliability	
Description	NESSUS Variable	Original Design	Samples (Error)
HIC	g_hic	57.7910%	100,000 (0.7%)
Chest acceleration	g_cg	92.2970%	100,000 (2.1%)
Chest deflection	g_chestd	99.9752%	2,000,000 (8.8%)
Femur axial load	g_femurl	46.4020%	100,000 (0.6%)
Footrest intrusion	g_fri	99.9623%	2,000,000 (7.1%)
Toepan deflection	g_tpd	100.0000%	2,000,000 (N/A)[1]
Brake pedal location	g_bpd	100.0000%	2,000,000 (N/A) [1]
Instrument panel def.	g_ipd	99.6870%	2,000,000 (2.5%)
Door aperture closure	g_dac	72.6750%	200,000 (0.7%)
Engine location	g_engd	99.6000%	2,000,000 (2.2%)

[1]NESSUS limit on samples and all samples resulted in a safe condition.

The probabilistic sensitivity factors computed are shown in Figure 10. These sensitivity factors define the parameters that can be most effectively modified to improve the reliability. From the figure, the nominal value of the yield strength of the small vehicle rail material can be most influential in increasing the reliability. The positive sign indicates that the reliability will increase if the yield strength is increased. The sign convention is opposite when the reliability is greater than 50%. The next most influential parameters are the nominal value of the front weld stiffness of the small vehicle rails followed by standard deviation (or scatter) of the small car yield strength. Several variables show negligible importance including the tire height and the foam properties for both vehicles.

The nominal value of the right car rail installation tolerance also shows some importance. This parameter is a center installation point and would not be considered for a design change without changes to the vehicle structure.

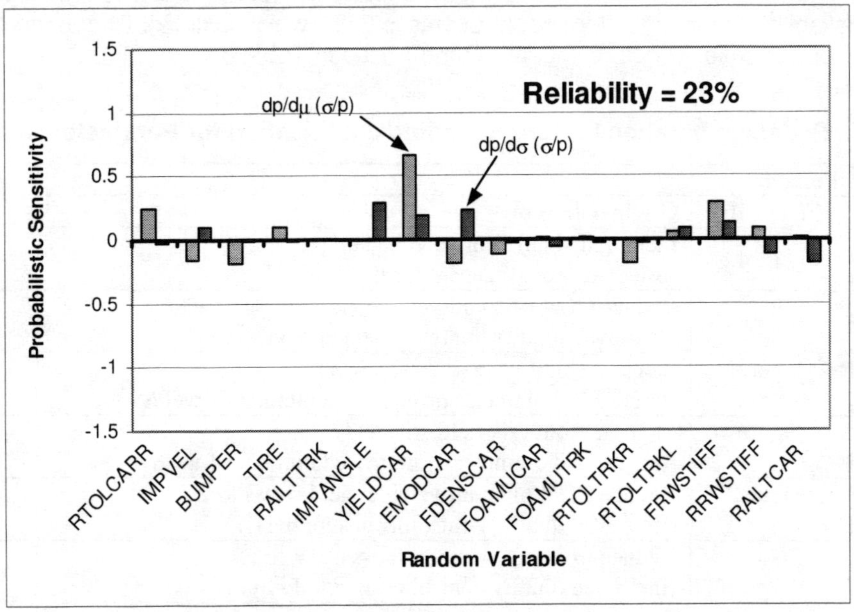

Figure 10. Probabilistic Sensitivity Factors for the Original Design.

Analysis Iterations:

The objective of the analysis is to provide a recommendation to improve the reliability of the small vehicle in a vehicle-to-vehicle frontal offset impact. The approach used is to rely on the probabilistic sensitivity factors to identify the dominant parameters (random variable mean and standard deviation) that will improve system reliability. Several restrictions were placed on the parameters that could be modified:

- Change only small car parameters.
- Parameters can only be changed by ±10% with exception of the weld stiffness. Changing the number of welds or weld diameter can increase/decrease the weld stiffness.

- Limit extrapolating the response surface models.

The following steps are performed on each parameters identified from the probabilistic sensitivity factors as candidates to improve the reliability:

- Modify the appropriated mean or standard deviation in the problem statement,
- Evaluate the reliability using Monte Carlo simulation,
- Evaluate the probabilistic sensitivity factors to find additional parameter candidates for reliability improvement if warranted, and
- Evaluate the acceptance criteria for event reliabilities once a model with acceptable system reliability is reached to identify the dominant events.

Table 7 lists the iteration history along with the justification for modifying each parameter. The table also lists the reference to the probabilistic sensitivity measures used to select the parameter to improve the reliability. The reliability improvements are shown in figure 11. The system reliability for the final design is 86% (1.5% error). The system reliability assuming independent events is 85%, which is a slightly conservative estimate for this case. These results are summarized in Table 8.

Table 7. Redesign Iteration History Including Justification for Parameter Changes.

Iter.	Reference		Reliability
0	Figure 10	Original design	23%
1	Figure 12	Increased small vehicle yield stress. (use a different material for the rails)	28%
2	Figure	Reduce COV of small vehicle yield stress to 4% (improved quality control from supplier)	30%
3	Figure	Increase small vehicle weld stiffness (larger weld diameter or increased number of welds)	40%
4	Figure	Increase front rail weld stiffness (larger weld diameter or increased number of welds)	48%
5	Figure	Reduce COV of small vehicle yield stress to 2% (improved quality control from supplier)	53%
6	Figure	Reduce front rail weld stiffness COV to 5%. (improved quality control of assembly line)	62%
7	Figure	Reduce rail thickness COV from 1.8% to 1% (improved quality control from supplier)	66%
8	Figure	Reduce front rail weld stiffness COV to 2%. (improved quality control of assembly line)	70%
9	Figure	Reduce COV of small vehicle yield stress to 1% (improved quality control from supplier)	79%
10	Figure	Reduce COV of small vehicle foam material to 10% (improved quality control from supplier)	83%
11	Figure	Tighten installation tolerances for bumper and rail (improved quality control of assembly line)	86%

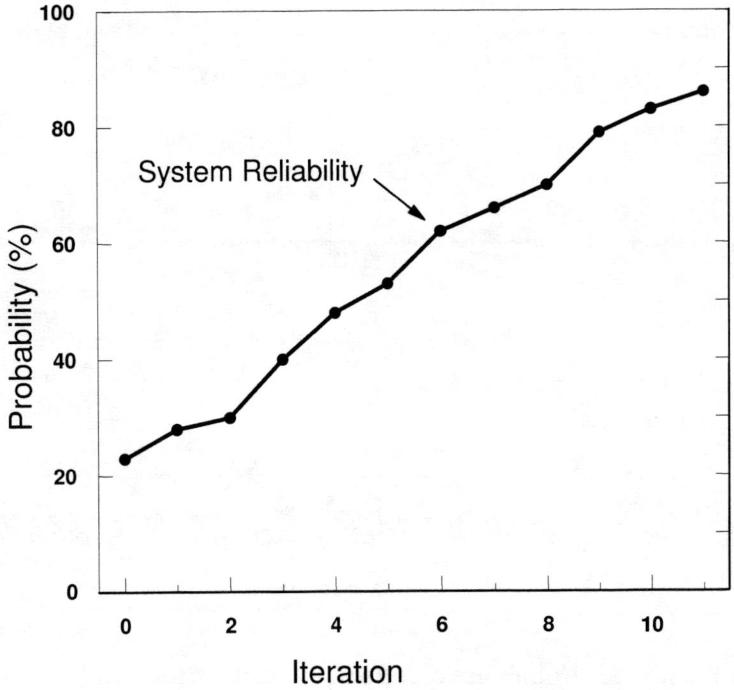

Figure 11. System Reliability Improvement

Table 8. System Reliability of the Final Design

System Reliability Approach	Reliability
NESSUS PFTA	86% (1.5% error)
Independent Events	85%

The parameters of the final iteration are listed Table 9. Finally, the event reliabilities were evaluated for the final configuration to identify the dominant acceptance criteria as listed in table 10. The dominant event for the original design was the femur axial load acceptance criteria. The femur axial load also shows the lowest reliability for the final design but increased from a reliability of 46% to 93%.

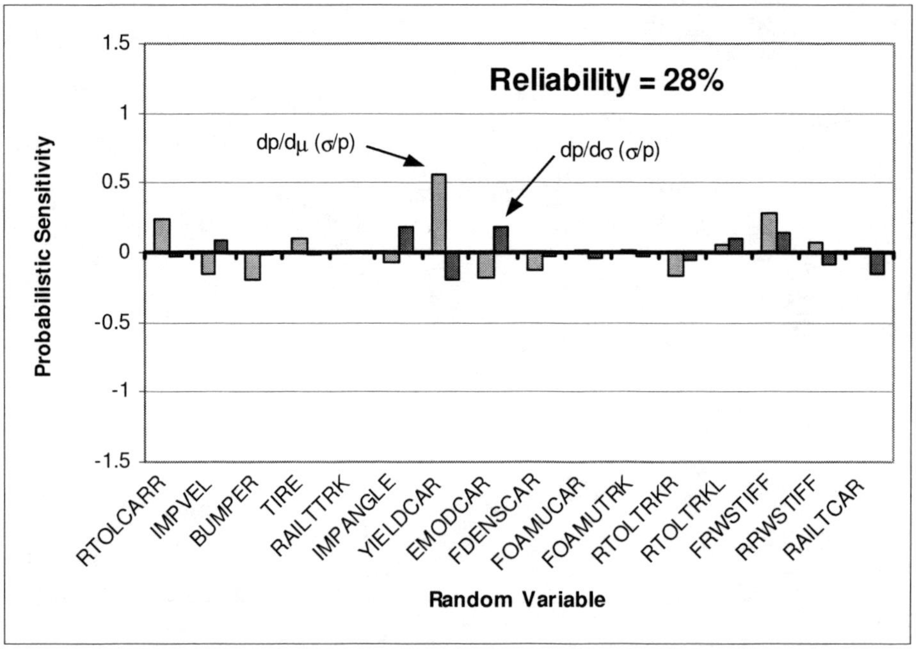

Figure 12. Probabilistic Sensitivity Factors for Iteration 1.

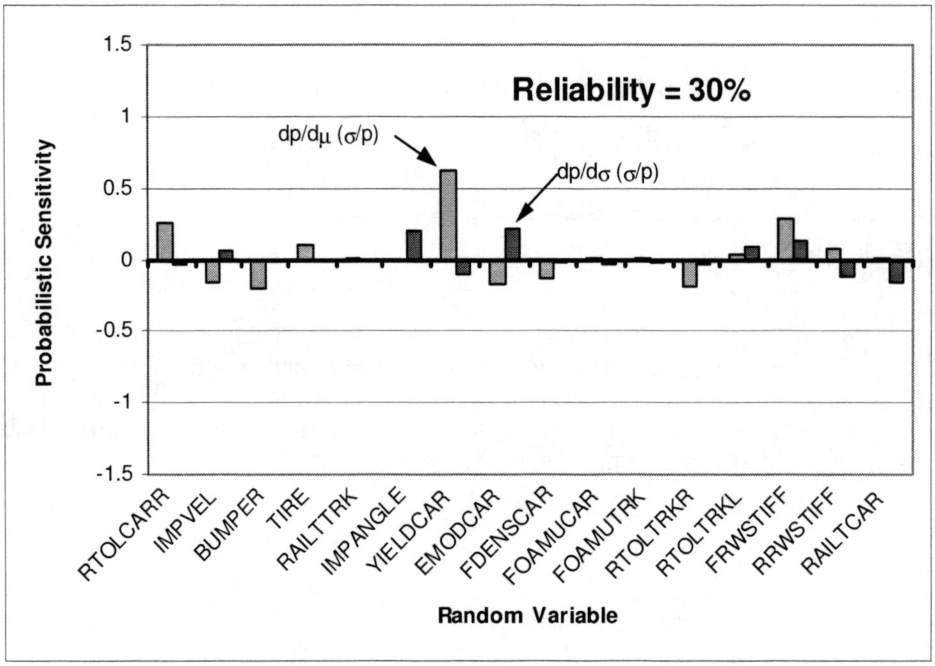

Figure 13. Probabilistic Sensitivity Factors for Iteration 2.

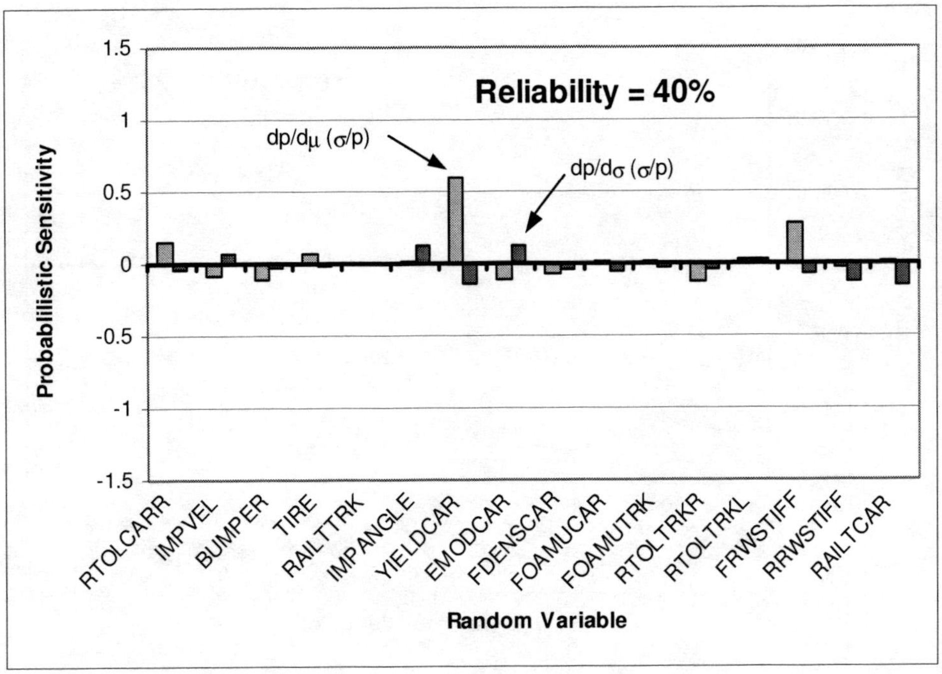

Figure 14. Probabilistic Sensitivity Factors for Iteration 3.

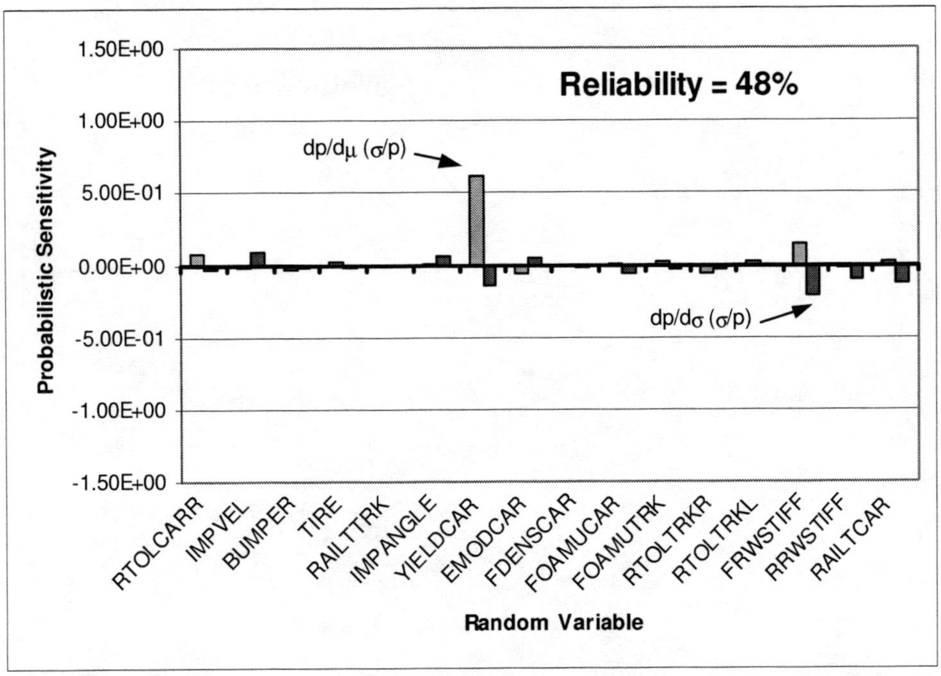

Figure 15. Probabilistic Sensitivity Factors for Iteration 4.

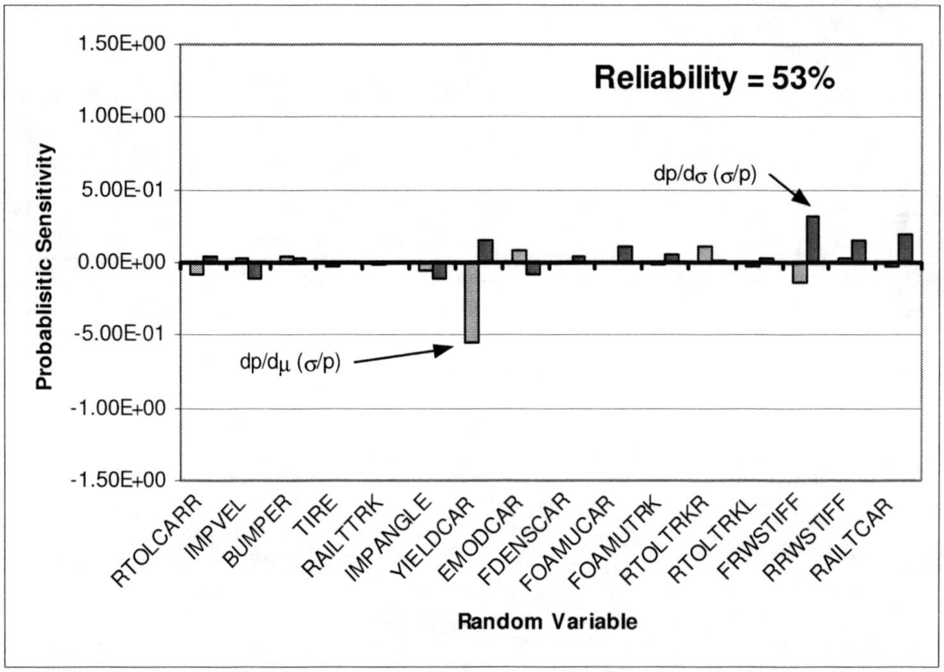

Figure 16. Probabilistic Sensitivity Factors for Iteration 5.

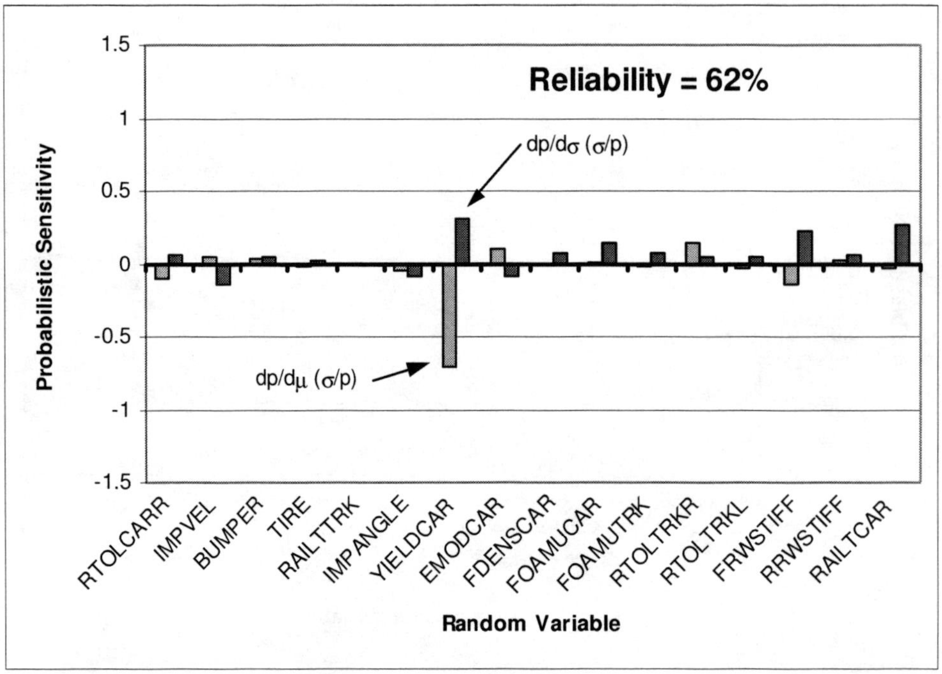

Figure 17. Probabilistic Sensitivity Factors for Iteration 6.

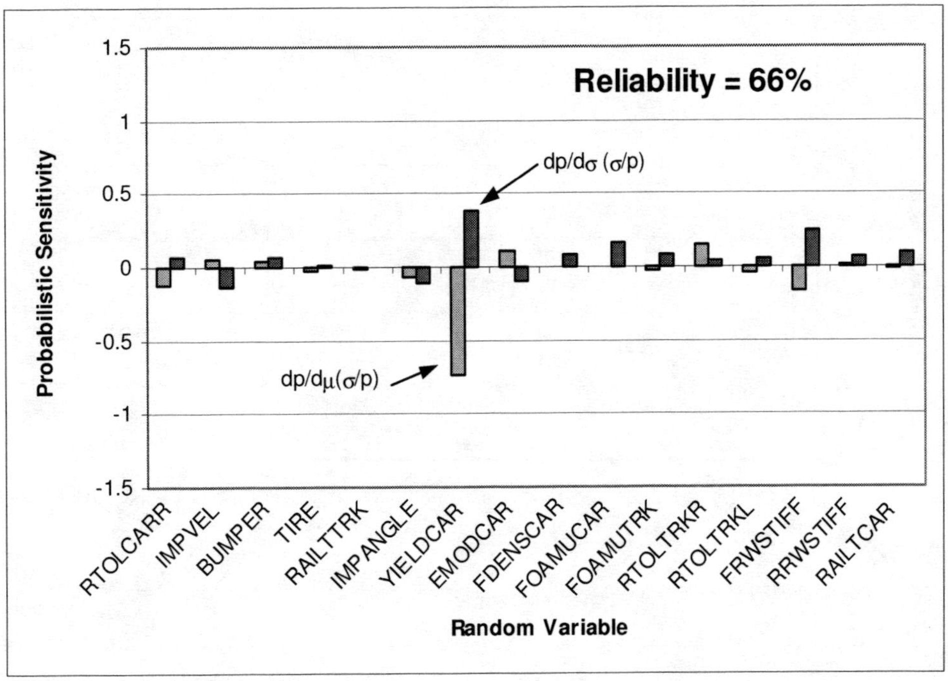

Figure 18. Probabilistic Sensitivity Factors for Iteration 7

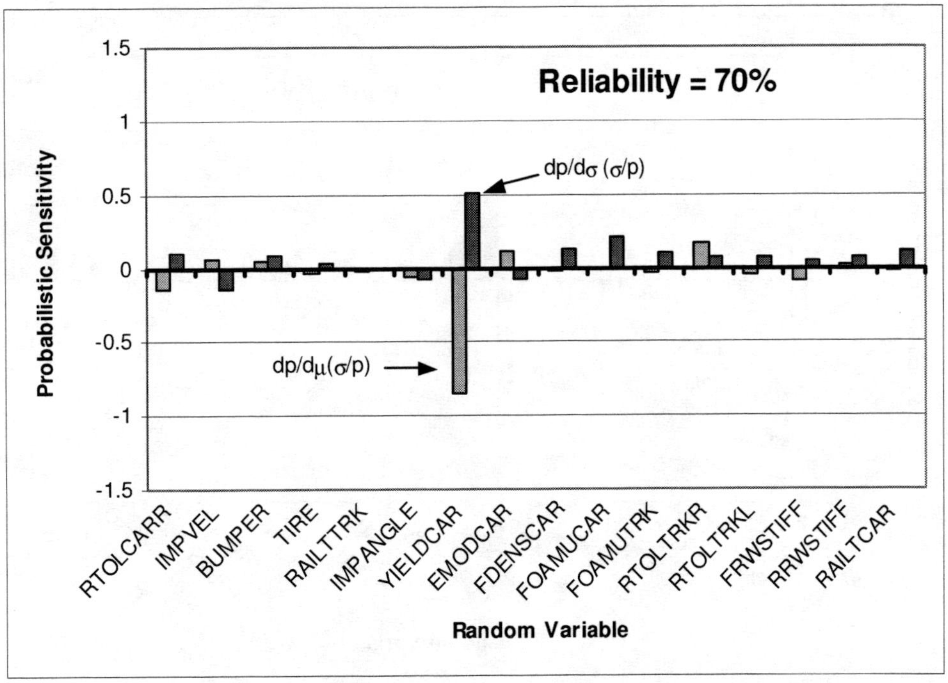

Figure 19. Probabilistic Sensitivity Factors for Iteration 8.

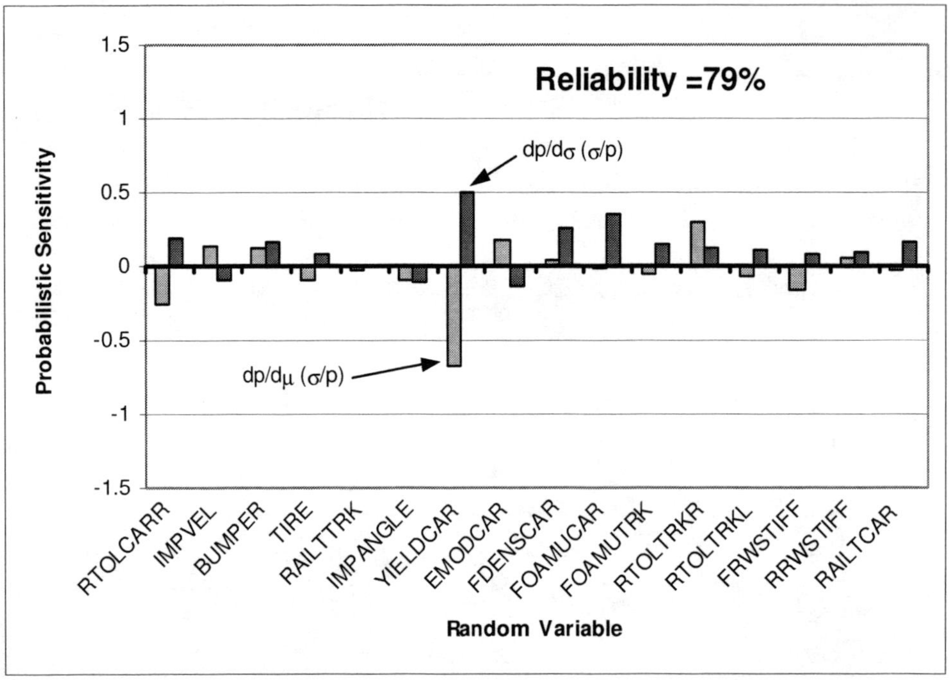

Figure 20. Probabilistic Sensitivity Factors for Iteration 9.

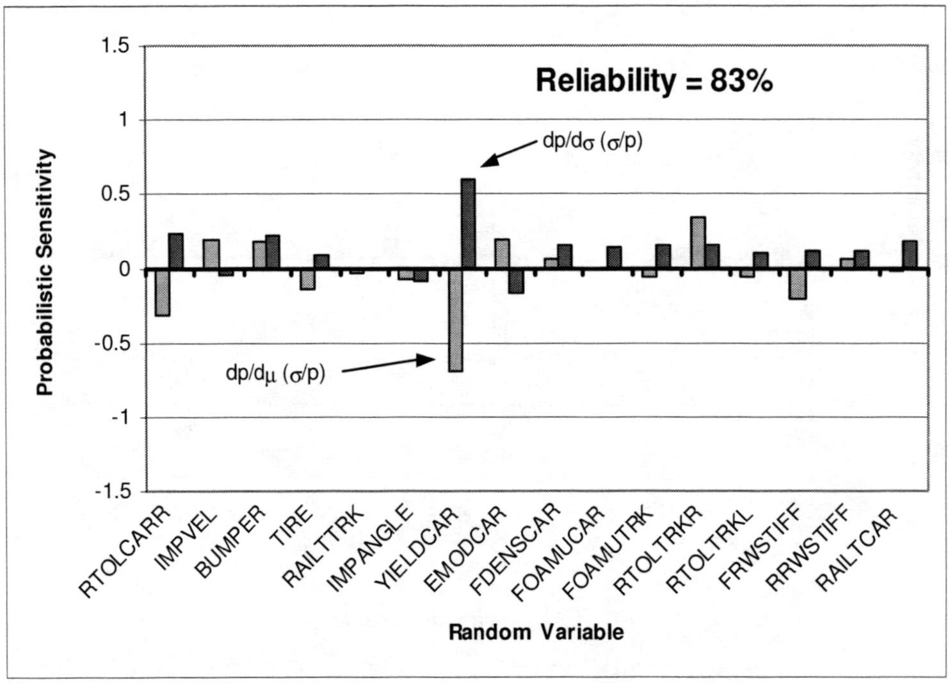

Figure 21. Probabilistic Sensitivity Factors for Iteration 10.

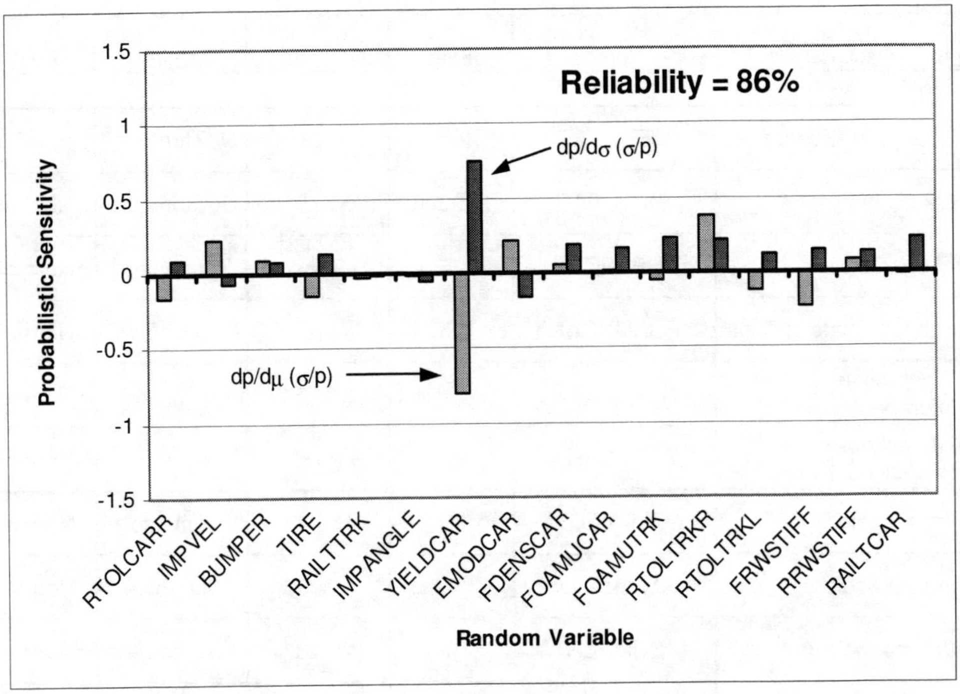

Figure 22. Probabilistic Sensitivity Factors for Iteration 11 (Final Design).

Discussions and Conclusions:

The stochastic crashworthiness model consists of ten acceptance criteria. The performance models for each acceptance criteria are defined in a stochastic framework. The LS-DYNA model is used to compute the structure response of the vehicle such as accelerations and displacements. These response quantities are used for the compartment intrusion performance measures and as input to the MADYMO program to compute occupant injury assessment measures. The LS-DYNA finite element model and the MADYMO model of a 50th percentile male Hybrid III dummy are integrated with a stochastic formulation to comprise the crashworthiness model. This model allows an automatic procedure to relate changes in the random variables to the LS-DYNA and MADYMO input and easily select model responses such as displacements used to define the acceptance criteria measures. The model allows the variation of any of the defined random variables.

Ten acceptance criteria were used to evaluate the crashworthiness of the vehicle-to-vehicle frontal offset impact. All acceptance criteria are evaluated for the small vehicle. The criteria include four occupant injury and six compartment intrusion measures.

Table 9. Random Variables (FINAL ITERATION)

	Description	Name	Mean	Standard Deviation	Units	Dist.
Small Car						
1	HSS yield strength	YIELDCAR	280	2.8	Newton/mm^2	Lognormal
2	Elastic modulus	EMODCAR	210000	10500.0000	Newton/mm^2	Normal
3	Foam density	FDENSCAR	8.00E-10	8.00E-11	Tonne/mm^3	Normal
4	Foam ultimate strength	FOAMUCAR	25	2.5	Newton/mm^2	Normal
5	Right rail assembly tolerance	RTOLCARR	0	2.1167	mm	Normal
6	Left rail assembly tolerance	RTOLCARL	0	4.23333	mm	Normal
7	Impact velocity	IMPVEL	15643	156.43	mm/sec	Normal
8	Impact angle	IMPANGL	0	3.3333	degrees	Normal
9	Bumper height	BUMPER	0	2.1167	mm	Normal
10	Tire height	TIRE	0	4.23333	mm	Normal
11	Front rail weld (effective plastic strain)	FRWSTIFF	1.6E-3	3.2E-5	-	Lognormal
12	Front left rail weld (effective plastic strain)	LRWSTIFF	1.33E-03	1.33E-04	-	Lognormal
13	Front right rail weld (effective plastic strain)	RRWSTIFF	1.463E-3	7.315E-05	-	Lognormal
14	Rail thickness	RAILTCAR	1.9	0.019	mm	Normal
Large Vehicle						
15	Foam density	FDENSTRK	4.95E-11	7.4184E-12	Tonne/mm^3	Normal
16	Foam ultimate strength	FOAMUTRK	210	31.5000	Newton/mm^2	Normal
17	Right rail assembly tolerance	RTOLTRKR	0	4.23333	mm	Normal
18	Left rail assembly tolerance	RTOLTRKL	0	4.23333	mm	Normal
19	Rail thickness	RAILTTRK	1.73	0.03114	mm	Normal
20	Overlap	OVERLAP	0	0	mm	Normal

Table 10. Reliability for each Acceptance Criteria (Original and Final Designs).

Acceptance Criteria		Reliability		Samples (Error)
Description	NESSUS Variable	Original Design	FinalDesign	Final Design
HIC	g_hic	57.7910%	94.0120%	100,000 (2.4%)
Chest acceleration	g_cg	92.2970%	98.8240%	200,000 (4.0%)
Chest deflection	g_chestd	99.9752%	99.9999%	2,000,000 (196%)[1]
Femur axial load	g_femurl	46.4020%	92.9330%	100,000 (2.2%)
Footrest intrusion	g_fri	99.9623%	100.0000%	2,000,000 (N/A)[2]
Toepan deflection	g_tpd	100.0000%	100.0000%	2,000,000 (N/A)[2]
Brake pedal location	g_bpd	100.0000%	100.0000%	2,000,000 (N/A)[2]
Instrument panel def.	g_ipd	99.6870%	99.9719%	2,000,000 (8.3%)
Door aperture closure	g_dac	72.6750%	98.7460%	200,000 (3.9%)
Engine location	g_engd	99.6000%	99.9997%	2,000,000 (88%)

[1]NESSUS limit on samples. The reliability is ±error*(1-Reliability) resulting in the prediction well within an order of magnitude of (1-Reliability).
[2]NESSUS limit on samples and all samples resulted in a safe condition.

The model consists of 20 random variables. These random variables include parameters that define key energy absorbing components of the vehicles such as material properties for bumpers and rails, test environment uncertainties such as impact velocity and angle, and manufacturing variations in the form of rail and bumper installation. Each of these random variables was characterized by a statistical distribution defined from manufacturing data, literature and/or expert opinion.

To facilitate the probabilistic analysis and the redesign analyses, a response surface model (RSM) was developed for each acceptance criteria performance model. The RSM approach aided in reducing the number of required LS-DYNA analyses and provided a fast running function that could be used for the design study tradeoff analysis.
The reliability for each acceptance criteria was computed using NESSUS to identify the dominant acceptance criteria of the original design. The femur axial load acceptance criteria event was found to have the lowest reliability (46%) followed by the HIC event (58%) and the door aperture closure event (73%). A system reliability analysis was used to include the contribution of all acceptance criteria to correctly quantify the vehicle reliability and identify important parameters.

The probabilistic fault tree analysis (PFTA) capability was used to develop a system reliability model. The model consists of the ten acceptance criteria defined by the RSM performance functions. The system reliability was 23% for the original design. The probabilistic fault tree analysis was also used to computes probabilistic sensitivity factors. These sensitivities are the derivatives of the probability of failure with respect to the random variable parameters (mean and standard deviation). The magnitude and direction of the sensitivities indicate the variables that contribute most to the reliability. For the original design it was determined that the yield stress of the small vehicle rail material contributes most to the reliability of the design.

A redesign approach was developed and performed using the probabilistic sensitivity factors. These sensitivities identify the most effective parameters to improve the reliability. A redesign using 11 iterations was performed resulting in a system reliability of 86% (original design reliability was 23%). The design changes include increasing the rail material yield strength and reducing the variation, reducing the variation of the bumper and rail installation tolerances, and increasing the rail weld stiffness and reducing its variation, and reducing the variation of the foam properties of the small vehicle. It should be noted that evaluating a new design requires very little computational expense since the performance functions are defined by response surface model.

A system reliability analysis is critical to correct evaluation of the vehicle performance especially for evaluating the probabilistic sensitivity factors at the system level for redesign analysis. Certain parameters such as stiffness/strength parameters can improve reliability for compartment intrusion performance measures but may be detrimental to the crash pulse attenuated to the vehicle occupant. The system model correctly accounts for events with common variables (correlated events) and thus correctly identifies the important variables on the system level.

A final evaluation of the reliability for each acceptance criteria in the new design was performed to identify the dominant criteria. The reliability of the original design was dominated by the femur load acceptance criteria (46%). The femur axial load also shows the lowest reliability for the final design but increased from a reliability of 46% to 93%.

Several limitations were identified during development and analysis of this model and warrant further investigation. Some of these limitations affect the performance models for each acceptance criteria:

1. Obtain LS-DYNA solutions to include the left rail stiffness as a random variable. The variation for this parameter did not yield complete LS-DYNA solutions and thus were not included in the performance models.
2. Verify solutions for response surface fit of important variables in the design regions. This step will require additional LS-DYNA analyses to verify the accuracy of the response surface in the vicinity of the reliability region for each event. Recreate the response surface if it is not accurate.
3. Obtain perturbed solutions for other important variables where a linear approximation was used.

These three items should be pursued prior to using these results for decision analysis.

The following recommendations are made to improve the stochastic crashworthiness model to aid in future assessments:

1. Develop a meta-model capability to automate the response surface generation procedure used in this effort.
2. Include additional meta-models such as Kriging to enhance the predictive models by more accurately modeling the response of highly nonlinear applications encountered in impact analysis.
3. Develop an algorithm to automatically change the perturbation size when a response evaluation fails. This capability would assist in using fast probability integration methods such as the advanced mean value (AMV) method for stochastic crashworthiness analysis. The AMV method has been shown to be more accurate and efficient than the response surface approach when perturbed solutions can be obtained.

REFERENCES:

1. Southwest Research Institute, "Probabilistic Structural Analysis Methods (PSAM) for Select Space Propulsion System Components," Final Report NASA Contract NAS3-24389, NASA Lewis Research Center, Cleveland, Ohio, 1995.

2. Riha, D.S, Thacker, B.H., Enright, M.P., Huyse, L. and Fitch, S.H.K., "Recent Advances of the NESSUS Probabilistic Analysis Software for Engineering Applications," Proc. AIAA/ASME/ASCE/AHS/ASC 42nd Structures, Structural Dynamics, and Materials (SDM) Conf., AIAA-2002-1268, Denver, CO, 22-25 April 2002.

3. Wu, Y. -T., Wirsching, P. H., "A New Algorithm for Structural Reliability Estimation," J. Engineering Mechanics, vol. 113, 1987, pp. 1319-1336

4. Torng, T.Y., Wu, Y.-T., and Millwater, H.R., "Structural System Reliability Calculation Using a Probabilistic Fault Tree Analysis Method," Proceedings 33rd SDM Conf., Part 2, pp. 603-613, April 1992.

5. Hassan , J.E., Nusholtz,G. and Forrest, M. D., "A stochastic approach for occupant crash simulation", SAE 2000-01-1597

6. LS-DYNA is a general-purpose transient dynamic finite element program. LS-DYNA is a registered trademark of Livermore Software Technology Corporation.

7. MADYMO combines multibody and finite element techniques to for occupant safety analysis. MADYO is a registered trademark of TNO.

Proceedings of IMECE'03
2003 ASME International Mechanical Engineering Congress
Washington, D.C., November 15–21, 2003

IMECE2003-55057

MODELING TIRE BLOW-OUT IN ROADSIDE HARDWARE SIMULATIONS USING LS-DYNA

Fabio Orengo,
Graduate Research Assistant,

Malcolm H. Ray, Ph. D.,
Associate Professor
Worcester Polytechnic Institute, 100 Institute Road, Worcester, MA 01605

Chuck A. Plaxico, Ph.D.
Senior Research Engineer
Battelle Memorial Institute, 505 King Avenue Columbus, Ohio 43201

ABSTRACT

Often when vehicles interact with roadside hardware like guardrails, bridge rails and curbs, the interaction between the roadside hardware and the tire causes the tire to lose its air seal and "blow-out". Once the seal between the rim and rubber tire is lost, the tire deflates. The behavior of the deflated tire is much different than the behavior of an inflated tire such that when this behavior is observed in real world crashes or in full-scale crash tests, the vehicle kinematics are strongly coupled to the behavior of the deflated tire. Accounting for this behavior in LS-DYNA models is crucial in many types of roadside hardware simulations since the forces generated by the deflated tire often introduce instability into the vehicle that can cause rollover or spinout. This paper will present a method for accounting for tire deflation during LS-DYNA simulations and will present examples of the use of this type of improved model.

INTRODUCTION

Finite element models of pneumatic tires have been developed by others to investigate stresses and deformations in order to evaluate the performance of commercial tires [1]. The objective of such analyses has lead, necessarily, to very sophisticated models that take into account ply orientation, ply overlapping and mechanical properties of different layers of the composite structure of the tire.

This approach is necessary and practical if the objective of the study is the design and analysis of the tire itself ignoring the many other components of the vehicle. On the other hand, this approach becomes computationally inefficient for analyses involving large-scale overall response of the vehicle during impact with other structures, such as roadside hardware, where the tires are only one aspect of many intricate parts that must be modeled.

In most cases, the simulation of vehicle impact into roadside hardware has been approached using relatively simple models, where the particular aspects of the tire structure were not reproduced. Typically, the tire has been modeled with isotropic membrane elements (thin shells) which may be effective when the dynamics of the tire are negligible. A more realistic model is needed when the tire interacts significantly with other elements of the roadside environment, for instance, in those conditions where the tire directly plays an important role in the vehicle kinematics and stability.

BASICS OF TIRE MECHANICS

A basic understanding of pneumatic tire mechanics is required in order to develop a simplified model that produces realistic behavior of the tire under the types of loading to which the tire will be subjected. The most common tire construction technology in use today is the radial construction, thus our attention will be focused on this type of pneumatic tires.

The tire plays an important role in the vehicle's dynamics since it must absorb road irregularities while providing a vibration-free motion on smooth roads. There are two conflicting requirements: flexibility is needed to absorb road bumps, while the need for a vibration-free ride requires a rigid round wheel. This conflict is further aggravated by the fact that a pneumatic tire is a doubly curved surface, geometrically a non-developable surface that must be deformed to provide a contact area on flat pavement.

Clark [1] identifies some of the basic requirements for the structure of a pneumatic tire:

- No appreciable change of size upon inflation,
- Ability to envelop obstacles without sustaining damage,
- Ability of part of this envelope to deform from a surface of double curvature to a plane surface, and
- Enough rigidity to develop substantial forces in any direction.

Inflated shell structures, having a surface of double curvature made solely of a single isotropic material, have always failed to meet the requirements stated above.

A pneumatic tire has several structural elements, the most important of which is the carcass, made up of many flexible filaments that have a high Young's modulus (e.g., synthetic polymer fibers or fine hard drawn steel) embedded in a matrix of low modulus polymeric material like natural or synthetic rubber. In order to build a demountable tire carcass that can be fitted on a wheel rim, the layers of high modulus filaments are wrapped around the bead coils made of strands of hard drawn steel wire.

Tension in the carcass cords is set up by inflation pressure that is resisted by the tension developed in the steel wire bead coils. The rubber that encases the steel wire coils is pressed against the rim by the inflation pressure; this reaction between rim flanges and tire bead enables traction and braking forces to be transmitted by friction.

Demounting of a conventional tire is made possible by a well in the rim base, also known as drop center. The shape of the rim is usually tapered with a five degree angle to the flange to pretension the tire bead upon inflation.

The third important structural element of the tire is the tread, the only part of the tire that contacts the road surface. The tread must be made of abrasion-resistant materials and must provide high friction with the road in order to transfer the traction, cornering and braking forces. In belted radial tires, the belt is essential to the functioning of the tire. Without it the radial ply of the carcass can become unstable.

MECHANISM OF LOAD CARRYING: INFINITE FLEXIBLE MEMBRANE

Consider an inflated toroidal structure made of an infinite flexible membrane such as a toy balloon mounted on a rigid rim. The air pressure puts the membrane in tension as it assumes a shape determined by equilibrium and compatibility conditions. The membrane tension is resisted by a reaction at the edge of the rim. If a flat plate is pressed against the membrane while the structure is supported by the rim, a reaction will develop between the membrane and the plate, proportional to the product of the area of contact and inflation pressure.

The question arises as to how the reaction is transferred to the rim. Considering that the air pressure is constant everywhere on the surface of the rim and that the surface over which the pressure acts is also constant, the resultant of the air pressure on the rim is zero whether the tire is loaded or not.

These simple considerations highlight the fact that the mechanism of load reaction from tire-rim interface is substantially different from the mechanism between the tire and the road surface.

A closer look at the geometry of the deformed shape of the tire shows that the curvature of the membrane side walls is greater in the region between the plate and the rim. Hence, because of the greater curvature, the membrane stresses in the region are lower than elsewhere on the membrane wall, since the tension is equal to the product of the inside pressure and the radius of curvature. Further, the local distortion of the membrane causes an increase of the angle between the side wall membrane and the rim cross section; this causes the decrease of the radially outward component of the tension in the membrane. Figures 1 and 2 illustrate this behavior.

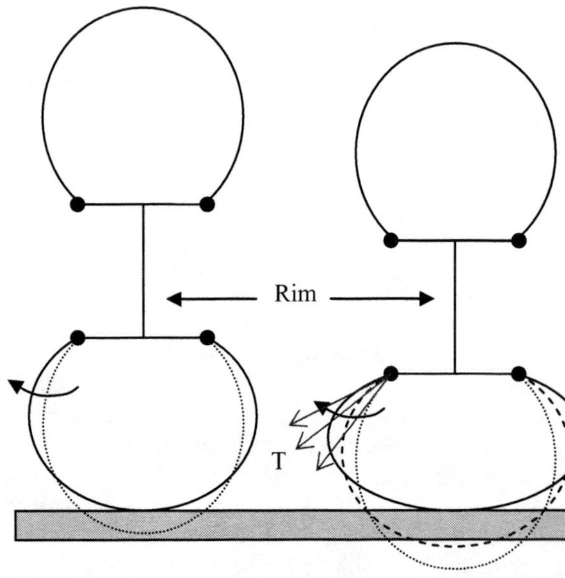

Figure 1: The deflection of the sidewalls reduces the radially outward components of the membrane.

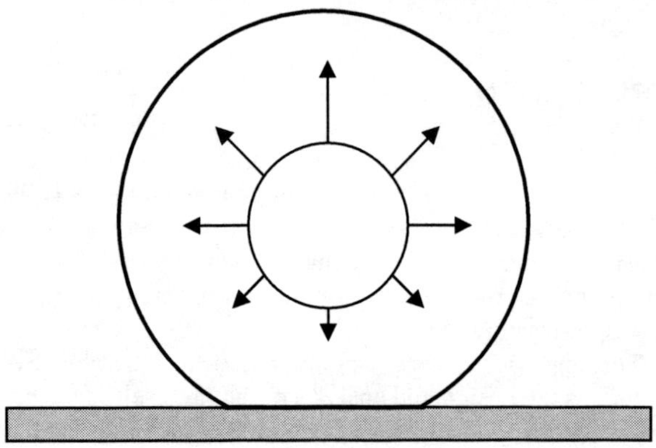

Figure 2: Distribution of the radially outward component of the membrane tension along the rim edge.

The net effect due to the reduced tension in the sidewalls and the reduction of the radially outward component in the region of the contact is that the rim "hangs" in the tension of the undeflected region. This system of load transmission is analogous to the one present in a spoked wheel where the hub hangs by the spokes in tension on the top of the wheel. Figure 3 illustrates the difference between a solid tire and a pneumatic tire in the load transfer mechanism.

Usually a tire has a considerable bending stiffness that supports a significant part of the load. Bending stresses will be present in the tire's walls especially where the radius of curvature decreases. High bending stresses will be present around the edge of the contact surface with the ground. This will lead to a non-uniform contact pressure distribution between the tire and the ground.

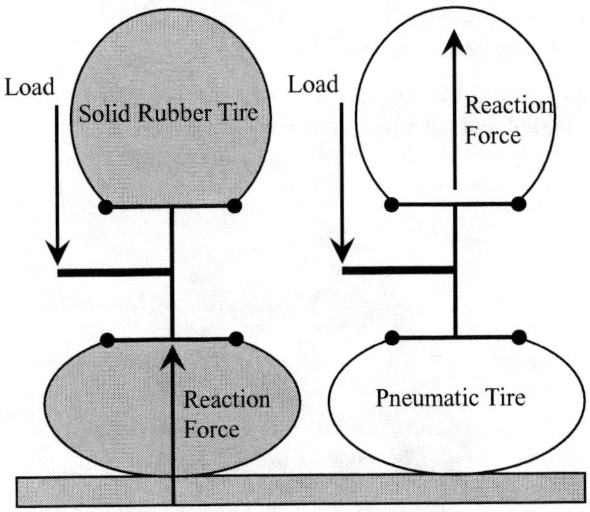

Figure 3: Load carrying mechanism comparison between a solid and an inflated tire.

MECHANISM OF LOAD CARRYING: TIRE STRUCTURE

When a tire is inflated, the cords in the reinforced rubber are in tension and the tire assumes a shape compatible with equilibrium and compatibility conditions. When the tire is loaded against a planar surface the rubber tread is compressed and the tire assumes a flat shape in the contact patch; if there were no tread, the tire inside surface would be flat. In this case, the tensions in the cords in the flat area are no longer related to the inflation pressure since the curvature is zero. This means that the tension in the cords is primarily transmitted from the adjacent free wall of the tire and locally modified around the edges due to the bending stresses.

In a commercial tire the bending stiffness of the different plies and the changing thickness of the rubber in the cross section cause the actual contact pressure with the roadway to be greater than the actual inflation pressure and to vary in the x and y directions in the contact patch. Consequently, the reaction load of the tire is greater than the product of the inflation pressure and the area of contact.

Clark [1] points out that those processes of load carrying seem to be linearly additive. The load transfer mechanism can be modeled as two different mechanisms acting in parallel. The first is analogous to the behavior of the infinite flexible membrane, where in the bottom part the tension in the cords is reduced and the local deformation causes a reduction of the radially outward component of the tension.

The second mechanism involves the tread. The tread can be thought of as a uniform belt around the whole periphery of the tire. This belt is tensioned under the inflation pressure, since it is curved except in the area where the pressure forces are transferred to the ground by compression without producing any resultant pressure on the belt (i.e., belt in this area has zero curvature). As a consequence, the absence of reaction forces in this area of the belt causes a force on the top sector of the tire to be unbalanced as shown in Figure 4. This resultant force pulls the bead coils against the base of the wheel rim, thus transmitting an upward force to the rim. Both mechanisms lead to the same type of load transfer mechanism with the rim suspended because of the tension in the shoulders of the undeformed upper part.

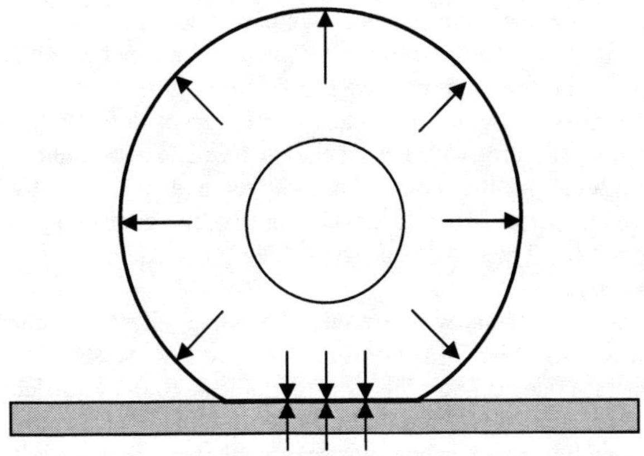

Figure 4: Load transfer mechanism from ground to wheel by effects of inflation.

ISOTROPIC SHELL TORUS MODELING TECHNIQUE

A tire model composed of isotropic shells with uniform thickness is an attractive modeling approach since it is simple and computationally efficient. Unfortunately, this modeling technique leads to very unrealistic tire behavior especially in cases where the tire undergoes large deformations. A simple isotropic shell model will perform adequately only if the inflation pressure is very high and the thicknesses of the side walls, as well as the thickness of the tread, are unrealistically thin. In such a simple model the bending stiffness becomes negligible and the tire works like an inflated balloon as described previously. Often the shells have to be so thin that the tire starts expanding, thereby increasing the external dimensions.

When the pressure is reduced and a more realistic thickness of the tire sidewall and tread is modeled, the elastic response of a double curvature surface becomes important.

Under small loads the tire model will behave linearly as it is loaded against the ground and its response will be somewhat overly stiff. As the load increases, the shell surface will begin reversing itself inside the tire assuming a lower deformation energy state. A simple finite element model of a loaded hemispherical shell as shown in Figure 5 illustrates this behavior, which is common for elastic panels with double curvature. Figure 5b shows the typical behavior of isotropic double curvature shell surfaces for large deformations. In the footprint area the deformed surface changes its concavity assuming a low energy deformation state. The deformed shape reduces the net footprint area. This behavior is not consistent with the real tire deformation.

LS-DYNA supports membrane elements that have no out-of-plane bending stiffness. The use of this type of element solves the problem of dimensional stability of the tire under inflation pressure, but, since it has no out-of-plane stiffness, the membrane is computationally unstable, especially where the curvature of the surface is small. Furthermore, when the membrane element undergoes large deformations, unrealistic pleats form where the flat geometry creates a possible hinge line. Since membrane element formulation does not account for bending, these large bending deformations are not associated with any deformation energy. The folding of the elements also causes mesh tangling that leads to permanent deformation of the tire (Figure 6). This behavior will cause numerical instabilities in the contact algorithms and will eventually terminate the run. To reduce the number of these tangled pleats, a very refined mesh is required, making this modeling approach impractical.

Shell elements with a reduced bending stiffness could be used to avoid the formation of hinge lines. Many static finite elements codes support this type of element. To our knowledge, however, LS-DYNA does not include this type of element in its element library. One possible solution is to superimpose two sets of elements: one layer of membrane elements and coincident layer of shell elements with both layers of elements sharing the same nodes. This will allow the analyst to impose different properties governing the membrane behavior of the model and the bending behavior. Of course this solution will result in twice the number of elements but it could lead to a better performing model than one using only shell or membrane elements.

Some preliminary analyses with this type of superimposed element model proved somewhat successful. However, unrealistic deformations of the tire in the area of contact (e.g., the formation of an "inverted hump" in the tire tread in the region of contact with the ground surface) suggested that the model needed further improvement in order to accommodate the type of loading that the tire model would be subjected to in non-tracking impacts.

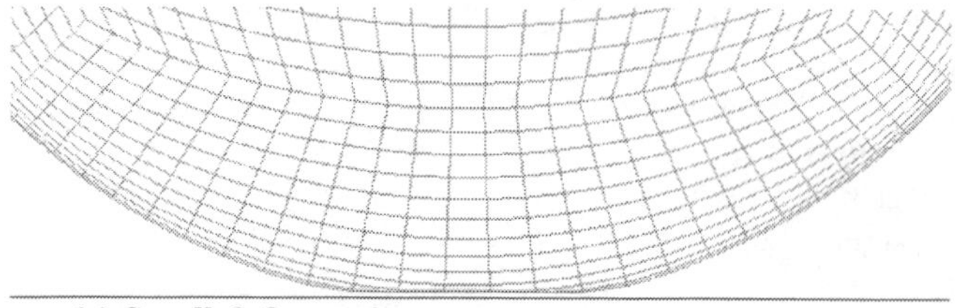

(a) Small deformations

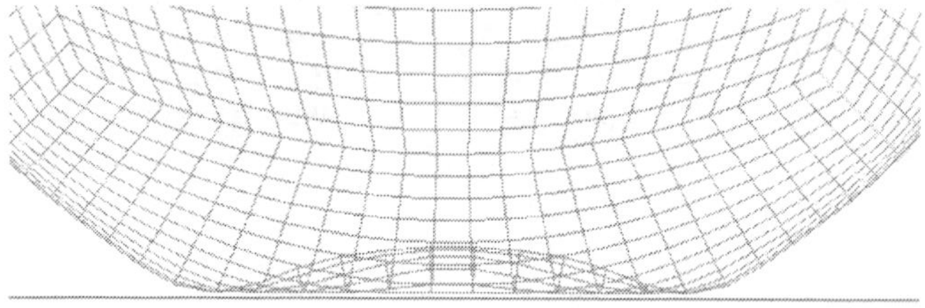

(b) Large deformations

Figure 5: Double curvature shell surface (Hemisphere) compressed against a planar surface.

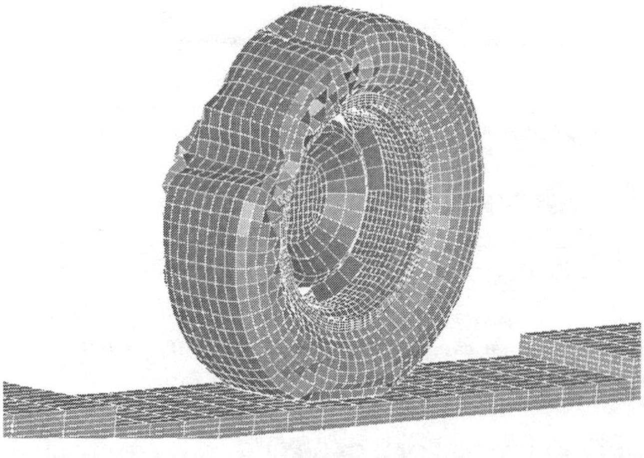

Figure 6: Membrane tire - permanent deformation due to mesh tangling.

ANISOTROPIC MODELLING TECHNIQUIE

The critical parts of the tire structure that affect the overall tire mechanics tire must be incorporated in the model.
These are:

- Bead coils,
- Radial fibers,
- Rubber sidewall,
- Under belt radial fibers,
- Steel belt, and
- Tread.

In addition to the structural elements, some non-structural elements were also included in the model. For example, a membrane between the two bead coils was introduced to describe a closed volume for defining an inflating air pressure. The rim also plays an important role since its geometry is the contact surface for the bead coils.

Bead Coils

The bead coils are generally made of hard drawn steel cable. This structural element can be effectively modeled using two rings of beam elements. In order to best transmit contact forces between the tire bead coils and the wheel rim, the bead coils were modeled using two rings of shell elements with elastic properties and a relatively high Young's modulus (100 MPa).

Radial Reinforcement

The radial reinforcement in the shoulders was modeled using beam elements with properties of steel equivalent to a distributed layer of radial steel fibers. The same approach was used to model the under-belt layers of radial plies. The element formulation adopted was the Belytschko-Schwer beam element.

Figure 7 shows the model of the radial reinforcements and of the bead coils.

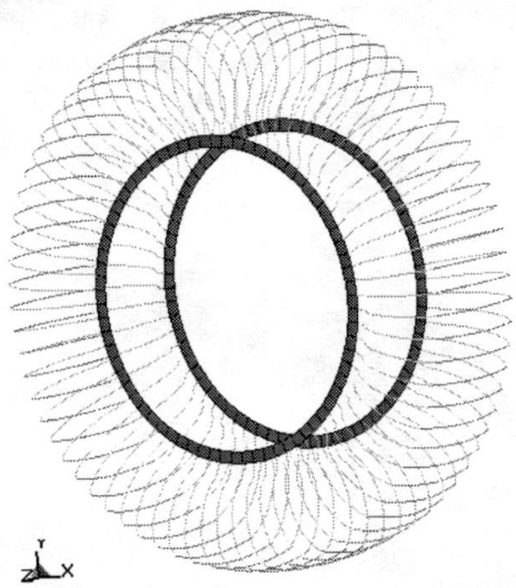

Figure 7: Finite element model of the radial reinforcement fibers and of the bead coils.

Shoulder Rubber Matrix

All the radial beams were interconnected with a set of Belytschko-Tsay shell elements with warping stiffness option turned on in LS-DYNA with the properties of rubber. The material properties were found in literature [1] and represent common values used for the shoulder rubber in tire manufacturing (E=4.5 MPa, υ=.49, ρ=1.1 kg/m^3). This element formulation proved to be very efficient in this complex application where high deformations, element warping, and a rotating reference frame provide a critical test for element stability.

Steel Belt

The steel belts were modeled with Belytschko-Tsay shell elements with uniform thickness and with properties of steel.

Tread

The geometry of the tread plays an important role in the performance and stability behavior of the model. Its primary role is to uniformly distribute the pressure load from the contact with the roadway to the belt and to the under-belt carcass reinforcing fibers to prevent the formation of an "inverted hump" that could result in local implosion of the tire periphery in the contact area. This requires the geometry of the external surface that undergoes contact with the roadway to be different from the belt geometry.

Thick-shell elements or solid elements should be used to accurately model the variable thickness in the tread geometry. Under-integrated thick-shell elements were used successfully in static loading simulations and for simulations where the tires roll on flat and smooth surfaces. Unfortunately, in a verification simulation involving the tire model rolling over a series of "pot-holes", hourglass modes were excited, causing instability of the analysis.

Since the tread rubber has a low Young's modulus relative to its mass density, it is feasible to model the tread with solid hexahedronal elements (i.e., even though the element dimensions would be relatively small, the critical time-step for analysis is not reduced). For computational efficiency, the tire tread was meshed using one point under-integrated elements with three elements through the thickness to avoid hourglass modes.

The use of these solid under-integrated elements coupled with stiffness hourglass control suppressed the hourglass modes without affecting the time step, resulting in a computationally efficient model. Due to the low modulus of the rubber tread, the automatic-single-surface contact with interior contact flag turned on was necessary to prevent negative volume of elements that may collapse during impact with the ground.

Figure 8 shows the complete model of the tire's structural parts.

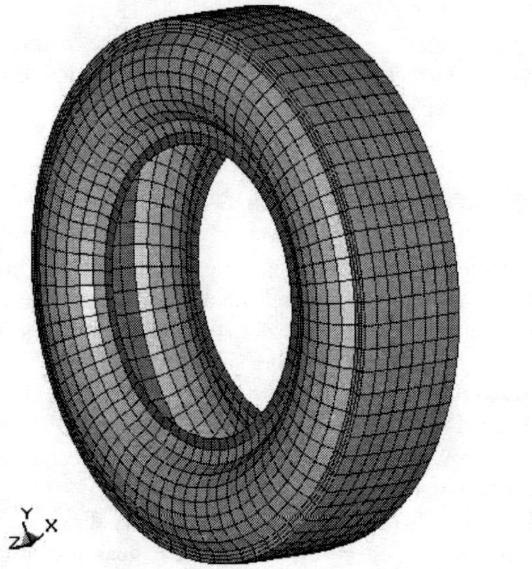

Figure 8: Model of all structural parts in the tire structures.

"Virtual" Elements

To inflate the tire with a gas under pressure an airbag must be defined. LS-DYNA defines this kind of simple pressure airbag only on a geometrically closed volume. For this reason a membrane has to be present between the bead coils to have a toroidal closed geometry. This membrane must be extremely flexible since it has no structural purpose. However, since it is involved in contact with the rim bed, its thickness must be reasonable to avoid contact instabilities (Figure 9).

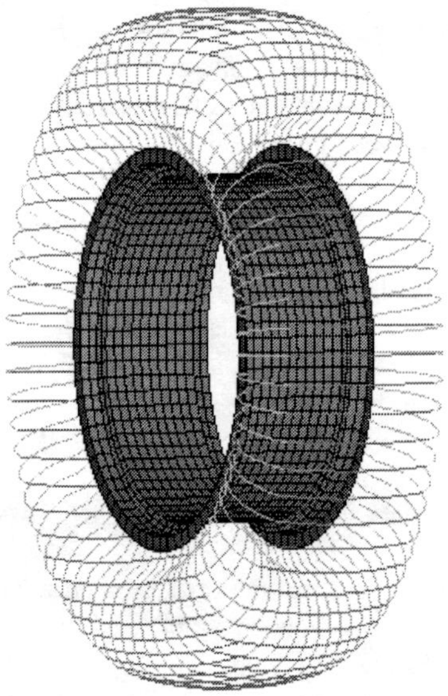

Figure 9: Tire model with the "virtual membrane" used to define a closed volume of the air-bag definition in LS-DYNA.

Material densities were, where necessary, mass-scaled to achieve the correct overall weight of the wheel and to maintain a time step greater than 3.5 microseconds.

MODELING TIRE DEBEDDING

When a vehicle leaves the roadway it is likely that the tires will interact with objects on the roadside such as curbs, guardrail posts, and other roadside appurtenances that may lead to deflation of the tire. Tire deflation in these cases is due to transverse overloading of the side shoulder of the tire that causes the bead coil to separate from the flange of the rim and slide toward the well in the rim base. When the coil falls off the edge of the well, the air seal is immediately lost and the tire deflates completely.

To model this failure mode an imaginary disk was included in the model attached to the rim base near the drop center curbside edge. This surface is used as a sensing switch, exploiting the functionality of the TERMINATION-_CONTACT card in LS-DYNA. This modeling option in LS-DYNA makes it possible to interrupt the analysis when the nodes on this virtual surface contact the bead coils. After the simulation is interrupted, a simple restart file can be created which allows the analyst to reset the inside pressure of the tire to zero and restart the simulation at the same point in time with the tire deflated.

In order for this modeling technique to work, the geometry of the wheel rim that interacts with the tire must be modeled accurately. Ignoring the geometry of the rim can lead to an overly stiff model. For example, if the rim base is modeled as a flat cylindrical surface, then as the bead coil is pushed inward during impact with a curb, the bead coil will tighten around the rim because of the non-circumferential displacement caused by friction. This leads to a self-locking mechanism that completely prevents any further movement of the bead coil. This self-locking mechanism is similar to the one that occurs when a rope ring is tightened around a pole; the rope can slide along the pole when the rope is not under tension but will lock when the rope is pulled tangentially along the pole. Thus, it is important for the tapered part of the rim base and the center drop to be accurately modeled if a debedding failure of the tire is of interest in the analysis. Figure 10 shows the finite element model of the rim and the sensing witch surface of the tire debedding.

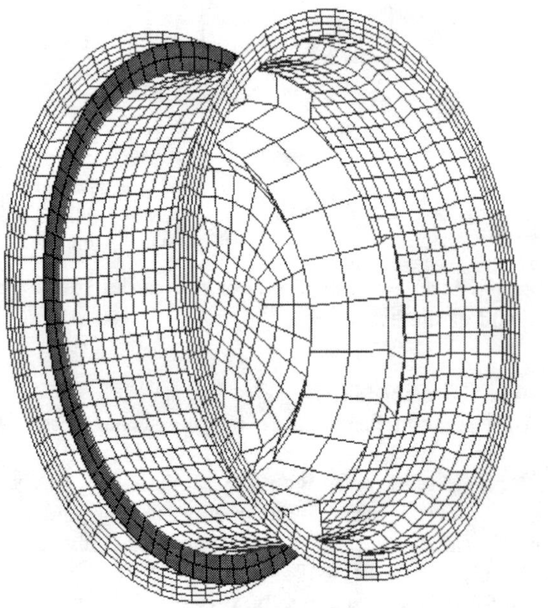

Figure 10: Model of the rim with the center well and the debedding sensor surface (fill color).

EXPERIMENT DESCRIPTIONS

Two experiments were conducted on the spare tire of a 1995 C2500 Chevrolet pickup truck mounting Uniroyal Laredo Lt 245/75 R16 tires in order to verify the utility of the model.

The first test, called a "debedding test," was carried out to calibrate the debedding sensing switch membrane. The entire wheel was mounted on an INSTRON 8803 testing machine. The wheel was bolted to a steel hub clamped to the bottom of the testing machine (Figure 11). A symmetrical "side pusher" was used to "blow out" the tire (Figure 12). For safety reasons and accuracy the test was conducted in displacement controlled mode with a compression speed of 25.4 mm per minute. Under this testing condition the seal between the rim and the tire was gently broken and the tire could gently "blow-out" without an "explosion."

The tire under testing conditions was inflated to 0.24 MPa (35 psi). At 78 mm the tire began to deflate and at 82 mm the seal was completely broken (Figure 13).

The deflection of the side pusher was neglected since the magnitude of this was negligible with respect to the magnitude of the deflection of the tire shoulder. Wood was used for the contact interface to avoid damaging the tire sidewalls by pinching the rubber.

The second test was intended to measure the global stiffness of a pneumatic wheel assembly with respect to a vertical displacement applied to the rim. The test was performed using the Instron 8803 testing machine. The rim was bolted to a steel hub clamped in the vise of the testing machine as shown in Figure 14. The tire was compressed against a rigid surface made of steel. The test was conducted in displacement control up to 13.345 kN load, which is greater than twice the static load of the truck.

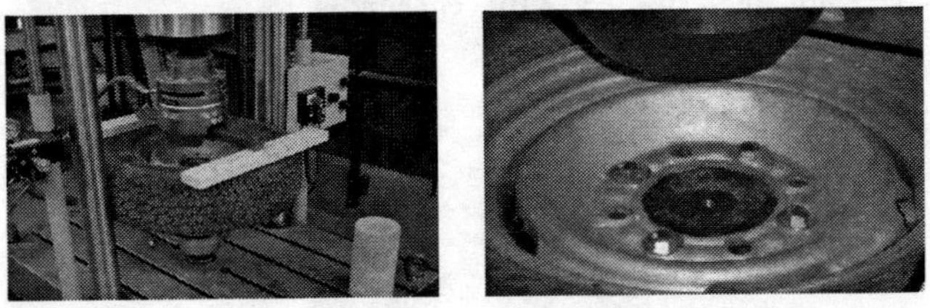

Figure 11: Debedding test setup and close up to the fixture and bolts.

Figure 12: Vertical top view of the pusher and tire under testing.

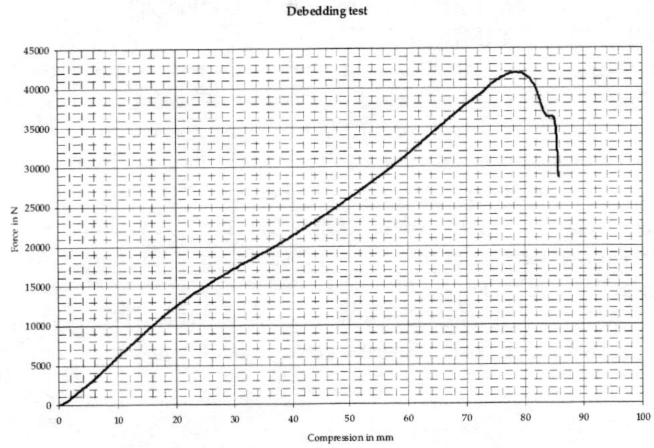

Figure 13: Force versus compression curve measured by the load cell of the Instron testing machine.

Figure 14: Steel hub and fixture used in the compression test, and test set-up.

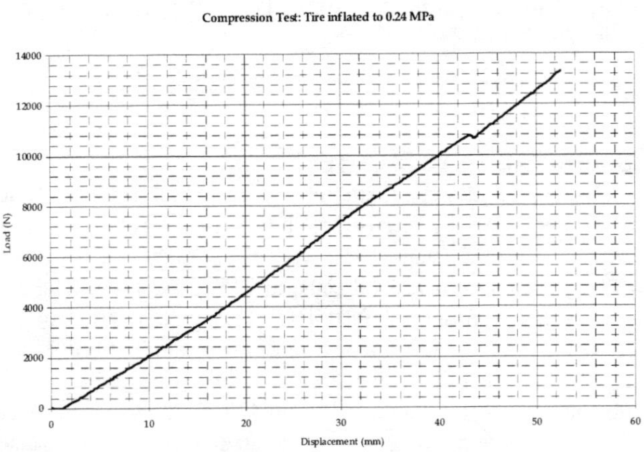

Figure 15: Load-deflection curve for the compression test.

COMPARISONS BETWEEN THE TESTS AND THE NUMERICAL SIMULATIONS

For the debedding test, the numerical model can be easily calibrated so the detection of the "blow-out" condition can be matched almost exactly with the experiment. The "blow-out" condition is based on the deflection of the tire bead coil with respect to the rim flanges and the center well.

Regarding the tire compression test, the simulation results are very close to the test result as shown in Figure 16 (for a deflection range between 0 and 32 mm), where the errors in the energy are negligible. Deflections of a correctly inflated tire rolling on a road surface are expected to be in that range.

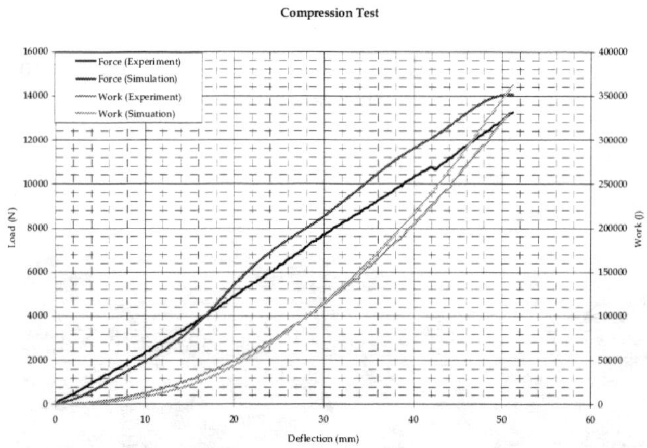

Figure 16: Comparison of the experimental results with the results obtained from the numerical simulation with LS-DYNA.

The deformed shape computed in the simulation compares very well to the deformation of the tire in the test as shown in Figure 18.

NUMERICAL STABILITY TEST

To test the numerical stability of the model, the tire model was "rolled" over a series of "pot-holes" made from a series of unequally spaced steps. The rim was loaded against the roadway with a vertical load equal to 7357.5 N and accelerated by a constant force of 9810 N. (Figure 19) This simple simulation is believed to be a severe testing simulation for any numerical instability since the tire undergoes large deformations as it compressed against the leading edges of the steps while the reference frame rotates with a continuously changing angular velocity.

Figure 17: Comparison between the test and the simulation of the deformed shape at different loads.

Figure 18: Image of the tire rolling over one of many steps.

FULL SCALE TESTS AND COMPARISON WITH LS-DYNA

A full-scale test program was conducted to study the interaction between tire and curb in non-tracking impact scenarios. The results of those tests indicated that the tire is most likely to fail due to the debedding of the bead coils rather than rupture of the tire structure. Of the sixteen tests that were conducted, four tests resulted in tire debedding and blow-out. The deflated tire in those cases did not suffer any major damage to the tire structure and the tires were simply reinflated and used in subsequent tests. Figure 19 shows a sequence of photographs of the tire failure event in one of the tests; frame C captures the "blow-out" of the tire and the tire's sudden deflation. From the analysis of video frames it was estimated that the tire was completely deflated in less than 60 milliseconds after initial impact with the curb.

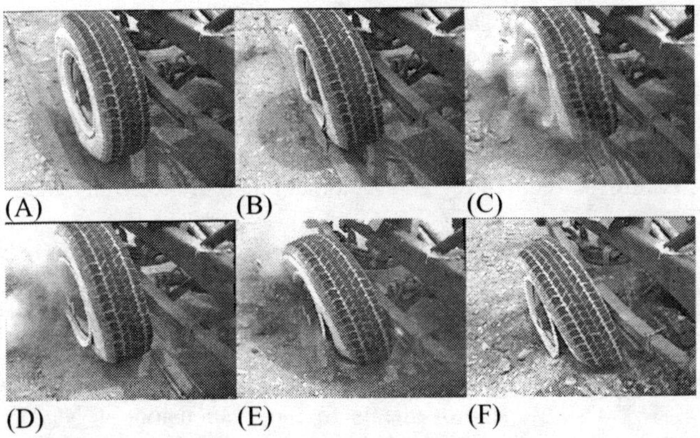

Figure 19: Sequence showing a tire debedding from full-scale test 603XD0135A.

The tire model developed and presented in this paper was implemented into the NCAC Chevrolet C2500 pickup model version 9 [4] [5] including the subsequent modification made on the suspension by Tiso [6]. The vehicle model was then used in analyses of non-tracking impacts into various roadside curbs to study the curb-tire interaction and vehicle stability [7]. In this type of analysis, the tire model plays a critical role in the overall kinematics of the vehicle and therefore the behavior of the tire must be properly accounted for.

Figure 20 shows the results of an LS-DYNA simulation of the truck model impacting a four-inch sloped curve (i.e. very similar to the test shown in Figure 19) which illustrate the failure sequence of the tire during impact. Frame C captures the "blow out" of the tire and the following frames show the behavior of the deflated tire as it interacts with the curb and the ground surface. The behavior of the tire in the simulation (Figure 20) compares very well to the behavior of the tire in the physical test (Figure 19).

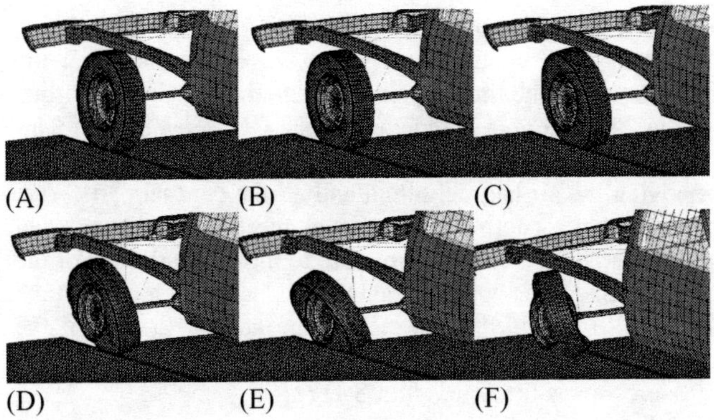

Figure 20: Sequence showing an LS-DYNA simulation result of a tire debedding with the new tire model.

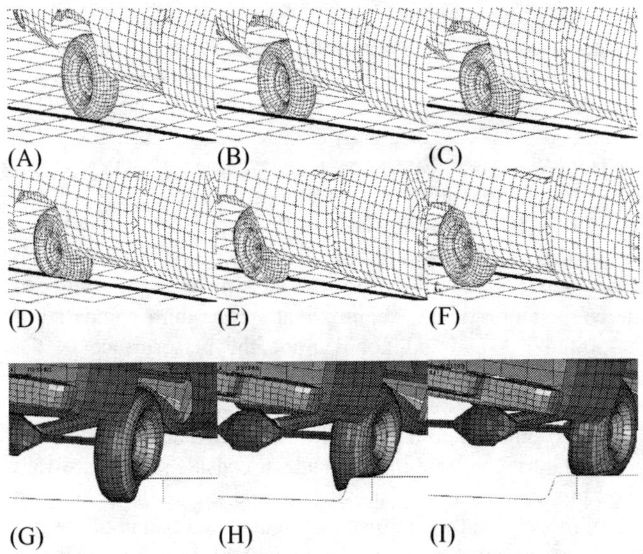

Figure 21: Sequence showing the tire finite element model performances in non-tracking impacts used in the original and WPI-modified NCAC C2500 Version 9 (reduced).

For comparative purposes, Figure 21 illustrates the behavior of the tires modeled with thin isotropic shell elements (i.e., as is typical in many reduced-element vehicle models) for the case of the pickup model impacting a six-inch sloped curb. In this particular model tire deflation is not accounted for. As the tire compresses against the curb, the mesh of the tire forms to the shape of the curb. Then, as the tire proceeds to traverse the backfill area behind the curb, it springs back resuming its original shape resulting in a roll displacement of the vehicle that counters any initial adverse roll angles caused by the initial impact with the curb. Consequently, the model erroneously predicts that rollover is highly unlikely. Frames (G), (H) and (I) clearly show the unrealistic behavior discussed above.

FURTHER IMPROVEMENTS

From the computer analyses it was determined that the pressure inside the tire did not vary significantly during normal rolling. Therefore a simpler approach to modeling the air pressure inside the tire was later adopted which involved removing the air bag definition and all the "virtual parts" (i.e. parts needed to define a closed volume for the air bag) and simply applying a constant pressure to the inside surface of the tire elements. This approach led to a reduction in analysis time and it also provided a means for the analyst to control the deflation rate of the tire since the pressure is defined using a pressure-versus-time curve in LS-DYNA.

The computational performance of the pneumatic tire model was strongly affected by the contact algorithms used. In a preliminary analysis the required computation time was considered unacceptable for general use of the model. After careful debugging of the model, it was determined that the contact algorithm that was being used to compute contact forces between the tire's bead coils and the wheel rim was requiring over 65% of the total CPU time in the analysis. This contact was originally defined using the AUTOMATIC_GENERAL contact option in LS-DYNA and was subsequently replaced with the AUTOMATIC_SINGLE_SURFACE contact option with the parameters SOFT=2 and EDGE=1 in order to activate the shell edge-to-edge contact algorithm. This simple change reduced the computation time by approximately one-third compared to the initial run and did not degrade the performance of the model. It is believed that the model's efficiency could be further improved by conducting a contact optimization study.

Two other important failure modes of pneumatic tire are possible during vehicle impact against roadside safety hardware (e.g., guardrails and median barriers) and should be included in future models: 1) the tire pinching against the flange of the rim and 2) rupture of the tire due to impact with an object such as a guardrail post.

CONCLUSIONS

The generic tire modeling methodology developed in this research is an effective means of modeling pneumatic tires for use in impact analyses with roadside hardware simulations. The simplicity of the tire model makes it relatively easy to maintain and to implement into existing vehicle models while not compromising the computational efficiency of the reduced-element models that are widely used in roadside safety research. However, further improvements to the model may be necessary if rupture of the tire structure is likely (e.g., tire snagging on guardrail posts during impact with guardrail system).

ACKNOWLEDGMENTS

This research was sponsored by NCHRP under project 22-17, "Recommended Guidelines for Curbs and Curb-Barrier Combinations," and by the FHWA under the Centers of Excellence program.

REFERENCES

[1] Samuel K. Clark, editor, Mechanics of Pneumatic Tires, U.S. Department of Transportation National Highway Traffic Administration, U.S. Government Printing Office, Washington, D.C., 1981, 931 pp.

[2] Thomas D. Gillespie, Fundamentals of Vehicle Dynamics, Society of Automotive Engineers Inc, 1992.

[3] Ross H. E. JR., Sicking D. L., Zimmer R.A., and Michie J.D., Recommended Procedure for the Safety Performance Evaluation of Highway Features. In NCHRP Report 350, Transportation Research Board, National Academy Press, Washington, D.C. 1993.

[4] Zaouk A. K., Bedewi N. E., Kan C., Schinke H.," Evaluation of a multi-purpose pickup truck model using full scale crash data with application to highway barrier impacts", Proceedings of the 29th International Symposium on Automotive Technology and Automation, Florence, Italy, June 1996. pp. 39-46.

[5] Zaouk A. K., Bedewi N. E., Kan C., Marzougui D.," Development and evaluation of a C-1500 pickup truck model for roadside hardware impact simulation", Proceedings of the FHWA Vehicle Crash Analysis Crash Conference, Mclean, VA, July 1996, pp. 1-31.

[6] Tiso P., Improvements to the suspension of NCAC C2500 Pickup truck finite element model, MS Thesis, Worcester Polytechnic Institute, 2001.

[7] Ray M., Plaxico C., Orengo F., Quarterly Progress Reports to the National Cooperative Highway Research Program (NCHRP) on Project 22-17 "Recommended Guidelines for Curbs and Curb-barrier Combinations", Report #12 and #13.

[8] Plaxico C., Ray M. Keeney T., "Modeling the tire rotation and steering in the C2500 pickup truck", Quarterly Report, FHWA Center of Excellence in LS-DYNA Modeling, The University of Iowa, 7 July 1998.

[9] Hallquist, J. O., "LS-DYNA Theoretical Manual", Livermore Software Technology Corporation, Livermore, CA, 1998.

[10] Hallquist, J. O., "LS-DYNA User's Manual", Livermore Software Technology Corporation, Livermore, CA, 1998.

Proceedings of IMECE'03
2003 ASME International Mechanical Engineering Congress
Washington, D.C., November 15–21, 2003

IMECE2003-41374

AN INTEGRATED APPROACH FOR NUMERICAL SIMULATION OF FULL VEHICLE BEHAVIOUR DURING TRANSIENT DYNAMIC MANEUVRES ON ARBITRARY OBSTACLES AND ROAD SURFACES

E. Duni, G. Monfrino, R. Saponaro, M. Caudano, F. Urbinati
FIAT Research Centre, Vehicle Division,
Strada Torino, 50 – 10043 Orbassano, ITALY
M. Spinelli, A. Pizzuto
FIAT Auto Spa
Corso Settembrini, 40 –10135 Torino, ITALY

ABSTRACT

This work describes a numerical methodology based on the finite element method used for the transient dynamic simulation of the full vehicle rolling on different kind of obstacles. Some issues related to the tire finite element model development and its validation, by numerical-experimental comparison, have been discussed. The strategy to combine the static simulations such as the tire inflating, the vehicle weight application and suspension pre loading, with transient dynamic analysis of the car rolling over the obstacle has been chosen. The methodology, based on integration of Abaqus Implicit and Explicit codes, has been successfully applied for the dynamic simulation of Fiat Punto car passing over comfort and pothole obstacle; it is in a good stage of development for a bump obstacle (misuse airbag test).

Keywords: vehicle dynamic, tire, obstacle, pothole, comfort.

INTRODUCTION

Both HW and SW last developments permit an increase in the complexity of numerical analyses in terms of problem size and typology of phenomena to be studied. In this new scenario, analysts try to extend virtual simulation to problems considered in the past, for procedural and size limits, numerically impossible to be evaluated and faced only by experimental approach. Some of these problems are related to non linear dynamic transient behavior of full vehicle FEM models passing over obstacles.

One critical point in facing such problems is related to the choice of the best way to combine the simulation of quasi-static and dynamic phenomena. The ideal numerical solution of this problem is to perform quasi static analysis (equilibrium prior to the dynamic simulation) using an implicit code and use the implicit results as initial condition of the dynamic analysis performed with an explicit code. As Abaqus code contains an implicit and an explicit solver, with the possibility to transfer information from the first to the second and vice-versa, it has been chosen as reference code for the present work.

The most critical issue in the transient dynamic simulation based on finite element method is related to the development of a reliable tire finite element model. The existing analytical and numerical tire models, integrated in a multi body approach, commonly used by car manufacturers, have some limitations. The main is that they need a continuous calibration based on data coming from experimental tests carried out on real component. The principal aim of tire finite element model developed in the present work is to describe the dynamic tire response in general conditions. A finite element model for a standard 165/65R passenger car tire with limited d.o.f. has been developed, based on information in terms of weight, stiffness, eigenmodes, usually available in the early development stage of a new car. A calibration of the model in terms of material proprieties and component thickness has been carried out for first. Then, three kinds of validations have been performed for a tire rolling with constrained axle height over a rectangular obstacle with the dimensions 100x25 mm: quasi-static, static and dynamic validation. In the numerical/experimental comparison of the results in terms of vertical and longitudinal load, the spin acceleration and the rolling radius variations a good agreement can be observed.

As case study in the present work a FIAT car of the "B" segment has been considered. The tire models, previously developed and validated, are combined with suspensions and car model. An implicit analysis has been carried out for first, in order to find the equilibrium condition of the car subjected to gravity load in presence of the road. Then, an Abaqus/Explicit analysis has been done as a restart from implicit one. The explicit analysis has been carried out for three different obstacle shapes: rectangular (100x25 mm), pothole obstacle and a bump (250x85mm) for misuse airbag tests. Three different velocity conditions are respectively applied: 50 km/h, 40 km/h and braking from 60 km/h. The dynamic vehicle response in terms of vertical and longitudinal forces, displacements at wheel center and strut house has been carried out. For the rectangular obstacle, a comparison between finite element and multi body results is reported. The stress and strain contour distributions give a complete view on real structural behavior of critical components

The Choice Of The Numerical Approach

The equations of equilibrium governing the non linear dynamic response of a system of finite elements is:

$$M\,{}^{t}\ddot{U}+C\,{}^{t}\dot{U}+{}^{t}K\,{}^{t}U={}^{t}R \qquad (1)$$

where: M, C, ${}^{t}K$ are respectively, the mass, damping and stiffness matrices;

${}^{t}R$, is the external load vector;

${}^{t}U$, ${}^{t}\dot{U}$, ${}^{t}\ddot{U}$ are the displacement, velocity and acceleration vectors of the finite element assemblage at time "t".

The procedures used for the solution of general systems of differential equations can be divided in: direct integration and mode superposition. In the direct integration of the Eq. (1) a numerical step-by-step procedure is used. The application of this method is based on two ideas:

- trying to satisfy the Eq. (1) only at discrete time intervals "Δt", instead of any time "t";

- assume the variation of the displacements, velocities and accelerations within each time interval "Δt". Obviously, the choice criteria on these assumption, determines the accuracy, stability and cost of the solution procedure.

Namely, in the solution of such problems, the choice stands between the use of explicit or implicit time integration method.

Explicit Method

In this method, known also as central difference method, the displacement equilibrium solution at time "t+Δt" is based on using the conditions of equilibrium at time "t". Such integration schema does not requires a factorisation of the stiffness matrix in the step by step solution which can be carried out on the element level and relatively little high speed storage is required. A second very important consideration in the use of the central difference scheme is that the integration method requires that the time step "Δt" is smaller than a critical value, Δt_{cr}, which can be calculated from the mass and stiffness properties of the complete element assemblage. More specifically, in order to obtain a valid solution (for the case with no damping):

$$\Delta t \le \Delta t_{cr} = \frac{2}{\omega_{\max}} \qquad (2)$$

where $\omega_{\max}$ is the highest frequency of the finite element assemblage with "n" degrees of freedom. An approximation to the stability limit is written as the smallest transit time of a dilatational wave across any of the elements in the mesh:

$$\Delta t_{cr} \approx \frac{L_{\min}}{c_d} \qquad (3)$$

where L_{min} is the smallest element dimension in the mesh and c_d is the dilatational wave speed in terms of effective Lamé's constants, λ and $G=2\mu$.

All integration schemes that require the use of a time step Δt smaller than a critical time step Δt_{cr}, such as the central difference method, are considered to be conditionally stable. If is used a time step larger than Δt_{cr} the integration is unstable. It means that any errors resulting from the numerical integration or round off in the computer grow and makes the response calculations worthless in most cases. Since the total cost of the analysis is approximately inversely proportional to the magnitude of time step, it results that if the time step can be "m" times as large, the cost would be reduced by a factor of "m". As it can be seen from Eq. (3), the Δt_{cr} is directly proportional to minimum element length and it is very important to find a compromise between total analysis cost reduction and stress accuracy requirements. It is particularly indicated for numerical simulation of short dynamic phenomena and commonly used in automotive industry for crash analysis.

A big advantage related to commercial numerical codes based on this method is the facility to define and manage the contact problems. In the Abaqus/Explicit code for example, two different contact algorithms are available: the kinematic and the penalty. In the case of kinematic enforcement of contact conditions, in each increment of the analysis the solver first advances the kinematic state of the model into a predicted configuration without considering the contact conditions. Then the solver determines which slave nodes in the predicted configuration penetrate the master surfaces. The depth of each slave node's penetration, the mass associated with it, and the time increment are used to calculate the resisting force required to oppose the penetration.

The penalty contact algorithm, instead, results in less stringent enforcement of contact constraints than the kinematic contact algorithm.

A negative issue related to such method is that it is not possible to be applied in the simulation of static phenomena.

Implicit Method

This method uses the equilibrium conditions at time t+Δt in order to find the displacement field solution at the same time. Using standard finite difference expressions to approximate the acceleration and velocity components in terms of displacement components, the solution for displacement field at time t+Δt, ${}^{t+\Delta t}U$, can be obtained from the following equilibrium equation:

$$M\,{}^{t+\Delta t}\ddot{U}+C\,{}^{t+\Delta t}\dot{U}+K\,{}^{t+\Delta t}U={}^{t+\Delta t}R\,. \qquad (5)$$

The effectiveness of these integration schemes, which are unconditionally stable derives from the fact that to obtain accuracy in the integration, the time step "Δt" can be selected without any requirement [such as defined by Eq. (3)] and in many cases 'Δt" can be orders of magnitudes larger than the Δt_{cr} defined by Eq. (3).

A big advantage of step-by-step solution scheme based on this integration method is that it can be used for solving both, static and dynamic problems, whereas the central difference method (explicit) solution could not be used if mass and damping effect are neglected.

Regarding contact definition in the Abaqus/Standard solver, based on the implicit integration schema, a pure master-slave contact algorithm is used: nodes on one surface (the slave) cannot penetrate the segments that make up the other surface (the master). Two different formulations for the contact can be used: small-sliding and finite-sliding. When using the small-sliding formulation, the relationship between the slave nodes and the master surface at the beginning of the simulation is established and maintained throughout the analysis. The finite sliding contact formulation requires to determine constantly which part of the master surface is in contact with each slave node. This is a very complex calculation, especially if both the contact bodies are deformable. At each slave node in the contact a constraint is applied. The non linear equation solving process, in the solvers based on this integration method, is expensive and if the equations are very non linear, it may be difficult to obtain a solution.

Trying to find out the right method to perform the transient dynamic analysis, referring to the previous comments about numerical integration methods, it may be advantageous to use different operators to integrate the response for different phenomena involved in the transient mission. The use of a combination of operators for the integration of static and dynamic responses raises the questions of which operators are to be chosen and how to couple them in order to take advantage from each one. As Abaqus code has both implicit and explicit solvers, with the possibility to transfer information between them by the *IMPORT capability, applied in both directions, it has been chosen as reference code for non-linear transient analysis of the car passing over the obstacle.

Tire Modelling

The most critical issue for the present work is related to the construction of an appropriate tire finite element model, able to transmit correctly transient dynamic loads to the hub. Tire manufacturers commonly use, in the development stage of their products, detailed FEM models that have an excessive computational cost in a full vehicle environment. It is therefore necessary to develop a general purpose FEM model with a restricted number of d.o.f. (in any case, much greater than a multibody model) having as a principal goal to describe the dynamic tire response in general conditions, and not to be used as a tire designing tool.

Tire Multibody Model

The simplest kinds of these models have a single point tire-road interface. For relatively large wavelengths the geometry of the road surface can serve directly as input to the model while for short wavelengths an effective road description that takes into account the enveloping properties of the tire must be used. The basic assumption is that the contact patch mainly reacts quasi-statically. Consequently, the effective excitation of the tire can be assessed from the quasi-static enveloping properties of the tire. The effective road surface corresponding to an obstacle is obtained from experiments with constant axle height. In any case, this model needs, for different obstacle shapes and tires, both quasi-static and dynamic validation.

Other models (like the Swift Tire by TNO) allow the evaluation of both harshness and handling vehicle performances, combining a 3D rigid ring approach (lateral deformation of the sidewall included), with the quasi-static enveloping properties of the tire and a description of combined slip forces through synthetic formulas (the so called Magic Formulas).

In other approaches the rigid ring becomes a flexible ring and the combined slip forces are obtained from a contact patch brush model.

Tire Finite Element Model

The development of the general purpose tire finite elements model with limited number of d.o.f. is based on 165/65 dimensions tire for two reasons: is mounted on the FIAT Punto car, chosen as test case for the present study; is a standard passenger car tire and a lot of information, both experimental and multibody simulation, is available.

The tire model, with section reported in Fig. 1, consists of two parts: the plies model and the rubber model. The plies are modelled using membrane elements arranged on the section perimeter according with their characteristics; in particular the material stiffness given to these elements is obtained, using the classical lamination theory, from the composed stiffness of the plies placed one upon another. The rubber parts of the tread and seat parts are modelled using solid continuum elements made of isotropic material.

As it can be seen in Fig. 1, two versions of the tire model were developed, in order to study the differences in the tire behaviour due to a solid model of the sidewall rubber. Furthermore a third model was developed starting from the second version of Fig. 1 to evaluate a circumferential mesh refinement influence in the near-obstacle contact patch.

The tire finite element models displayed in Fig. 1 are:

- *Model 1*, only membrane elements with equivalent material and thickness properties on the side;

- *Model 2*, the same as model 1 with the addition of solid elements for modelling the rubber part on the side;

- *Model 3*, different from model 2 for the presence of a finer circumferential mesh in the near obstacle contact zone.

Static and Modal Validation. This process includes the calibration of the rubber behaviour and of the membrane elements thickness, based on the following general data, usually available in the early development stage of a new car: weight, vertical and lateral stiffness and first in plane eigenmodes.

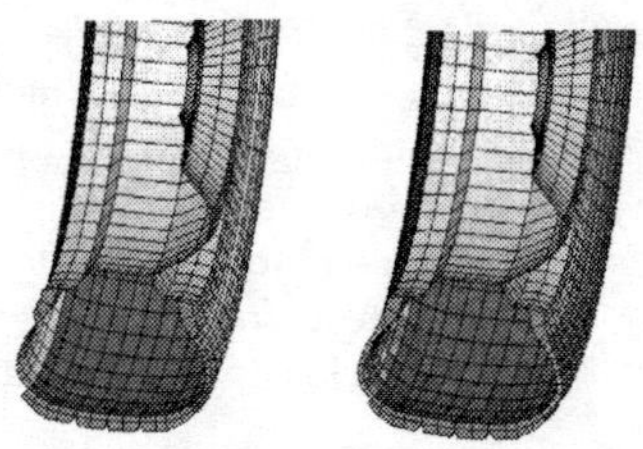

FIGURE 1

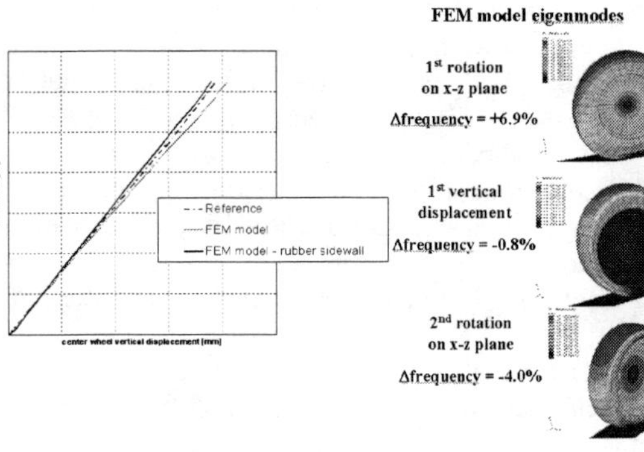

FIGURE 2

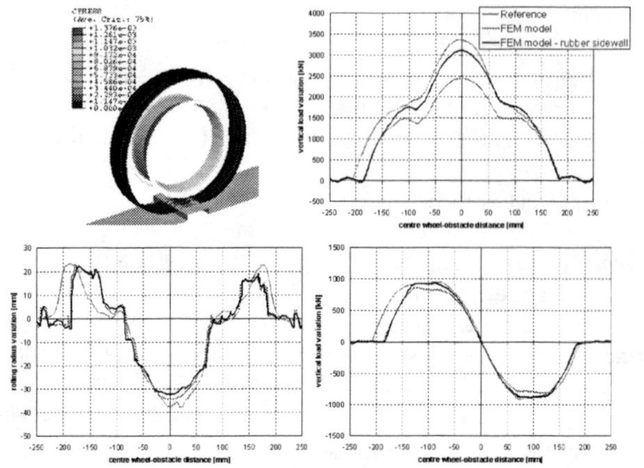

FIGURE 3

In order to perform the static validation of the model, Abaqus/Standard code has been used and a step-by-step procedure has been carried out. The numerical procedure can be summarised as it follows: -tire seating on the rim; -inflating, performed simulating the air inside the tire as a distributed load on the plies (it is possible to add the air behaviour simulation using an appropriate modelling strategy); -loading, performed using the interaction between tire and ground to achieve the prescribed vertical load at the hub. The loaded configuration of the wheel, previously obtained, is the condition used to perform the modal analysis; to reproduce the test conditions the rim is completely fixed at ground except for the d.o.f. of rotation around the wheel axis.

A comparison between experimental and numerical results carried out for the tire *model 1* and 2 as previously described, in terms of radial stiffness and first three in plan eigenmodes, is reported in the Fig. 2. Analyzing these results, the following considerations can be drawn:

- independently from sidewall modelling strategy, there is a good agreement between experimental and numerical values in terms of tire radial stiffness;

- the numerical values of the tire three first in plane eigenvalues overlap quite well with respective experimental one; indeed, the first numerical eigenvalue is 6.9 % higher than experimental one, the 2nd numerical eigenvalue is 0.8 % lower than the experimental while the 3rd numerical value is 4% lower than the respective experimental value.

Quasi-Static Validation. The test facility has a long flat steel road surface with the tire rolling with the wheel centre fixed in longitudinal, vertical and lateral directions allowing the experiments to be done with constrained axle height. Different obstacle shapes with different heights can be mounted on the road surface. The effective road surface velocity is very low in order not excite tire dynamics and to give the air spring system time to settle. For the tire under consideration in the present work, a rectangular obstacle with 25x100mm dimensions has been examined.

From the numerical point of view, first, a static analysis has been carried out to achieve the equilibrium conditions of the tire in the presence of a 312 kg constant vertical load. Than, starting from this equilibrium condition and fixing the distance between hub and ground, the wheel rolling maneuver over the obstacle has been simulated. The validation is carried out in terms of vertical and longitudinal load and rolling radius variations with respect to the longitudinal distance between wheel and obstacle centre (the zero distance means that the wheel centre coincides with the obstacle centre). A comparison of numerical and experimental results is reported in Fig. 3, for *model 1* and 2 described before. From an overview of all these results, the following considerations can be drawn:

- in terms of rolling radius and longitudinal load variations, there is a good agreement between numerical and experimental results, independently from the finite element used;

- there are some differences between numerical and measured results in terms of peak vertical load variation;

- adding the solid model of rubber on the side, the agreement between numerical and experimental results in vertical load increases.

Dynamic Validation. A specific test stand can be used. It is mounted on top of a 2.5 m diameter drum and is designed to measure tire dynamic behaviour in the frequency range 0-100 Hz. The vertical height of the wheel axle can be adjusted to load the tire while during the experiments the wheel axle's vertical, longitudinal and lateral motions are constrained. The reaction forces of the tire, both longitudinal and vertical, are measured.

The numeric simulation of tire rolling dynamic manoeuvre has been carried out based on Abaqus/Explicit code. In this case, a flat road profile provided with a rectangular obstacle of 25x100 dimensions is used instead of the bench test drum. The initial conditions for this explicit analysis have been from the static loading condition previously carried out in Abaqus/Standard (implicit solver). On the reference node of road profile, modelled as a rigid body, a boundary condition of type velocity in the longitudinal direction has been imposed while all the other d.o.f. are constrained to the ground. The simulations were carried out at the velocity 40 km/h with the axle height corresponding to initial vertical load of 312kg.

Different numerical simulations have been carried out in order to investigate the influence of following parameters on dynamic behaviour of the tire rolling on the obstacle: the finite element dimensions, material damping and contact properties. Regarding the damping, a general damping effect provided by Rayleigh damping has been introduced.

For the chosen vertical load condition and longitudinal velocity, the vertical force (ΔFz), the longitudinal force (ΔFx) and the spin acceleration (ΔΩ) variations are plotted versus time in the Fig. 4. In the same figures the multibody simulation results are reported too. In the Fig. 5 the contact pressure distribution between tire and road before, during and after the obstacle are reported.

From a global overview of these results, the following considerations can be outlined:

- as shown in Fig. 6, for damping factor calibrated for 1st in plane frequency and defined for different fraction of critical damping, ξ, ranging from 2% up to 7%, no significant change in dynamic tire response is verified. For ξ > 9% instead, the tire response, especially in terms of longitudinal force amplitude is different with respect to the responses corresponding to smaller values of critical damping.

- independently from the finite element model used, there is a good agreement between multi body and the finite element results in terms of spin acceleration (ΔΩ) and longitudinal force (ΔFx) variations.

- regarding the vertical force variation (ΔFz) comparison:
  - there is the same shape of vertical force variation in multibody and finite element simulation; indeed, the two peak shape of vertical force variation of the tire overcoming the obstacle, is present in both curves reported in the Fig. 4;
  - the vertical force peak value obtained by the finite element simulation, in the hypothesis of simplified finite element model, results to be 14% lower than the one obtained by multibody simulation. This value decreases in 8% by introducing the solid elements for sidewall rubber modelling.
  - in the hypothesis of *model 3* (rubber solid elements on the side and fine mesh in the contact zone) the response obtained by the finite element simulation is almost equal to the multibody simulation (see Fig. 4).

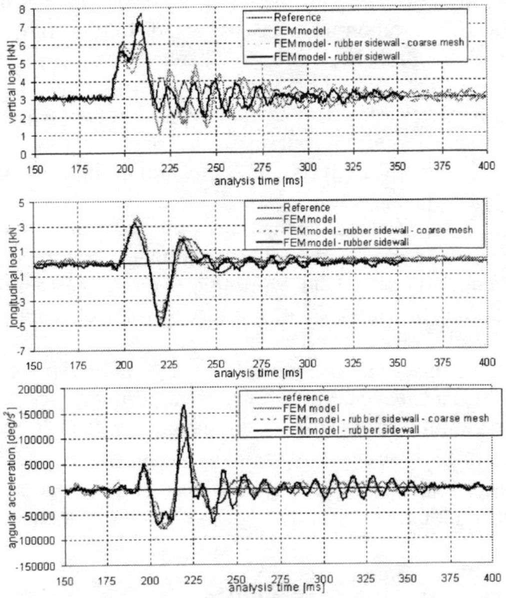

FIGURE 4

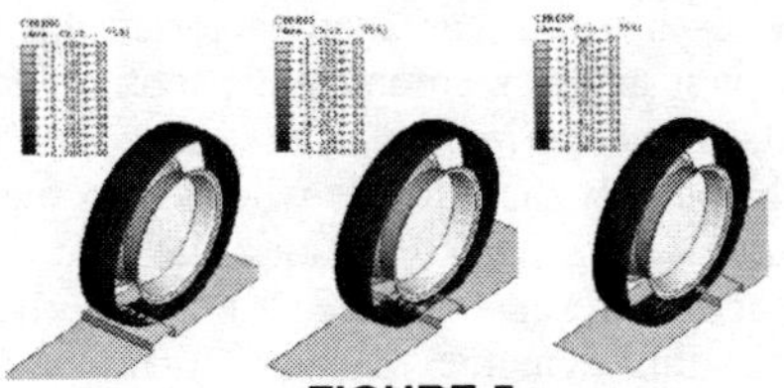
FIGURE 5

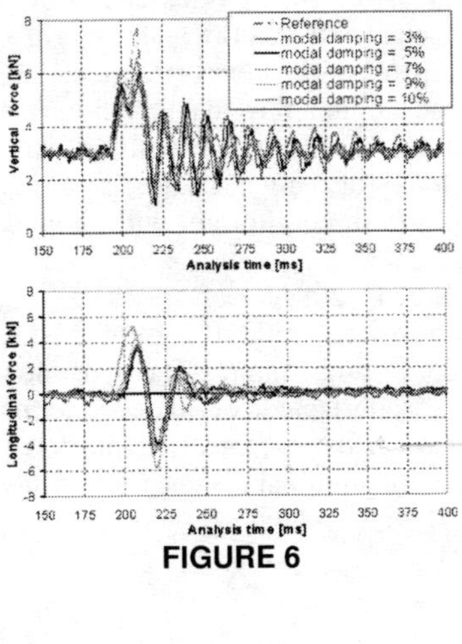

FIGURE 6

Transient Dynamic Analysis Of The Vehicle Passing Over The Obstacle

Until now, the two most important aspects related to the transient dynamic analysis of the vehicle in the presence of road profile have been faced: developing of a tire finite element model of reduced d.o.f. and the choice of the approach to apply in order to combine static and dynamic phenomena, based on the integration of Abaqus implicit and explicit solvers.

Regarding the full vehicle transient dynamic procedure, in the Abaqus/Standard code the following phases are simulated using a step by step approach: the tire seating and inflation, performed for the four wheels of the model as described in the single tire model; the gravity load enforcement and consequent suspensions pre-loading; the contact enforcement between the wheels and the ground; the equilibrium between vehicle and the ground under the gravity field previously applied.

During the methodology development phase of the explicit analysis, in order to overcome the difficulties related to the problem dimensions, two different vehicle models have been

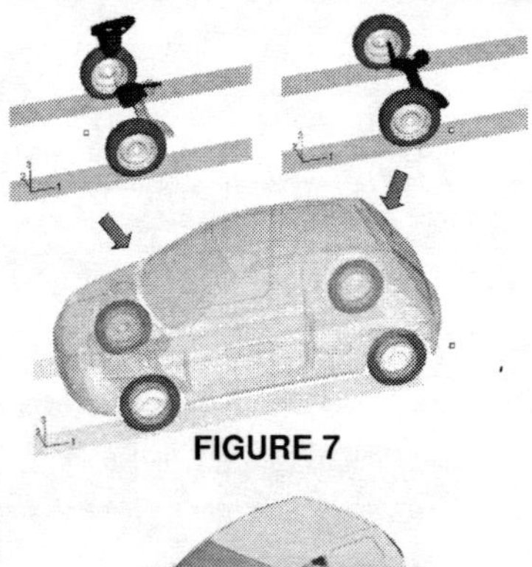
FIGURE 7

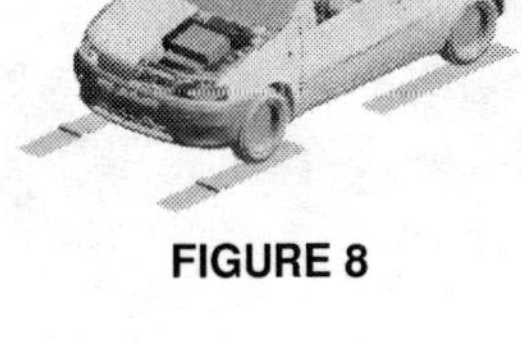
FIGURE 8

carried out:

- *the semi-flexible model* (Fig. 7) created by transferring implicit equilibrium results (stress and strain information) only for some significant elements of the model: the tires, the subframes, the strut-houses, whose behavior is of particular interest in these kind of maneuvers, while all the other components are modeled as rigid bodies;

- *the complete model* (Fig. 8) where all components are modeled as *deformable* bodies and whose initial stress/strain condition is transferred from previous implicit analysis.

The Comfort Obstacle

The comfort obstacle has a rectangular shape with 100x25 mm dimensions as reported in Fig. 9. The experimental test is carried out with the vehicle running with velocity of 50 km/h. The Fiat PUNTO car has been considered as test case in the present study. The dynamic analysis consists in reproducing experimental conditions: a relative vehicle-ground velocity of 50km/h, and consequent wheels angular velocity, with gravity load applied to full vehicle model, in order to guarantee the continuity with the static equilibrium.

Due to the large problem dimensions the complete model analysis is limited to 60 ms necessary for the front wheel to overcome the obstacle, while for the semi-flexible model the analysis time is expanded to a total of 400 ms.

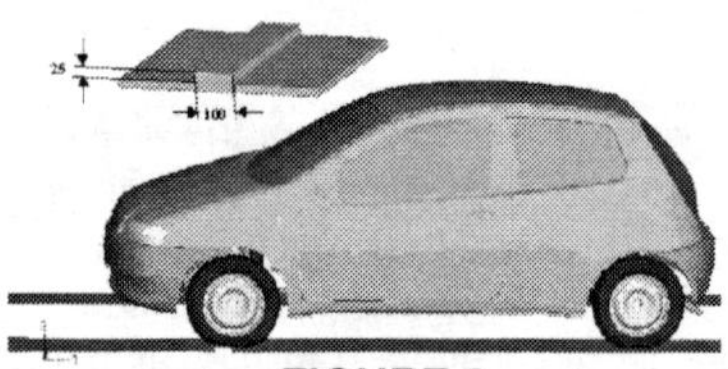

FIGURE 9

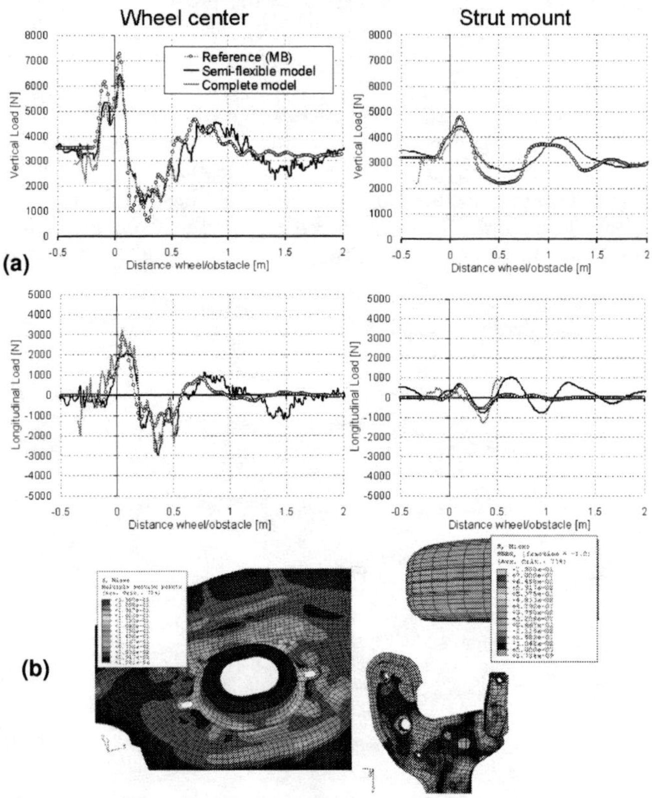

FIGURE 10

The full car behavior evaluation is carried out in terms of wheel center and strut mount vertical and longitudinal load variation plotted versus the longitudinal distance between wheel and the obstacle center (the zero distance means that the wheel center coincides with the obstacle center). A comparison of the finite element results, obtained from both semi-flexible and complete models, with the multibody one is reported in the Fig. 10-a. The mises stress distribution for strut house in the correspondence of the obstacle center, is reported too. From an overview of all these results, the following considerations can be outlined:

- in terms of wheel center longitudinal force, the finite element simulation results correspond rather well with multibody one. Leaving the obstacle, a higher than multibody amplitude vibration can be observed in the finite element simulation, probably due to different ways to introduce damping effects.

- in terms of wheel center vertical force, while the two peak shape force distribution in the first phase of obstacle is verified in both analysis (multibody and finite element), the maximum peak value obtained from multibody simulation results to be 15% higher than one obtained from finite element analysis. It is important to underline that there are some small differences in static load distribution among the multibody and the finite element model as confirmed from different values of static loads before the obstacle.

- the mises stress contour (Fig. 10-b) corresponding to the most critical instance of whole dynamic maneuver shows that the maximum mises stress value is smaller than material yield stress (a high strength material is used with σ_y=320Mpa and σ_R=360Mpa). Consequently, no plastic deformation involves the strut mount material during this dynamic maneuver.

The Pothole Obstacle

The pothole obstacle, which has a length of 2700mm and a depth/height of 114mm, as reported in the Fig. 11, is commonly used from car manufacturers for the structural evaluation of suspension components, subframe and strut house attachments. The experimental test is carried out with a velocity of 40km/h and only the left side wheels pass over the obstacle, while the right side wheels rolls on a flat road. The same finite element model representative of Fiat Punto car has been used as test case for this dynamic maneuver. The tire finite element model is the same as one previously used with the exception of the fine mesh size, which is extended in order to cover the ramp. The contact between tire and rim is also introduced.

As described for the comfort obstacle, the explicit dynamic has to reproduce the physical condition of the test; so the relative vehicle-ground velocity of 40km/h and the initial spinning velocity of the wheels are imposed appropriately. The vehicle model is in equilibrium with ground under the gravity loading.

In the Fig. 11-a a plot of the car rolling over the pothole obstacle is reported with a deformed shape of the tire in correspondence of the "ramp". The dynamic behavior of the car rolling over this obstacle is evaluated in terms of:

- the vertical displacement of the left wheel center with respect to the longitudinal distance between wheel and obstacle ramp (Fig. 11-b);

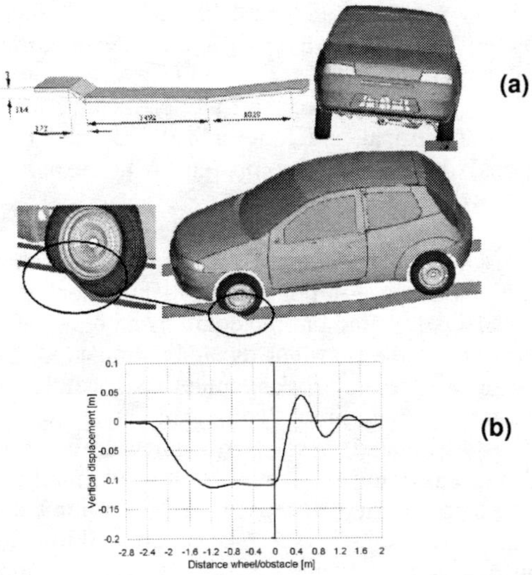

FIGURE 11

- longitudinal and vertical load variation, in correspondence of left wheel center and strut mount, with respect to the longitudinal distance between wheel and obstacle ramp (Fig. 12-a);

- equivalent mises stress and plastic equivalent deformation contours (Fig. 12-b).

Analyzing all the results reported in these figures, the following considerations can be pointed out:

- the peak value of both longitudinal and vertical left wheel center load is quite the same and equal to 24000N; in terms of vertical load that means an amplification of the load due to the dynamic maneuver which is 6 times the static load applied to the single front wheel

- due to the wheel acceleration and energy absorbed by shock absorber, the peak value of strut mount vertical load is 10500N dramatically reduced with respect to the load passing through wheel center.

- the mises stress contour shows that the stress level present in the strut mount is greater than material yield stress and consequently, plastic deformations are present in the component (Fig. 12-b).

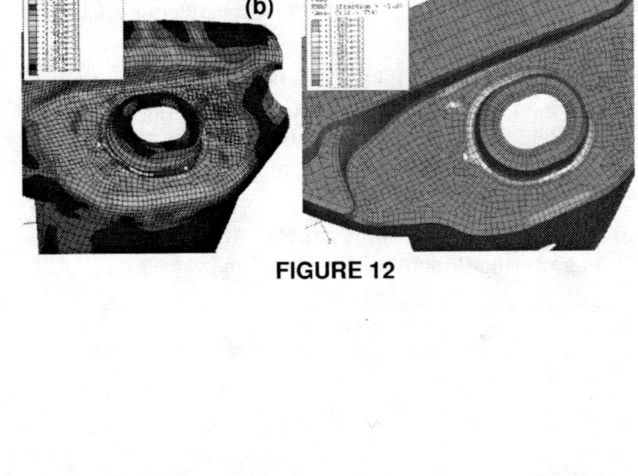

FIGURE 12

Method Development: Misuse Airbag

This simulation is an attempt to use the reduced finite element tire model in extremely non linear conditions. This test consists of a passing over a smoothed big rectangular obstacle (Fig 13, dimensions 85x250mm) while the front wheels are braking. The basic idea is to evaluate if the modeling strategy

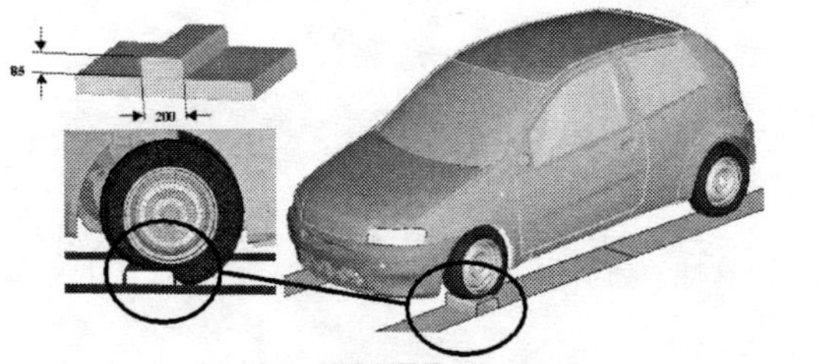

FIGURE 13

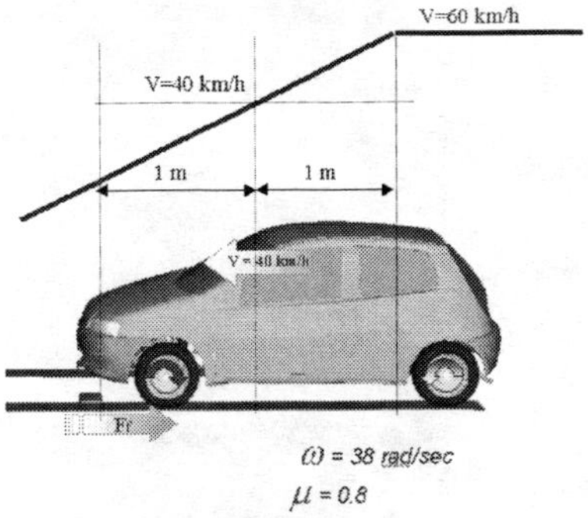

FIGURE 14

developed for the previous obstacles is appropriate in a test involving an impact between tires and obstacle. So the model used for this analysis is the same described in the previous paragraph (including the internal contact between tire and rim in the wheels) with the exception of four friction elements added to the center of the wheels to simulate the brakes effect. The experimental test is performed starting from a constant velocity of 60km/h and putting on the brakes starting from a distance of two meters before the obstacle (Fig. 14). This braking condition is maintained during the entire transient dynamic maneuver.

The simulation follows the same sequence with the variation that the braking phase is shorten from two to one meter to limit the analysis cost. Therefore the initial velocity imposed to the vehicle is evaluated applying an appropriate braking law. Moreover, it is important in this analysis to consider the load transfer between rear and front axis of the vehicle due to the brake action. Due to these considerations, the initial conditions imposed to the model in the explicit analysis are: the initial velocity of 40km/h coupled to the braking force applied to the front wheels (fig. 14); the gravity load applied to all the vehicle model, in order to guarantee the continuity with the static equilibrium.

The semi-flexible model has been used for this particular transient dynamic analysis. As for the car considered as test case in the present study no experimental data are available at the moment, the car behavior evaluation can be done considering the accelerations at some significant points of the

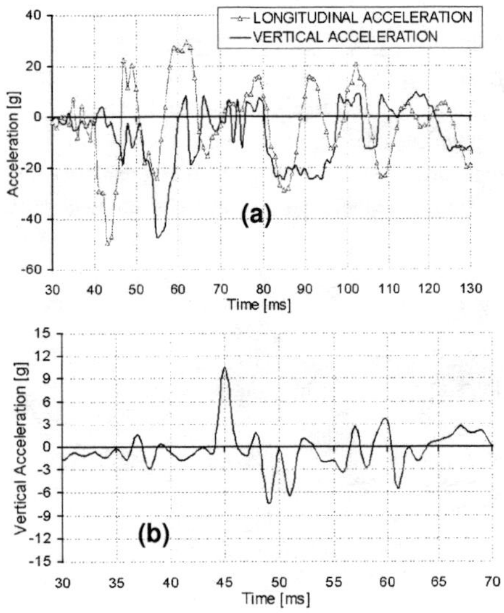

FIGURE 15

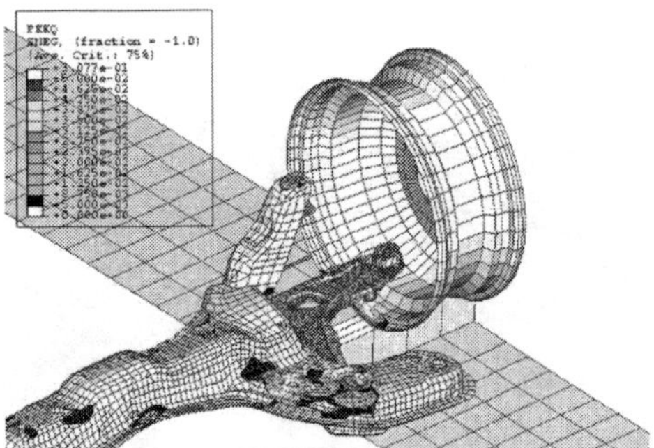

FIGURE 16

car such as wheel center (fig. 15-a), strut mount (fig. 15-b) and accelerometer location.

Furthermore the contour plastic deformation, reported in the fig. 16 for subframe and wheel rim, can be used for a more exhaustive comprehension of the internal behavior of the structures.

An important remark based on these results can be summarized as it follows: the integrated approach presented in this paper can be also successfully used for numerical simulation of some extreme maneuvers involving the highly non linear contact properties of the tire with the obstacle characterized by the impact and the slipping of the tire (instead of rolling).

Conclusions

An integrated CAE methodology, based on integration of Abaqus Implicit and Explicit codes, able to evaluate the transient dynamic response of the full vehicle rolling on the road in the presence of the obstacle, has been presented. Assuming as a case study the Fiat Punto car, the methodology has been successfully applied allowing to predict accurately the loads entering in the structure and the consequent damage caused to it.

The most critical point of the methodology regards the development of the tire finite element model, of limited degrees of freedom, which doesn't require a continuos calibration with experimental data like multibody tire models. A sensibility analysis of mesh size and modeling technique adopted on the tire response rolling on the obstacle has been carried out. The tire validation process is composed of a static, quasi-static and dynamic evaluation where the finite element results, expressed in terms of vertical and longitudinal load and rolling radius variations, have been compared with experimental and multibody data.

Another important issue encountered regards the best way to perform the quasi static (tire inflating, vehicle weight application, suspension pre loading) simulations needed in the overall analysis to find out the equilibrium between vehicle and ground prior to the transient dynamic simulation. The strategy chosen in the present work, applied successfully for dynamic simulation of the Fiat Punto car, completed with tires (previously validated) passing over three different obstacles, is based on performing the static simulations in Abaqus/Implicit code (a step by step approach has been implemented) and importing the final stress/strain information into Abaqus/Explicit code where the dynamic simulation takes place. Even if there are actually in Abaqus code some limitations regarding the element types that can be imported from Implicit to Explicit code and vice-versa, the present work has demonstrated that such kind of approach can be successfully applied in automotive industry for virtual evaluation of the components particularly stressed during obstacles dynamic maneuvers.

ACKNOWLEDGMENTS

The authors would like to thank FIAT-Auto Company, that financially supported this work.

REFERENCES

1. Abaqus/Standard Users' Manual, 2002, Version 6.3, Hibbitt, Karlsson & Sorensen, Inc.
2. Abaqus/Explicit User's Manual, 2002, Version 6.3, Hibbitt, Karlsson & Sorensen, Inc.
3. Bakker, E., Pacejka, H.B. and Lidner, L., 1989, "A new tire model with an application in vehicle dynamics studies", SAE paper 890087.
4. Bakker, E., Nyborg, L. and Pacejka, H.B., 1987, "Tyre modelling for use in vehicle dynamics studies", SAE paper 870421.
5. Bathe, K. J, 1982, "Finite Element Procedures in Engineering Analysis", Prentice-Hall, Inc., Englewood Cliffs, New Jersey 07632.
6. Dixon, J. C., 1991,"Tires, suspensions and handling", Cambridge University Press, England.
7. Genta, G., 1993, "Meccanica dell'Autoveicolo", Levrotto & Bella, Italy.
8. Morelli, A., 1999, "Progetto dell'Autoveicolo", CELID, Italy.
9. Pacejka, H.B. and Bakker, E., 1993,"The magic formula tyre model", 1st International Colloquium on Tyre Models for Vehicle Dynamics Analysis, Delft, The Netherlands, October 21-22, 1991, Vehicle System Dynamics, **21** supplement, pp. 1-18.
10. Reimpell, J., and Stoll, H., 1996, "Automotive Chassis – Engineering Principles", Arnold, England.

Proceedings of IMECE'03
2003 ASME International Mechanical Engineering Congress
Washington, D.C., November 15–21, 2003

IMECE2003-41767

Meshless Analyses of Fracture, Plasticity and Impact

A Eskandarian[1], Y. Chen[1], M Oskard[2] and J. D. Lee[1]

[1] School of Engineering and Applied Science
The George Washington University
Washington, DC 20052, U. S. A

[2] Turner-Fairbank Highway Research Center
Federal Highway Administration
U. S. Department of Transportation
McLean, Virginia 22101, U. S. A.

Abstract

The governing equations for rate-independent large strain plasticity are formulated in the framework of meshless method. The numerical procedures, including return mapping algorithm, to obtain the solutions of boundary-value problems in computational plasticity are outlined. The crack growth process in elastic-plastic solid under plane strain conditions is analyzed. The large strain plastic response of material under high-speed impact is simulated. Numerical results are presented and discussed.

Keywords: meshless method, crack growth, plasticity, impact

1. Introduction

The finite element method has been the dominant technique in computational mechanics in the past several decades, and has made significant contributions to the crashworthiness analysis and design. An alternative method for this purpose is the rapid growing meshless methods, which constructs the approximation of field variables entirely in terms of nodes. This method eliminates the need of a highly structured mesh as required in finite element method, is therefore less susceptible to mesh distortion due to large material deformation, and has advantages in dealing with moving discontinuities or for problems involving considerable meshing or remeshing as in structural optimizations [1-7].

Fracture mechanics deals with the rupture of solids in the presence of cracks. Linear elastic fracture mechanics (LEFM) was originally developed to describe crack growth and fracture under essentially elastic conditions. However, such conditions are met only for plane strain fracture of high strength metallic materials and for fracture of intrinsically brittle materials like glasses, ceramics, rocks, etc. There are many important classes of materials that are too ductile to permit description of their behavior by LEFM. It is observed that cracked specimens made of ductile materials subjected to monotonically and slowly increasing load show a considerable amount of crack tip plasticity and non-negligible amount of slow and stable crack growth prior to the onset of fast fracture. In this work the slow crack growth process in elastic-plastic solid is formulated as a moving boundary value problem in plasticity and analyzed by meshless method.

The underlying structure of the finite element method that originates from their reliance on a mesh is not well suited to the treatment of extreme mesh distortion. Meshless methods require no explicit mesh in computation and therefore avoid mesh distortion difficulties in large deformation analysis. In this work the large strain plastic impact/contact problem is formulated and the dynamic large strain plastic response of material under impact is simulated.

2. Formulation

In this work, we begin with two basic assumptions in classical plasticity. The first one states that the strain can be decomposed into an elastic part and a plastic part:

$$E_{KL} = E^e_{KL} + E^p_{KL} \ , \tag{1}$$

where $\boldsymbol{E}$ is the Lagrangian strain tensor. In case of small strain, $\boldsymbol{E}$ is reduced to $\boldsymbol{e}\ \{\varepsilon_{ij} \bullet (u_{i,j} + u_{j,i})/2\}$, the symmetric part of the displacement gradients. To be discussed later, the evolution of $\boldsymbol{E}^p$ is defined through a flow rule, eq. (1) is considered as a definition of the elastic strain $\boldsymbol{E}^e = \boldsymbol{E} - \boldsymbol{E}^p$ or $\boldsymbol{e}^e \equiv \boldsymbol{e} - \boldsymbol{e}^p$ in case of small strain. The second assumption relates the 2nd order Piola-Kirchhoff stress tensor $\boldsymbol{T}$, which will be reduced to the Cauchy stress tensor $\boldsymbol{t}$, to the elastic strain $\boldsymbol{E}^e$ as

$$T_{KL} = A_{KLMN}(E_{MN} - E^p_{MN}) \quad \text{or} \quad t_{ij} = A_{ijkl}\varepsilon^e = A_{ijkl}(\varepsilon_{kl} - \varepsilon^p_{kl}) \ . \tag{2}$$

The essential feature that characterizes plastic flow is the notion of irreversibility. This basic property is built into the following formulation. The elastic domain E is defined to be the set that contains all the admissible states ($\boldsymbol{T,P}$) in the stress space as

$$\mathrm{E} \equiv \{(\boldsymbol{T,P}) \ | \ f(\boldsymbol{T,P}) < 0\} \ , \tag{3}$$

where f is a scalar-valued yield function and $\boldsymbol{P}$ is a set of internal variables. The boundary of E, denoted by ∂E, defined as

$$\partial\mathrm{E} \equiv \{(\boldsymbol{T,P}) \ | \ f(\boldsymbol{T,P}) = 0\} \ , \tag{4}$$

is called the yield surface in stress space. The states outside of ∂E, i.e., $f(\boldsymbol{T,P}) > 0$, are nonadmissible. The equations of evolution for $\boldsymbol{E}^p$ and $\boldsymbol{P}$, called the flow rule and the hardening law respectively, can be written as

$$\dot{\boldsymbol{E}}^p = \gamma \boldsymbol{r}(\boldsymbol{T,P}) \ , \tag{5}$$

$$\dot{\boldsymbol{P}} = -\gamma \boldsymbol{h}(\boldsymbol{T,P}) \ , \tag{6}$$

where $\boldsymbol{r}$ and $\boldsymbol{h}$ are prescribed constitutive functions which define the direction of plastic flow and the type of hardening; the parameter γ is a nonnegative function, called the consistency parameter, and is assumed to obey the following Kuhn-Tucker conditions:

$$\gamma \geq 0 \quad , \quad f(\boldsymbol{T},\boldsymbol{P}) \leq 0 \quad , \quad \gamma f(\boldsymbol{T},\boldsymbol{P}) = 0 \quad . \tag{7}$$

In this work, we use a special and classical model for metal plasticity, the J_2 flow theory for 3-D and plane strain problems. First, we define the deviatoric stress tensor as

$$S_{KL} \equiv T_{KL} - \tfrac{1}{3} T_{MM} \delta_{KL} \quad . \tag{8}$$

It is seen that $S_{MM} \equiv 0$. A set of internal variables, typical for metal plasticity, is chosen to be

$$\boldsymbol{P} = \{\boldsymbol{\beta}, \bar{\beta}\} \quad , \tag{9}$$

where $\boldsymbol{\beta}$, often named as the back stress, defines the center of the von Mises yield surface in the stress space and is a second order symmetric tensor with $\beta_{ii} = 0$ and $\bar{\beta}$ defines the isotropic hardening of the von Mises yield surface. The J_2 model has the following yield condition, flow rule and hardening law [8,9]:

$$f(\boldsymbol{T},\boldsymbol{P}) = \sqrt{\xi_{ij}\xi_{ij}} - \sqrt{2/3}(\sigma_Y + cH\bar{\beta}) \equiv \|?\| - \sqrt{2/3}(\sigma_Y + cH\bar{\beta}) \quad , \tag{10}$$

$$\dot{\boldsymbol{E}}^p = \gamma \frac{?}{\|?\|} \quad , \tag{11}$$

$$\dot{\boldsymbol{\beta}} = \gamma \tfrac{2}{3}(1-c)H \frac{?}{\|?\|} \quad , \tag{12}$$

$$\dot{\bar{\beta}} = \gamma \sqrt{2/3} \quad , \tag{13}$$

where $? \equiv \boldsymbol{S} - \boldsymbol{\beta}$; σ_Y is the von Mises strength; H is a constant representing the slope of the stress-strain relation in plastic loading; c is a constant with $0 \leq c \leq 1$. It is noticed that: (1) $c = 1$ implies $\boldsymbol{\beta} = 0$ which is the case of isotropic hardening; and (2) $c = 0$ implies that $\bar{\beta}$ does not affect the size of the yield surface - it is the case of kinematic hardening.

The central difference method is used in this work for dynamic problems. The geometric and material nonlinearities are included in the analysis of high-speed impact problems. The equations of motion, at time step n, may be written as [10]:

$$\boldsymbol{Ma}^n = \boldsymbol{F}^n = \boldsymbol{F}^{ext}(\boldsymbol{d}^n, t^n) - \boldsymbol{F}^{\text{int}}(\boldsymbol{d}^n, t^n) \quad , \tag{14}$$

subjected to

$$G_I(\boldsymbol{d}^n) = 0 \quad , \quad I = 1,2,3,.....n_c \quad , \tag{15}$$

where $\boldsymbol{a}$ and $\boldsymbol{d}$ are the acceleration and displacement vector, respectively; $\boldsymbol{F}^{ext}$ represents the boundary conditions of specified loading; $G_I(\boldsymbol{d}^n) = 0$ are the essential boundary conditions; $\boldsymbol{F}^{\text{int}}$ indicates that the internal nodal forces depend on the displacements. Actually, nodal forces depend on stresses, which depend on the strain and strain rates through constitutive equations and through return mapping algorithm in the case of plasticity, which in turn depend on the displacements and their derivatives. The procedures and the flowchart for the explicit time integration can be found in Belytschko et al [10]. However, it is noticed that this work is based on meshless method and the way to impose the essential boundary conditions at each time step is very different from that in finite element method. The boundary conditions during impact are rigorously imposed.

3. Slow Crack Growth

During the process of crack growth, even in the case of two-dimensional self-similar crack growth, the crack size becomes a monotonically increasing but unknown variable. Precisely speaking, one more unknown variable corresponds to the need of one more governing equation for the system. In order to establish the governing equation for the crack size, various kinds of relations between the crack size and other fracture parameters have been proposed and tested [11,12]. Here we take the experimental curve relating the applied load with the incremental crack size as input and calculate the displacements, strains, stresses, internal variables, and plastic energy as output.

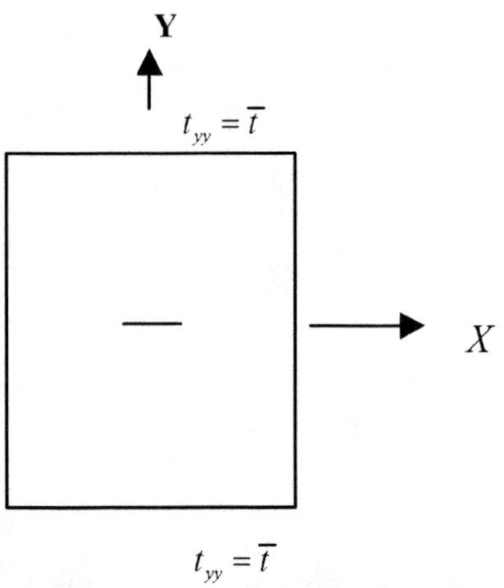

Fig. 1 Center-cracked specimen

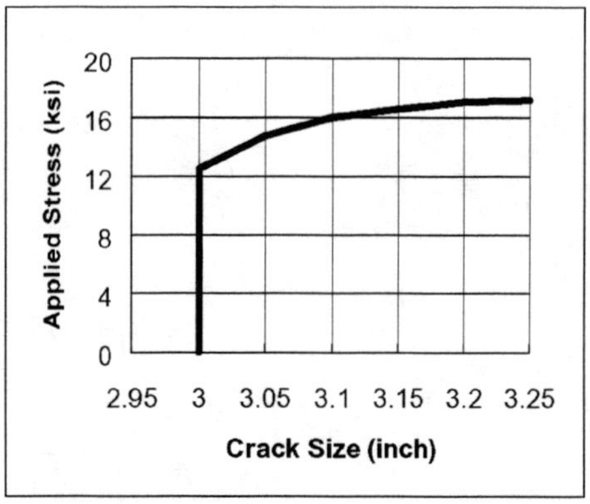

Fig. 2 Applied stress ($t_{yy} = \bar{t}$) vs. crack size (a)

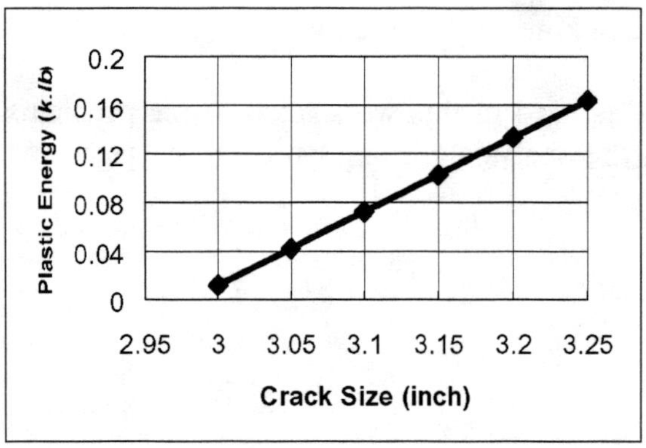

Fig. 3 Plastic energy vs crack size

A rectangular plate of length $2L$, width $2W$, and thickness B, with a centered line crack of initial crack size $2a^o$ subjected to symmetric boundary conditions is shown in Fig.1. The specimen is made of 2024-T3 aluminum alloy. A realistic experimental curve relating the applied load $t_{yy} = \bar{t}$ and the crack size a is shown in Fig.2. The J_2 flow theory is used to describe the material behavior of the cracked specimen. Its elastoplastic behavior during a monotonically increasing load (cf. Fig.2) has been simulated. The relevant input data are:

$$L = 6.0 \text{ in.} \quad , \quad W = 6.0 \text{ in.} \quad , \quad a^o = 3.0 \text{ in.} \quad , \quad B = 0.062 \text{ in.}$$

$$E = 10300 \; ksi \quad , \quad \upsilon = 0.33 \quad , \quad \sigma_Y = 55 \; ksi \quad , \quad H = 250 \; ksi$$

and, for the case of plane strain, a combined isotropic/kinematic hardening rule, $c = 0.5$, is assumed. The applied loading $t_{yy} = \bar{t}$ increases from 12.5 *ksi* at $a^o = 3.0$ in. to 17.1 *ksi* at $a =$ 3.25 in.

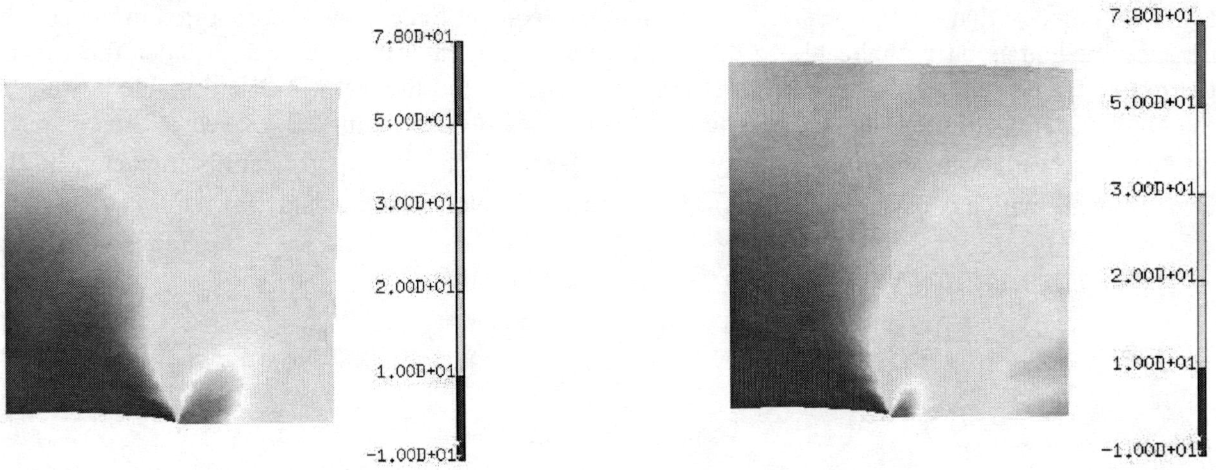

(a) at initial crack size $a = 3.0$ inch (b) crack size grows to be $a = 3.25$ inch

Fig.4 Stress distribution (t_{yy}) (plane stress case)

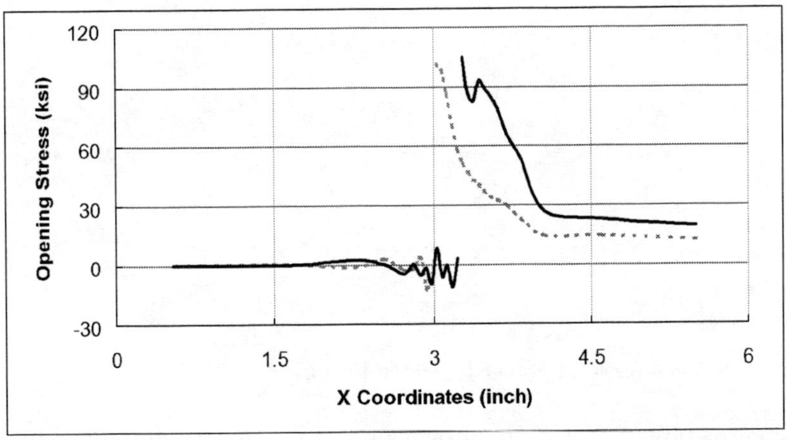

Fig. 5 Stress distribution (t_{yy}) along the line of crack (plane strain case)

solid line: a = 3.25 inch, dot line: a = 3.0 inch

Some of the numerical results are shown in Fig. 3 to Fig.5. For plane stress case the plastic energy (= plastic energy density integrated over the area of the specimen) is shown as a function of the crack size in Fig.3. The stress distributions of t_{yy} at $a = a^o = 3.00$ inch and $a = 3.25$ inch for plane stress are shown in Fig.4. For plane strain case the opening stresses t_{yy} at $a = 3.00$ in. and at $a = 3.25$ in. are plotted for $y \cong 0.125$ in., $x \in (0,\ W)$ in Fig.5.

4. Impact/Contact

High-speed impact is a dynamic large-strain elastic-plastic problem. It is also a contact problem between two material bodies. During the process of contact, penetration is prohibited but separation is allowed. In fact, separation begins as soon as the pair of points in contact has the tendency to move apart. It is noted that in meshless method the displacement, as well as the velocity and the acceleration, of a point is not equal to the nodal value associated with that point; rather, it is a combination of nodal values of points in the neighborhood of that point; however the instantaneous contact force being incorporated in the central difference method is only applicable to the point in question. In this work, we consider the impact between two bodies: one is a rigid cylinder (radius =0.4 meter) and the other is a deformable beam (0.2 meter by 2 meters), as shown in Fig.6. There are two cases: (1) the cylinder is fixed in space and is impacted by a beam with an initial velocity $v^o = 50m/\sec$, (2) the fixed end beam is impacted by the cylinder moving with a constant velocity $v^c = 10m/\sec$. The relevant input data are

$$E = 0.21*10^{12}\ N/m^2\ ,\quad \upsilon = 0.28\ ,\quad \sigma_Y = 0.2*10^9\ N/m^2\ ,$$
$$H = 0.21*10^9\ N/m^2\ ,\quad c = 1\ ,\quad \rho^o = 0.785*10^4 K_g/m^3\ .$$

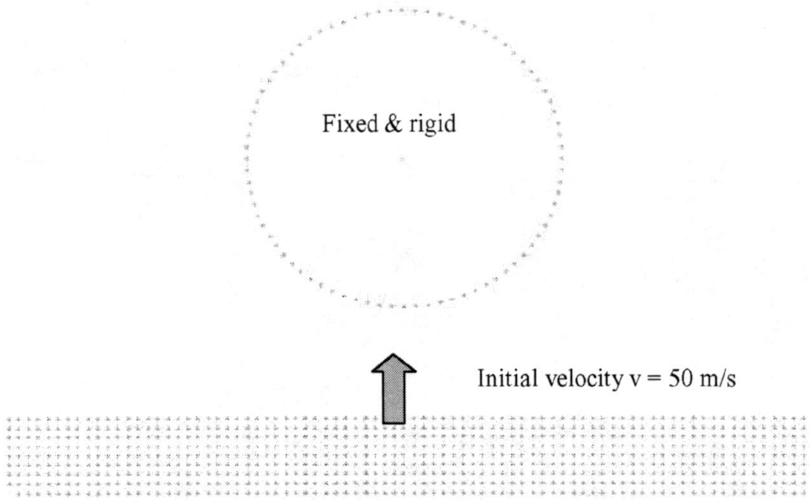

Fig.6 A rigid cylinder impacted by a beam

Some of the numerical results of case 1 are shown in Fig.7. They are the von Mises stress distribution during the contact/separation process. Fig.8 are snapshots of case 2. They are von Mises stress and normal stress σ_{yy} (normal to the contact surface), respectively.

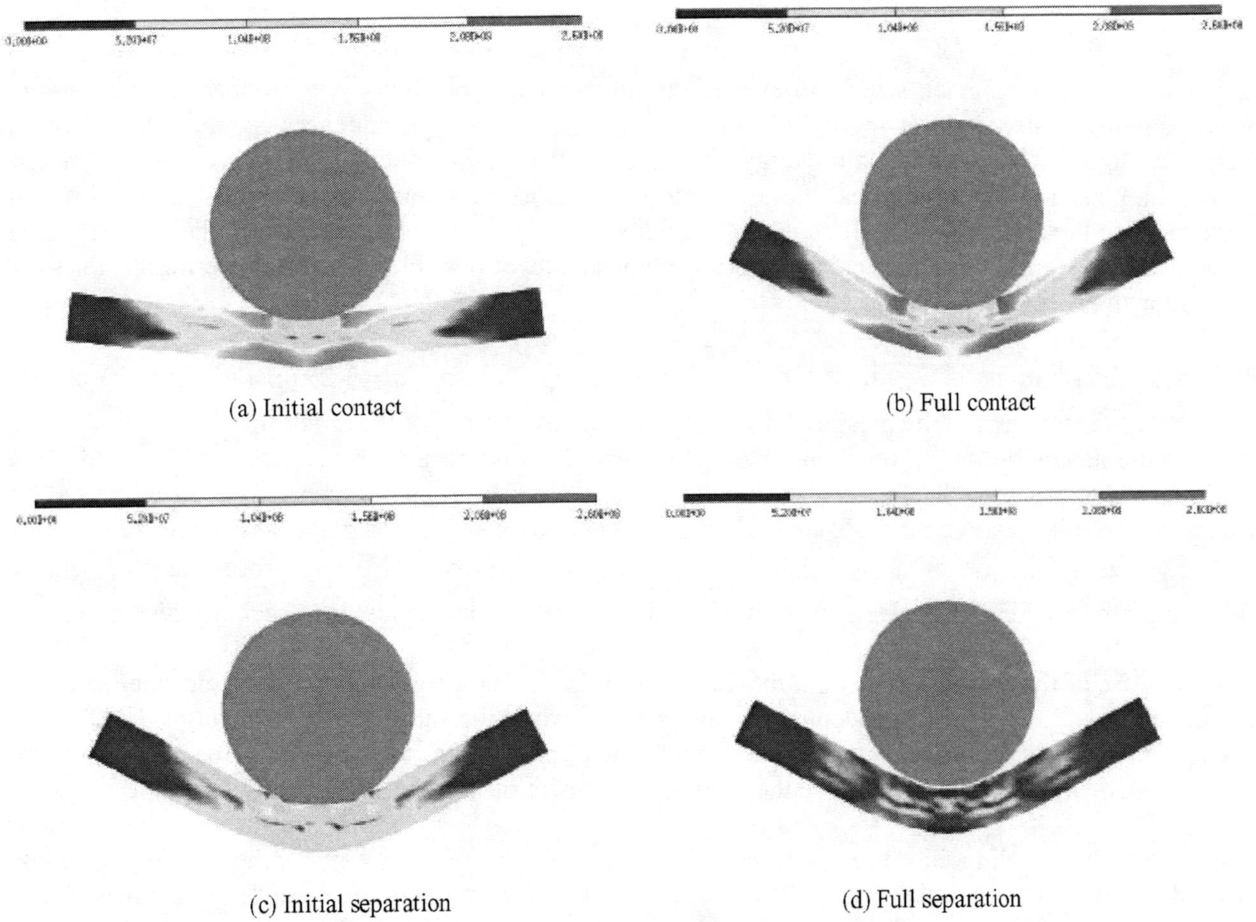

(a) Initial contact

(b) Full contact

(c) Initial separation

(d) Full separation

Fig.7 Von Mises stress distribution during impact/contact/separation (impact with initial velocity)

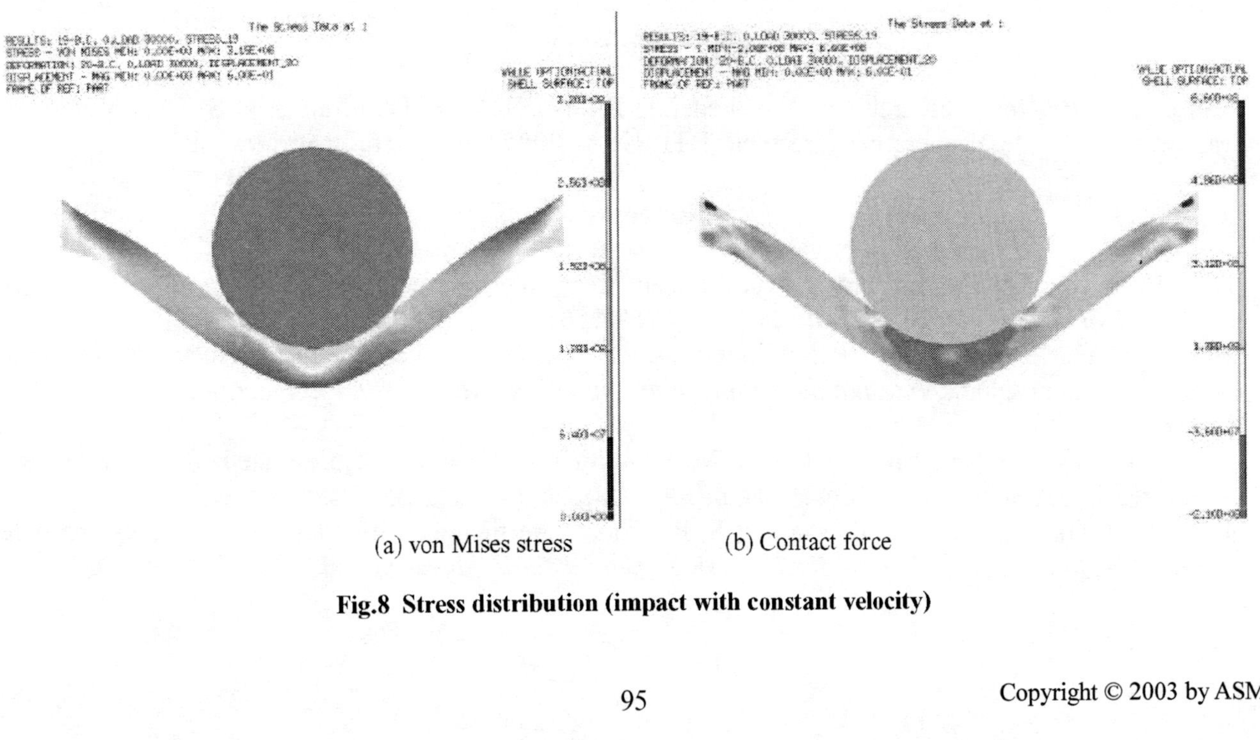

(a) von Mises stress

(b) Contact force

Fig.8 Stress distribution (impact with constant velocity)

5. Discussions

In fracture mechanics, crack size is an ever-increasing variable, plasticity is a dissipative process, and plastic energy is also an ever-increasing quantity. These are the consequences of the second law of thermodynamics. Therefore, plastic energy should be a monotonic function of crack size and, in this sense, Fig.3 has a profound physical meaning. Moreover, since the input data is a realistic experimental curve relating the applied load and the varying crack size (Fig.2), the relation between plastic energy and crack size being linear is of particular importance and makes it possible to serve as a criterion for slow crack growth process [11,12].

The stress distributions of t_{yy} at $a = a^o = 3.00$ in. (cf. Fig. 4a) and $a = 3.25$ in. (cf. Fig. 4b) clearly show that, as crack grows, the plastic region and the stresses in front of the crack tip increase and in the wake of the advancing crack tip an unloading process is on-going and the stresses are reduced. This indicates that crack growth process as a moving boundary value problem has been solved properly and accurately. For illustrative purpose, we use the same input data, Fig.2, for plane strain case. The stress t_{yy} plotted as a function of x in Fig.8 also indicates that the plastic loading process and the elastic unloading process are taking places in front of and in the wake of the advancing crack tip, respectively.

The first case of the impact problem, a moving beam impacts on a rigid and fixed cylinder, aims at the simulation of a moving vehicle encountering an obstacle, while the second case, a moving rigid cylinder impacts on a fixed end beam, belongs to a general category of metal forming problems. A much more realistic but difficult problem would be the analysis of impact process between two deformable bodies.

Impact/contact modeling has been a research subject pursued by many researchers using meshless methods as well as finite element methods. The meshless method started to capture the interest of a broader community of researchers only several years ago, and now it becomes a growing and evolving field. Meshless method has overwhelming advantages in modeling vehicle impact/crash process involving large deformations. It is showing that this is a very rich area to be explored, and has great promise for many very challenging computational problems in the field of transportation safety.

Acknowledgement

Sponsorship for this research by the Federal Highway Administration of US Department of Transportation, under cooperative agreement DTFH98-C-00054, is gratefully acknowledged.

References

[1] T. Belytschko, Y. Y. Lu and L. Gu, Element-free Galerkin methods, *International Journal for Numerical methods in Engineering* **37** (1994) 229-256.

[2] S. Jun, W. K. Liu and T. Belytschko, Explicit reproducing kernel particle methods for large deformation problems, *International Journal for Numerical Methods in Engineering*, **41** (1998) 137-166.

[3] T. Belytschko, Y. Krongauz, D. Organ, M. Fleming and P. Krysl, Meshless methods: An overview and recent developments, *Comput. Methods Appl. Mech. Engng.* **139** (1966) 3-47.

[4] T. Belytschko, Y. Guo, W. K. Liu and S. P. Xiao, A unified stability analysis of meshless particle methods, *International Journal for Numerical methods in Engineering,* **48** (2000) 1359-166.

[5] J. Chen, C. Wu and T. Belytschko, Regularization of material instability by meshfree approximations with intrinsic scales, *International Journal for Numerical methods in Engineering,* **47** (2000) 1303-1322.
[6] Y. Chen, J. D. Lee and A. Eskandarian, Dynamic meshless method applied to nonlocal cracked problems, *Theoretical and Applied Fracture Mechanics*, **38** (2002) 293-300.
[7] Y. Chen, A. Eskandarian, M. Oskard and J. D. Lee, Meshless analysis of plasticity with application to crack growth problems, *Theoretical and Applied Fracture Mechanics,* to appear.
[8] J. C. Simo and T. J. R. Hughes, Computational Inelasticity, Springer, New York, 1988.
[9] R. D. Krieg and S. W. Key, Implementation of a time dependent plasticity theory into structural computer programs, in: J. A. Stricklin, K. J. Saczalski (eds.), Constitutive Equations in Viscoplasticity: Computational and Engineering Aspects, AMD-20, ASME, New York, (1976)
[10] T. Belytschko, W. K. Liu and B Moran, Nonlinear Finite Elements for Continua and Structures, John Wiley & Sons, Chichester, 2000.
[11] J. D. Lee, H. Liebowitz and K. Y. Lee, The quest of a universal fracture law governing the process of slow crack growth, Engineering Fracture Mechanics 55 (1996) 61-83.
[12] K. Y. Lee, H. Liebowitz and J. D. Lee, Finite element analysis of the slow crack growth process in mixed mode fracture, Engineering Fracture Mechanics 56 (1997) 551-577.

Proceedings of IMECE'03
2003 ASME International Mechanical Engineering Congress
Washington, D.C., November 15–21, 2003

IMECE2003-42277

FEEDBACK CONTROL OF BRAKE SYSTEM ON RAILWAY VEHICLE CONSIDERING NON-LINEAR PROPERTY AND DEAD TIME

Masanobu Nankyo
Vehicle Control Technology Division,
Railway Technical Research Institute,
2-8-38 Hikari-cho Kokubunji ct.,
Tokyo 185-8540 JAPAN

Tadashi Ishihara
Department of Education
Fukushima Univ., Fukushima, JAPAN

Hikaru Inooka
Graduate School of Information Sciences
Tohoku Univ., Sendai, JAPAN

ABSTRACT

Increase of deceleration in high-speed and high-density train operation degrades riding comfort and frequently causes wheel skids. This requires an introduction of the control engineering to upgrade the control performance of brake systems on rolling stocks. We are now studying a control method for mechanical brakes that uses friction and pneumatic pressure, including non-linear elements as the basis of brake force. Furthermore, the system itself has certain "dead time", which is not ignorable and makes controlling difficult. One of our targets is to develop a brake control device that can control deceleration in accordance with a decelerating pattern that optimizes riding comfort and prevents wheel skids. In this paper, a design method of the controller for the deceleration tracking control and the system compensating the dead time are proposed. Finally, the effects of them are confirmed through computer simulations and experimental results.

1. INTRODUCTION

Brake performance of railway vehicles is represented by an index called "deceleration" which is a negative acceleration during braking. Vehicles equipped with the electric controlled pneumatic brake system are designed to apply the braking forces assigned to the brake handle notch according as planned deceleration. Signaling system is also designed based on the deceleration of each train, and frequently brake performance tests are carried out on the service tracks in order to confirm if the actual deceleration of the train can satisfy the specification. Thus the deceleration is a significant physical index for the brake systems.

Actual deceleration of a train under braking is a result of friction force, which is generated by a brake shoe pressed on a wheel tread with an air cylinder. So the value of deceleration is affected by the non-linear property of friction coefficients and the compressibility of air, it fall off from expectation so often. Excess deceleration causes wheel skidding, flat spots of tier and results in worse riding comfort. On the contrary, the lack of deceleration leads to braking distance expansion, which is serious problem on safety.

We are studying a control method of pneumatic brake system which makes the actual deceleration on the track to be desirable by using the feedback control theory to solve the problems above mentioned. Controlling of brake cylinder pressure according to the error between actual and reference deceleration is expected to reduce the error itself and to increase robustness against fluctuation of parameters or disturbance. In addition, pneumatic system involves essentially transmission delay, that is to say, dead time, which influences the stability of a feedback control system significantly. We tried to introduce Smith's method which is effective on compensating of the dead time and upgrading of the response. In this paper, modeling of each composing elements of a pneumatic brake system is introduced at first, and the design of control system considering both the non-linear property of each element and the dead time compensation is mentioned. Effect of this control method is confirmed with numerical simulations and experimental results.

2. MATHEMATICAL MODELS OF THE ELEMENTS CONSISTING PNEUMATIC BRAKE SYSTEM

Focusing on one of wheels on a vehicle, the pneumatic brake system can be represented as the Single-Wheel-Model shown in Fig.1. Define two time domain functions, that is, desired deceleration $\beta_{ref}(t)$ as input and actual deceleration $\beta(t)$

as output. The relation between input and output in the brake system is given by:

$$\beta(t) = k_0 \frac{\mu}{\mu'} \frac{p}{p_{ref}} \beta_{ref}(t) \tag{1}$$

Here k_0 is a constant including modeling error, p is pressure in the brake cylinder, p_{ref} is desired pressure, μ is the actual friction coefficient between the wheel tread and the brake shoe, μ' is the constant called average friction coefficient that is used as setting value in the control device. In Eq.(1), p / p_{ref} describes pneumatic property in the cylinder, and μ/μ' dose friction property.

2.1 Model of Brake Cylinder Pressure

In the beginning, we will examine the pneumatic property of the brake cylinder. During braking, the brake shoe is pressed forward the wheel tread by the brake cylinder. Under such condition, it can be supposed that there is no movement of the cylinder rod stroke, so the brake cylinder can be considered as a fixed volumetric reservoir. Therefore, let us introduce the model that the reservoir (volume: V, initial pressure: p_0, temperature: θ) is filled by the compressed air of which pressure is p_{ref} through the orifice (the effective area: a) as shown in Fig.2 in order to describing pneumatic property of compressed air in the brake cylinder. Suppose the time domain function $p(t)$ as the cylinder pressure, and the time origin t=0 is set at the moment when filling is started. Generally, the flow velocity around the orifice will change according to the pressure ratio $p(t)$ / p_{ref} [1]. The pressure ratio has a critical point where the velocity turns from the sound speed to the subsonic speed. As the pressure ratio is increased up to 1.0, the subsonic speed passing after the critical point is reduced. According to the properties above mentioned, $p(t)$ is represented by the following expressions according to the ratio of the pressure inside to that outside of the cylinder:

(a) $p_{ref} > 1.893 p_0$

$[p_0 < p(t) < 0.528\, p_{ref}]$

$$p(t) = p_{ref} \frac{t}{t_n} + p_0 \tag{2}$$

$[0.528\, p_{ref} \le p(t) \le p_{ref}]$

$$p(t) = \frac{p_{ref}}{2} \sin\left(\frac{t - t_c}{t_n} + 0.028\right) + \frac{p_{ref}}{2} \tag{3}$$

(b) $p_{ref} \le 1.893 p_0$

$$p(t) = \frac{p_{ref}}{2} \sin\left(\frac{2t}{t_n} + \sin^{-1}\left(\frac{2p_0}{p_{ref}} - 1\right)\right) + \frac{p_{ref}}{2} \tag{4}$$

where:

$$t_n = \frac{V}{185 a p_s \gamma} \sqrt{\frac{273}{\theta}} \tag{5}$$

$$t_c = \left(0.528 - \frac{p_0}{p_{ref}}\right) t_n \tag{6}$$

and p_s is atmospheric pressure, γ is ratio of specific heat, θ is absolute temperature in the cylinder. Line (a) in Fig.3 shows the case of (a), and line (b) does the case of (b).

Time domain response of the brake cylinder pressure described by Eq. (2), (3), (4) corresponds to the step response the input value of which is p_{ref}. It is efficient and conventional to approximate the step response of pressure by the transient response of first order lag element, because its mathematical structure is simple and many kinds of design techniques for control systems which have it as a plant have been proposed. However, in our case, Eq. (2),(3),(4) show that the time period while the pressure rises up to 63.2% level of the reference value is changed according to input value, i.e. the reference pressure p_{ref}, although simple first order lag element has constant rise time. Therefore it is required to consider the time constant as a variable which changes according to the input value for more accurate approximation of pressure property by first order lag structure. Let us call a variable T defined by the following equation the equivalent time constant:

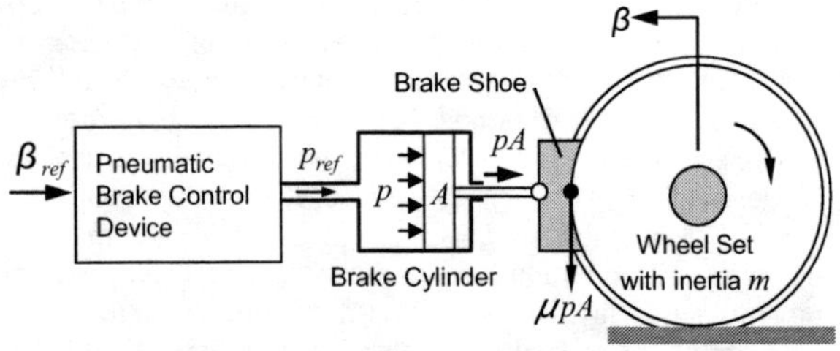

Fig.1 Pneumatic Brake System

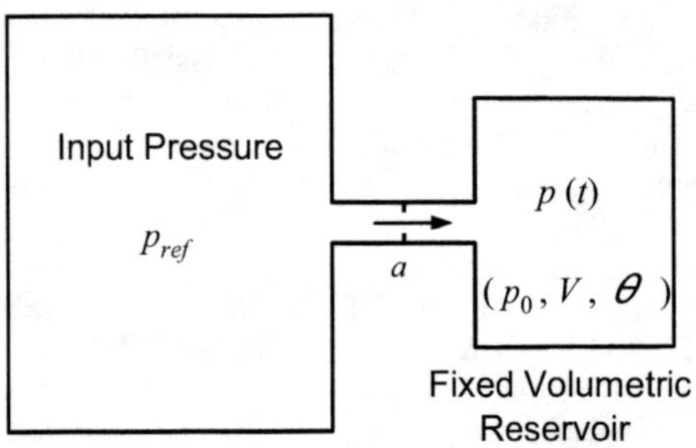

Fig2.Model of Brake Cylinder

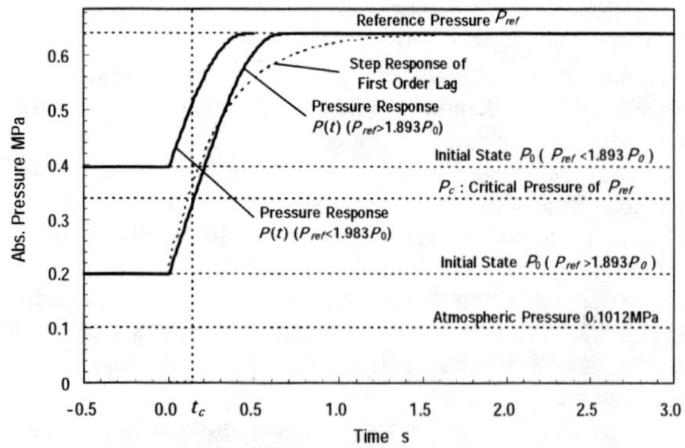

Fig.3 Step Response of Cylinder Pressure

$$p(T) = 0.632(p_{ref} - p_0) + p_0 \tag{7}$$

Actual value of T can be obtained easily with instrumentation and the maximum T comes from the maximum pressure on the specification of the train set concerned.

2.2 Dead Time

There are two kind of dead time (time delay) in the system. One is attributed to the transmission of pneumatic pressure by the sound speed, another is the time for calculation of deceleration. Firstly, the transmission speed of pressure follows the equation for the speed of sound u given as:

$$u = \sqrt{\gamma g R \theta} \tag{8}$$

Here g is acceleration due to gravity, and R is the gas constant. In our case, we consider that θ is assumed to be alike and constant, and the piping distance from the pneumatic control device to a brake cylinder is not varied. Therefore u is constant and also the transmission delay of pressure can be treated as a constant value. In Japanese typical EMU, it is estimated to be around 50msec on the average.

It is required to obtain actual values of deceleration of the vehicle on real time for the deceleration control. Most of EMUs have the speed sensor on its wheel axle, and so numerical differentiation method can be applied as a real time estimator of deceleration from measured speed data. In most cases, the signal from the speed sensors contains noise and the results of simple differentiation are significantly affected by noise. To reduce the effect of it, smoothing operation by a certain order polynomial approximation must be accompanied [2]. Therefore the system is obliged to contain the calculation delay equivalent to the order of smoothing. The real-time algorithm for differentiation using a Kalman filter or H_∞ filter is being studied. However it has been noticed that there remains some problems for application as mentioned by Nakazawa et al [3].

To simplify the model and the system, we assume that these delays studied above are equivalent each other. Finally, the following transfer function $G_{BC}(s)$ is obtained as a mathematical model of the brake cylinder pressure dynamics:

$$G_{BC}(s) = \frac{1}{1 + T s} e^{-Ls} \tag{9}$$

Here, T is treated as a variable parameter within $0 \leq T \leq T_{max}$, L denotes the equivalent summation of dead times above.

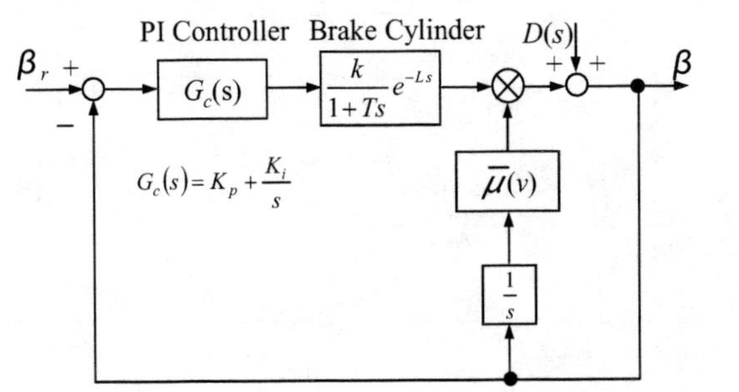

Fig.4 Brock Diagram of Deceleration Control

2.3 Friction Coefficient

Friction force generated from the contact between a brake shoe and a wheel tread is the basis of decelerating force on the vehicle. It has been said that the friction coefficient is typically seemed that its value is increasing as relative speed between them is decreasing. It can be called "negative ascending property." Based on this property, many kinds of mathematical model for the friction coefficient between them have been so far proposed such as hyperbolic function and exponential et al. We used the exponential function of the traveling speed of a vehicle, which is equivalent to the friction speed, because it can expressed well the experimental results which we have got so far. It is given by:

$$\mu(v) = a_1 \, exp(-a_2 v) + a_3 \tag{10}$$

Here v is vehicle speed, and constants set of a_1,a_2,a_3 should be identified from the measured results. Suppose that the constant term a_3 equals to μ' in Eq.(1), following model normalized by μ' is used for the system design instead of Eq.(10).

$$\bar{\mu}(v) = \frac{a_1}{\mu'} exp(-a_2 v) + 1 \tag{11}$$

2.4 Controlled Object and Controller

Through the above examinations, the following mathematical model of a brake system on railway vehicles is obtained as a controlled object:

$$G_p(s) = \frac{k_0 \bar{\mu}(v)}{1 + Ts} e^{-Ls} \tag{12}$$

Here $\bar{\mu}(v)$ in Eq.(12) should be handled as a variable parameter the range of which is decided from Eq.(11) and vehicle speed v, although Eq.(12) is expressed in the form of a transfer function. Using this model, the feedback control system of brake deceleration is represented as a block diagram shown in Fig.4, in which the PI (proportional plus integral) controller is applied as a controller.

3. DESIGN OF THE CONTROLLER

First, the specifications of deceleration control system are proposed to design the controller. In the actual operation of trains, the deceleration command from a driver or an onboard signaling device changes in "stepwise" so often that step response shape should be specified to keep riding comfort at certain level. Therefore the time response is set as a specification so that step responses are overdamping without overshoot nor undershoot. The tracking performance is also set as a specification so that the tracking error between actual and reference decelerations should be within a certain range (e.g. 0.5% of the reference).

3.1 Controller for Overdamping Response

Removing the delayed term e^{-Ls} from Eq.(12) and supposing k_0=1, the following nominal transfer function is obtained:

$$\overline{G}_p = \frac{\overline{\mu}(v)}{1+Ts} \quad (13)$$

PI controller is given as:

$$G_c = K_p + \frac{K_i}{s} \quad (14)$$

Here K_p is the proportional gain and K_i is integral gain. Consequently, the closed loop transfer function between the reference β_{ref} and the output β is given as:

$$G_o = \frac{G_c \overline{G}_p}{1+G_c\overline{G}_p} = \frac{\overline{\mu}(v)(K_p s + K_i)}{Ts^2 + (1+K_p\overline{\mu}(v))s + K_i\overline{\mu}(v)} \quad (15)$$

In the case where $K_p=0$, the necessary and sufficient condition satisfying that the step response of the system given by Eq.(15) becomes overdamping is that two poles are stable and real. In the case where $K_p \neq 0$, the condition for the same situation is given by:

$$4T\overline{\mu}K_i \le (1+\overline{\mu}K_p)^2 \quad (16)$$

It is noted that Eq.(16) is not sufficient but necessary. It can be confirmed easily that the necessary and sufficient condition for that the step response of the transfer function with two stable poles in the real number given as $G(s)=K(s+\gamma)\ /\ (s+\alpha)(s+\beta)$ $(0<\alpha\le\beta)$ becomes overdamping is $\gamma\ge\alpha$. By applying this condition with Eq.(16) to Eq.(15), the necessary and sufficient condition for the overdamping step response is obtained as follows:

$$\left\{K_i \le \frac{K_p}{T} \text{ AND } K_i \le \frac{\overline{\mu}K_p^2 + K_p}{2T}\right\} \text{ OR } K_i \ge \frac{\overline{\mu}K_p^2 + K_p}{2T} \quad (17)$$

Inequalities (17) can be expressed as a subspace under the curved surface which is formed in the coordinates system (K_p, K_i, v) by In.(17) as shown in Fig.5. Two surfaces in Fig.5 are plotted for each case of T=0.4 and T=0.8. They show that the surface moves downward when the equivalent time constant is increased. In addition, the surface is distorted by the effect of speed domain function $\overline{\mu}(v)$. Therefore, using Fig.5, we can design a required controller visually by drawing a segment parallel to v-axis considering a non-linear property of the friction coefficient. The segment which gives parameter set of (K_p, K_i) satisfying In.(17) is obviously under the surface with no intersection of it.

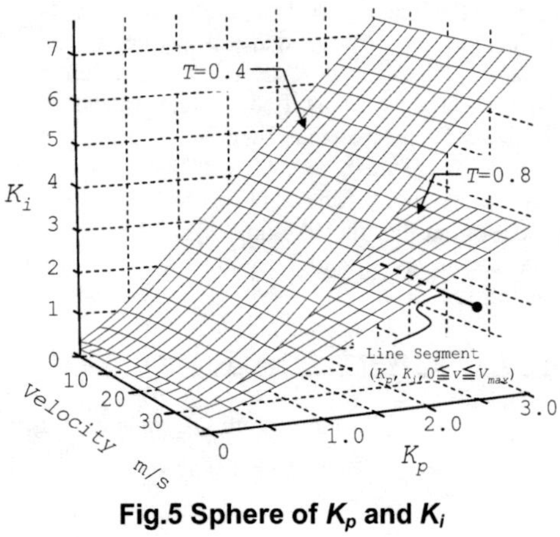

Fig.5 Sphere of K_p and K_i

3.2 Confirmation of tracking performance

It is clear that the steady state position error of the nominal system given by Eq.(13),(14) is zero by using the final value theorem. However, the control action is quitted when the a train or a vehicle stops (v=0) because our model is based on the brake system. Therefore the position error of the response should be considered within finite time range from starting point of brake to stopping. The friction coefficient in Eq.(11) has a great influence on the tracking performance. However it is difficult to obtain analytically the time response due to its non-linear property. Therefore, for the present, we use a computer simulation tool like MATLAB/Simulink to obtain and evaluate the time response numerically for design.

3.3 Numerical example

Using Step Response Method [6], suppose that a value set of controller parameter is obtained as a guide:

$$K_p = 2.4,\ K_i = 0.8 \quad (18)$$

The line segment by them can be placed under the T = 0.8 surface in Fig.5, so the step response becomes overdamping. According to the computer simulation of the system shown in Fig. 3 with controller given by Eq.(18), the results are plotted as Fig.6-a. Here the reference input is 1.0m/s^2 height step signal, initial speed of braking application is 50km/h, and the dead time is assumed L=0. As can be seen from this curve, the error from the reference is relatively large from start to end, though it is overdamping. Therefore it is difficult to say that the controller is sufficient for tracking, and it is required to adjust K_i by using Fig. 5 in order to reduce the tracking error. T=0.8 surface in Fig.5 is indicating the upper limitation of K_i, and it gives K_i < 3.0 when K_P = 2.4. For example, the K_i=2.0 response curve is plotted as Fig.6-b. It is clear that the response has been improved in tracking performance.

4. COMPENSATION OF THE DEAD TIME BY THE SMITH PREDICTOR

In the previous section, the method and procedure to

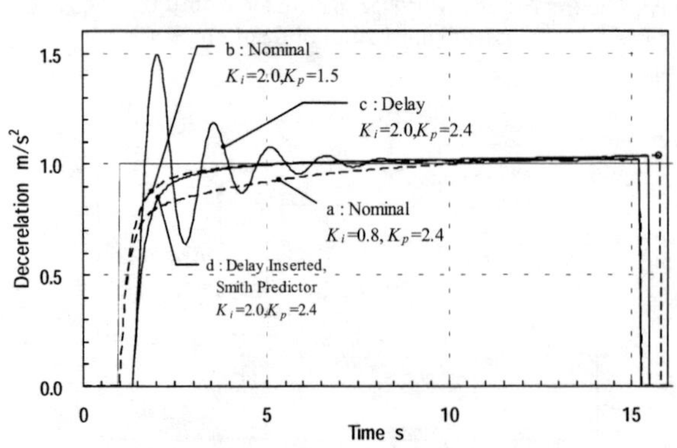

Fig.6 Simulation Results

design the controller for nominal model of brake cylinder is developed without taking account of the dead time element. However, by considering the dead time shown in Eq.(12), the response becomes oscillating trajectory as shown in Fig.6-c. Although this response is damped oscillation which converges around the reference deceleration until stop, it is not practical for railway vehicle braking due to several overshoots and undershoots. Therefore we introduce the Smith predictor as a compensator for the dead time [7]. The Smith predictor was traditionally used to compensate the time delayed system [8]. The closed loop system including the Smith predictor is shown in Fig.7, which predicts both of the genuine output and that through the dead time. As the characteristic equation of this system has no dead time term, we can apply the design method for the plant with no dead time, such as PID, kinds of compensation methods and so on. After rebuilding the simulation model of the system with the Smith predictor, almost the same curve as Fig.6-b can be obtained as shown in Fig.6-d. It is noticed that there is a time delay L between rising of the reference input and the response output.

4.1 Confirmation of Robustness

As well known, by applying the Smith predictor, there is possibility that the model mismatch or the error of delay resulting from parameter changing of the controlled object affects the stability of the system [9]. Let us remember that our model of brake cylinder represented by Eq.(9) has a variable parameter as equivalent time constant T, and so the confirmation of robustness against it is required.

For simple calculation, suppose that the pressure in the brake cylinder ascends linearly with constant speed (the speed

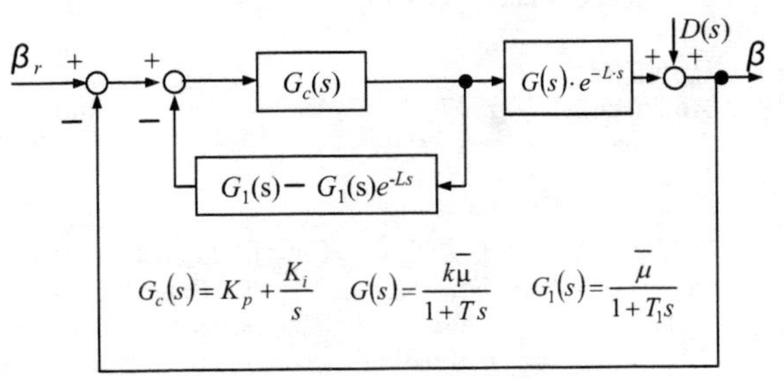

Fig.7 Closed Loop with Smith Predictor

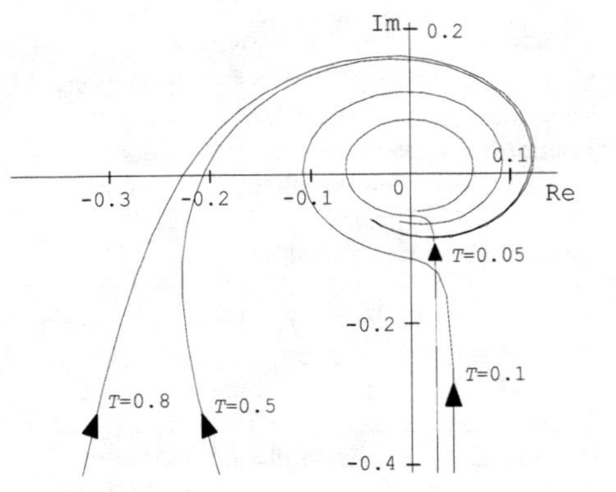

Fig.8 Nyquist Prot of Variable Parameter System

of sound) even after passing the critical point in the model shown in Fig.2. The equivalent time constant when the cylinder is pressurized from atmospheric pressure p_s to maximum working pressure p_{max} is assigned to T_1 in $G_1(s)$ included in the Smith predictor as follows:

$$T_1 = 0.6321\frac{t_n}{p_{max}}\left(p_{max} - p_s\right) \tag{19}$$

The time constant T in the controlled object model $G(s)$ is also obtained using the initial pressure p_0 and the reference pressure p_{ref} similarly as follows:

$$T = 0.6321\frac{t_n}{p_{max}}\left(p_{ref} - p_0\right) \tag{20}$$

Then, the following ratio is obtained:

$$\frac{T}{T_1} = \frac{p_{ref} - p_0}{p_{max} - p_s} \tag{21}$$

Equation(21) shows that T is in direct proportion to the input amplitude ($p_{ref}-p_0$). Suppose $k=T/T_1$ in Eq.(11), the controlled object is represented as follows:

$$G(s) = \frac{T}{T_1}\cdot\frac{\bar{\mu}}{1+Ts} \tag{22}$$

Using symbols in Eq.(15), the following Open-loop transfer function is obtained:

$$G_0(s) = \frac{G_c(s)G(s)\cdot e^{-Ls}}{1+G_c(s)G_1(s)\left(1-e^{-Ls}\right)} \tag{23}$$

The Nyquist loci by Eq.(23) are shown in Fig.8, in which the equivalent time constant T in $G(s)$ is reduced gradually from T=0.8 to T=0.05, and the other parameters are the same as the numerical examples shown in the previous section. It was found that the diameter of this vortical locus becomes small and the locus is moving apart from the point -1 ± j0 as T decreases

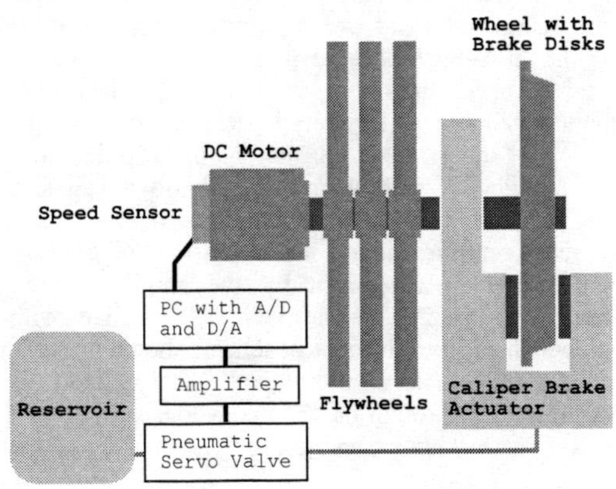

Fig.9 Experimental Setup

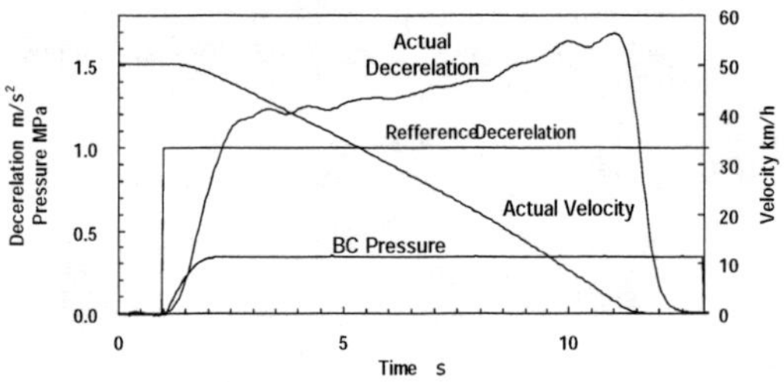

(a) Conventional Open Loop

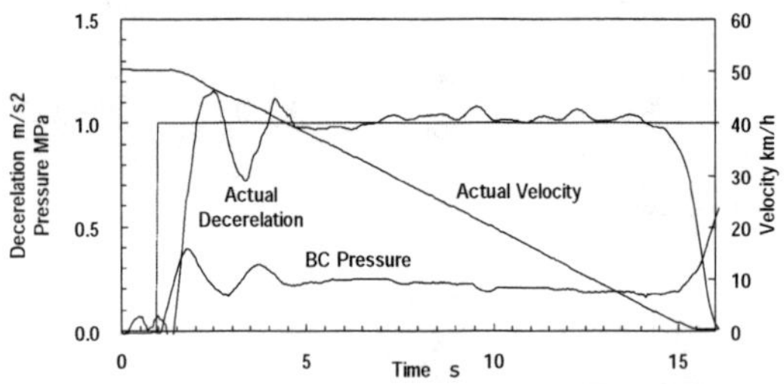

(b) Feedback control without Smith Predictor

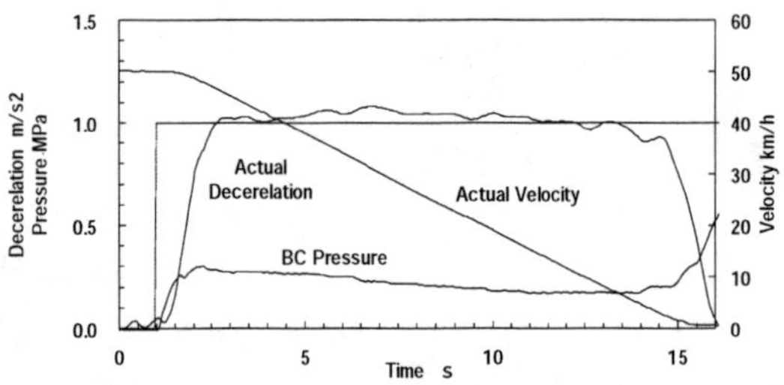

(c) Feedback control with Smith Predictor

Fig.10 Experimental Results

as shown in Fig.8. This indicates that the robustness of our system against a variable time constant will be ensured if the equivalent time constant T_1 in the Smith predictor is defined in accordance with the maximum working pressure.

4. EXPERIMENTAL RESULTS

To verify the concept mentioned so far, some experiments using a dynamo test stand were performed. Our experimental setup is shown in Fig.9. The dynamo test stand has an actual wheel with cheek disks and an axle set with fly wheels corresponding to 1/8 of one electric vehicle's inertia, which is driven by a DC motor placed at the end of the axle, and has a caliper equipment as a braking actuator. So it can simulate decelerating states similar to an actual train. In our experiment, the pressure controller device set, consisting of a PC and a pneumatic servo valve driven by the amplifier and D/A converter, was installed. The PC was installed with the program which realized the control scheme shown in Fig.7 with real time calculation at 50ms interval. The axle speed was sampled in the PC through the A/D converter and deceleration of the axle was calculated using a polynomial approximation method [2].

Fig.10 shows experimental results using the same parameters and controller in the numerical example. Fig.10-(a) is the result with an open loop system, which is conventional in Japanese EMU. In this trajectory, the error from the reference trajectory is getting large as the speed is reduced. In Fig.10-(b), feedback system with the controller designed by Fig.5 was applied. However the trajectory has several overshoot like as the simulation result shown as Fig.6-c. The reason is considered to be the existence of dead time above mentioned. Finally, we have got Fig.10-(c), where the Smith predictor is applied with the same controller as the case of Fig.10-(b). It is clear that this trajectory shows a smooth rising and the good performance of tracking to a reference trajectory with the proposed control system.

5. CONCLUSION

Tracking controls of train deceleration with pneumatic brake system considering its non-linear property and dynamics are studied in this research. Mathematical models for a brake cylinder and friction coefficients are built and the design method for the controller using a 3D diagram is proposed. Finally, the Smith predictor is introduced for a delay compensator. The effects of it are confirmed through computer simulations and experiments on tracking control under a step form input. The tracking error can be reduced significantly with the proposed control system design. For practical use, it is required to study the methods for structuring this controller as an actual vehicle brake system, and also necessary to confirm the behavior of response when wheel slips occurred.

ACKNOWLEDGMENTS

The authors are grateful to the member of brake control research group in Railway Technical Research Institute for a lot of good suggestions and discussions.

REFERENCES

[1] Andersen, B., 1967, The Analysis and Control of Pneumatic Systems, JOHN WILEY & SONS, INC., New York, pp.17-28.

[2] Savitzky, A., Golay, M.j.E., 1964, "Smoothing and Differentiation of Data by Simplified Least Squares Procedures," Analytical Chemistry, 36, 8, pp.1627-1639.

[3] Nakazawa, S., Ishihara, T., Inooka, H., "Real-Time Algorithm for Estimating Jerk Signals from Noisy Acceleration Data," Proc ICMIT, (2001), pp.423-428.

[4] Shimizu, T., Inooka, H., 1986, "Estimation of Solid Friction Using a Kalman Filter (1st Report, Simulation Studies for a Specified Friction Model)", Trans. JSME. Mech. Eng., C53-498, pp.2608-2612 (in Japanese).

[5] Suzuki, H., Miura, R., 1967, "The Jerking Motion in Servomechanisms, Considering Negative Slope of Friction Characteristics," J. Inst. Elect. Engrs. Jpn. 87, 950, pp. 2261-2268 (in Japanese).

[6] Ziegler, J. G., Nichols, N. B., 1942, "Optimum Settings for Automatic Controllers", Trans. ASME., 64, pp. 759-768.

[7] Smith, O. J. M., 1959, "A controller to overcome dead-time", ISA Trans., 6, 2, pp. 28-33.

[8] Schneider, D. M., 1988, "Control of Processes with Time Delays," IEEE Trans. Autom. Control, 24, pp. 186-191.

[9] Palmor, Z., 1980, "Stability Properties of Smith Dead-Time Compensator," Int. J. Control, 32, 6, pp. 937-949.

Proceedings of IMECE'03
2003 ASME International Mechanical Engineering Congress
Washington, D.C., November 15–21, 2003

IMECE2003-41214

ADVANCES IN ACTUATION SYSTEMS TO IMPROVE VEHICLE PERFORMANCE AND SAFETY

John S. Brader, Nathan R. Trevett, and David N. Rocheleau

University of South Carolina
Department of Mechanical Engineering
Advanced Actuators Research Group
300 Main Street
Columbia, SC 29208
Phone: 803-777-5648 Fax: 803-777-0106
brader@engr.sc.edu

ABSTRACT

Actuation and modeling for drive-by-wire applications are discussed. Previous work in advanced actuation for two automotive systems, active suspension and camless engines, is surveyed, outlining the major advancements throughout the previous decade. More recent research in these areas is discussed and focuses on recent improvements to system modeling and design. Specifically, a four-corner suspension model and a piezoelectric piloted hydraulic actuator for engine valve actuation are introduced. The four-corner suspension model addresses the three-dimensional parameters associated with active suspension design and improves upon the accepted quarter-car model. The piezoelectric based camless engine actuator is introduced as the next generation of camless engine actuation systems and addresses control issues through the relationship between input voltage and piezoelectric displacement.

INTRODUCTION

Automobiles at their onset were nearly entirely mechanical; however, throughout a century of development more and more systems have come to be controlled electronically. The advantages of electronically controlled components are that they often offer a faster response, provide increase control and variability through computer-controlled interfaces, and are not as prone to wear as mechanical devices. Furthermore, by eliminating frictional losses associated with mechanical members, electronic controls offer higher efficiencies.

X-by-Wire is the generic term used when relatively inaccurate mechanical systems are replaced with more precise electronic sensors and actuators. This X-by-Wire trend has been evident in the automotive industry for years, and morphing the phrase "fly-by-wire" (circa 1972) borrowed from the aerospace industries, has now become known as "drive-by-wire" in the automotive industry. The trend can be seen in the implementation of computer controlled fuel injectors to replace the carburetor and the development of electronically controlled ABS brakes. Drive-by-Wire is not a new program to implement, but a phrase capturing the existing trend of development and pointing in the direction of future advancements.

One of the major components of the Drive-by-Wire evolution is a family of advanced actuators. Actuators will soon be used to apply force to the brake pads when stopping and increasing the throttle when accelerating. They will be used to turn the wheels in cornering giving a new meaning to the term "power steering." These advanced actuators will altogether change the automotive world.

This paper will focus on two areas under development that are the most in need of advanced actuators and will make the most impact on the automotive field. The first is active suspension—a new form of ride control that will improve comfort and handling while adding a new level of safety. The second is the camless engine—an alternate means of engine valve actuation that will improve every aspect of engine performance from power delivery to fuel consumption.

NOMENCLATURE

CLE – Camless Engine
EVT – Electronic Valve Timing
ICE – Internal Combustion Engine
VCT – Variable Camshaft Timing
VTB – Virtual Test Bed
VVT – Variable Valve Timing
c_s – Damping coefficient of suspension

c_t – Damping coefficient of tire
k_s – Spring constant of suspension
k_t – Spring constant of tire
m_b – Mass of the body
m_t – Mass of the tire
x_b – Position of body
x_r – Position of road
x_s – Length of suspension
x_t – Length of tire
x_w – Position of wheel

ACTIVE SUSPENSION

Background

Automotive suspensions are a compromise between comfort and control. Traditional suspension systems are comprised of springs and dampers that store and release energy. Spring rates and damping coefficients are adjusted to set the balance between these two opposing criteria. Using conventional means, a single valued compromise is selected and set during the development of the vehicle. Whether it is a suspension system for a Cadillac limousine or a Ferrari sports car, some balance between comfort and control is chosen. As opposed to this passive, classical suspension, active suspension is the name given to an approach that allows this compromise to vary while driving. Active suspension can be divided into two primary areas: semi-active and fully active as described below.

The first level of advanced suspension is semi-active suspension. In this system one or more parameters of the suspension are automatically varied based on sensor inputs. The adjustments capable of on-the-fly manipulation are the ride height of the vehicle, the stiffness of the damper, and the spring rate. The adjustments are accomplished by changing either the pressure in the components or some internal geometry. For example, hydraulic dampers can be adjusted by using fast acting valves to change the orifice size between two fluid flow compartments. Pneumatic springs can be adjusted by changing the available volume the fluid can fill. Overall ride height can be controlled by the pressure in the chambers of the shock absorber unit. These are just a few examples of how pumps, reservoirs, and valves can be used to adjust and vary the parameters of a semi-active suspension.

Fully active suspensions are quite simple in concept but challenging in implementation. They consists of four actuators, one at each wheel, along with a full suite of sensors to monitor motion and force characteristics which are then fed back to a computer-controlled algorithm whose output is used to control the individual actuators. Fully active suspensions have the ability to detect bumps and then quickly compensate for them before the bumps are felt in the passenger compartment. They are also capable of reducing or eliminating the roll, pitch, and yaw of a vehicle felt during cornering, hard braking and accelerating. Cars with fully active suspensions can actually "lean" into turns to improve handling. A fully active suspension is a goal of suspension designers; however, the power requirements and cost considerations are currently high.

The price of improved comfort and control is paid in the increased weight, power consumption, and cost of active suspension systems. Sensors are needed to monitor several parameters: road modulations, steering, braking, speed, vehicle position and acceleration in three dimensions. Complex logic is required to account for all possible driving conditions. Fully active suspensions need actuators capable of large displacements (several inches) in microsecond time intervals under loads of several thousand pounds. The tires are capable of absorbing frequencies above 20-40 Hz, but the actuators must eliminate the rest [1].

Semi-active suspensions are becoming a reality in production vehicles. Volvo has included electronic controlled dampers designed by Ohlins Racing and Tenneco Automotive in their Four-C system. This Continuously Controlled Chassis Concept is one of the most advanced active suspensions systems available [2]. The Corvette along with other GM vehicles now offer magneto-restrictive shock absorbers.

Citroen has developed one of the first production level active suspension systems. It is being used on the Infiniti Q45 (Model G50 series). Pressure in the shock absorber can range from 0 to 1400 psi and is aided by spring supports [2]. Automotive industries and universities are continuing to work on advanced actuators to bring the potential of active suspension into reality.

Automotive Dynamics Model

The first step to develop or evaluate an effective active suspension is to understand the principles that govern vehicle dynamics. This can best be done through the modeling of a traditional suspension system.

An accurate model of a physical system is an invaluable tool. Models allow testing to be performed before physical systems are ever constructed. This drastically reduces cost, decreases development time, and improves the overall performance of the final product. Creating a model begins with an understanding of the physical system. Physical systems can be characterized using mathematical terms. The dynamics of an automobile is no exception.

A common tool used for suspension evaluation is a quarter-car model. This model examines one corner of the vehicle and its response to road disturbances. A schematic of a quarter-car model is shown in Fig. 1. This model is made up of two spring-mass-dampers placed on top of each other.

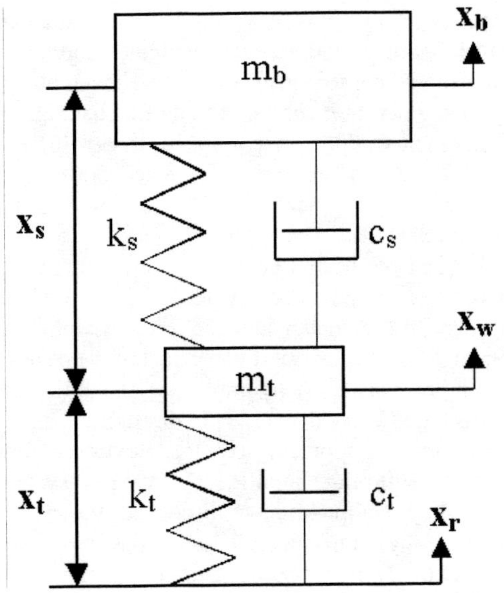

Figure 1: Quarter-Car Model [3]

The quarter-car model can be mathematically represented using Eq. (1) and Eq. (2) [4].

$$m_b \cdot \frac{d^2x_b}{dt^2} + c_s \cdot \frac{dx_b}{dt} + k_s \cdot x_b = c_s \cdot \frac{dx_w}{dt} + k_s \cdot x_w \qquad \text{Equation (1)}$$

$$m_t \cdot \frac{d^2x_w}{dt^2} + c_s \cdot \frac{dx_w}{dt} + (k_s + k_t) \cdot x_w = c_s \cdot \frac{dx_b}{dt} + k_s \cdot x_b + k_t \cdot x_r \qquad \text{Equation (2)}$$

A quarter-car model is a good means of simplistic analysis; however, it only captures a portion of the information needed to make an informed decision about a suspension system. The quarter-car model is confined to operate along a single axis. It is not capable of capturing the complexities, such as roll and pitch, of an actual vehicle.

Roll (side-to-side angle) and pitch (front-to-back angle) are the two parameters often negatively affected by improving the ride of a vehicle. The softer and more comfortable the ride becomes, the more severe the roll and pitch. Considering this, any model intended to optimize the performance of a suspension system must be able to observe not only responses to standard bumps and potholes but also to turns, accelerations, and other disturbances that tend to cause roll and pitch.

A full-car or four-corner model is one that can analyze the motions of an actual vehicle. A four-corner model allows each of the four tires complete freedom to encounter various bumps or accelerations. The four tires link through the suspension system to influence the motion of the car body. A four-corner model is a complex series of mathematical equations; and thus, it is difficult to develop a complete four-corner model. At this level of complexity each model has its own intricacies so the discussion will now move away from general models and universal rules to the specific model developed for this work.

A full-car model, partially shown in Fig. 2, has been developed at the University of South Carolina by the Advanced Actuators Research Group. The model can be quickly configured to represent any four-wheeled vehicle. The suspension subsystem can be removed and replaced with any form of active, semi-active or passive suspension system model. The model displays the performance of the vehicle as well as the power consumption of the suspension system.

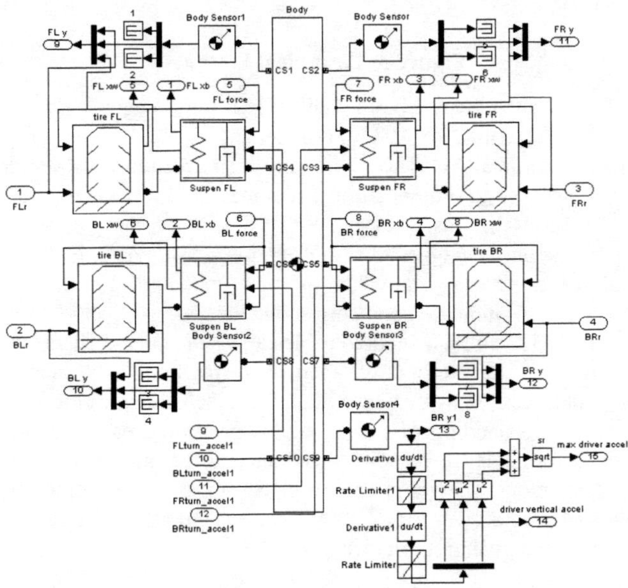

Figure 2: Full-Car Model

The full-car model has been developed using Matlab®/Simulink®. This model also utilizes the modeling power of SimMechanics®. SimMechanics takes physical bodies and constraints and formulates the appropriate equations of motion [5].

This model enables evaluation of any conceivable suspension system. Any system that can be mathematically characterized can be placed into the model. Furthermore, the course or track that the car follows is completely adjustable as are all of the parameters of the vehicle and suspension system. The model outputs real-time positions of any desired component, net accelerations experienced by the driver, and the power consumed by the active suspension. A screen shot of graphical outputs can be seen in Fig. 3.

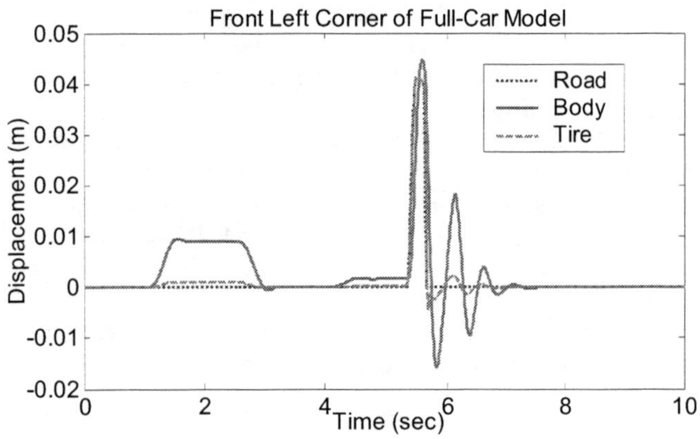

Figure 3: Graphical Outputs

The model is being validating using a Legends racecar [5/8-scale replica of 1937 Ford Sedan] at the University of South Carolina [6]. The car is fully instrumented with a suite of sensors and computational devices. Wireless network connections are used to relay the data to a laptop, which converts and analyzes the information. The data is being compared to its equivalent virtual model. Preliminary comparisons indicate good correspondence.

The full-car model enables a systems approach for the evaluations of automotive active suspension. Numerous actuators can be evaluated without ever being constructed. A good understanding of the principles governing each actuator enables a model to be constructed and placed into the existing full-car model. Criteria can be placed on the most important attributes in a given situation and the system can be optimized to meet the given requirements.

ENGINE VALVE ACTUATION

Background

Variable valve timing (VVT) efforts have been underway for several years and have taken primarily two distinct routes: mechanical VVT and camless valve trains. These enhancements to automotive engines are in response to the compromises associated with using a cam lobe to control the motion of engine valves. Primarily, the fixed profile of a single cam results in an engine that optimally runs at only a single rpm value. Furthermore, the single cam lobe results in a compromise between power and fuel efficiency.

Recognizing this compromise, automobile manufacturers have been attempting to provide vehicles capable of cylinder deactivation, variable valve timing (VVT), or variable camshaft timing (VCT). These new designs are mostly mechanical in nature. Although they do provide an increased level of sophistication, most are still limited to discrete valve timing changes over a limited range.

Early in the development of variable engines, Cadillac introduced its V-8-6-4 engine [7], which had the ability to deactivate cylinders. The engine changed the number of active cylinders by adjusting the position of the rocker-arm fulcrum, essentially preventing the rocker arm from displacing the engine valves. The Cadillac V-8-6-4 was the standard engine for all 1981 Cadillac models, but the engine experienced a short production run. Due to consumer complaints about the engine response and operation, especially when changing from one mode to another, Cadillac discontinued its variable engine. As an update to the short lived Cadillac V-8-6-4, GM will introduce its "Displacement on Demand" engines in some of their 2004 models [8].

Instead of the cylinder operating changes offered by the new GM engine, Honda introduced its VTEC engines to address the need for greater levels of engine sophistication [9]. This design incorporates multiple cam lobes and rocker-arms for each engine valve. A hydraulic spool valve connects the rocker arms together by driving a pin through the units.

As another approach to VVT, Lexus has developed a "variable valve timing, intelligent" (VVT,i) system for its engines. They claim to have produced the next level of sophistication by introducing continuously variable valve actuation [10]. The on-board computer monitors the engine demands and continuously adjusts the timing and overlap of the intake and exhaust valves.

Regardless of the VVT technology differences among the leading automotive manufacturers, the prime similarity of a camshaft remains. Therefore, limitations continue, since the timing is still a function of engine speed. These limitations have initiated research into camless engine technology.

Originally, camless engines (CLE) were developed for use as a design aid to automotive engine manufacturers. The use of a camless engine allowed the engineer to experiment with valve timing as a means of designing cam profiles. These early units were not limited by size or power consumption constraints. Instead they were solely developed for laboratory use as a design tool.

Aside from laboratory use, history shows that the idea of a camless internal combustion engine had its origins as early as 1899, when designs of variable valve timing surfaced [11]. It was suggested that independent control of valve actuation could result in increased engine power. More recently, however, the focus of increased power has broadened to include energy savings, pollution reduction, and reliability.

To provide the benefits listed above, researchers throughout the previous decade have been proposing, prototyping, and testing new versions of valve actuation for the internal combustion engine. Their designs have taken on a variety of forms, from electro-pneumatic to electro-hydraulic and more recently electromagnetic and piezoelectric piloted systems have surfaced.

A comprehensive project using solenoid control of pneumatic actuators was completed in 1991 [11]. This research included the development of the actuators, a 16 bit microprocessor for control, and comparative testing between a standard Ford 1.9 liter, spark ignition, port fuel injected four cylinder engine and the same engine modified for camless actuation. The actuators used during the research required an off-engine power source because an engine mounted compressor was not feasible. As Gould et al. state, their work cannot be considered feasible for implementation due to the high power requirements of the actuator.

In 1996 the next generation of camless engine was completed at the Ford Research Laboratory by, principally, Michael Schechter and Michael Levin [12]. They introduced a new concept of the hydraulic pendulum actuator. The use of a hydraulic pendulum decreases the system's energy consumption by converting the kinetic energy of a closing valve into potential energy stored in the pressurized fluid. This reduces the energy required for pumping the hydraulic fluid.

The hydraulic pendulum also allows for the solenoid-based-system to slow valve velocity. This results in soft seating the valve and is a favorable attribute of the updated system. Another benefit is the ability to vary the opening and closing velocity of the valve. This allows for increased variation to engine valve parameters.

Through the use of a hydraulic pendulum, a complete four cylinder ICE was produced and found some success. However, the system is complicated and requires multiple components. The use of a hydraulic pendulum requires two solenoids and two check valves per engine valve and both a high pressure and low pressure hydraulic fluid supply. (Schechter et al. state that two solenoids can run a pair of valves as long as the pair is synchronized. However, this detracts from the concept of fully independent valve control.)

The camless engine developed by Ford and described above was then enhanced at the University of Illinois at Urbana-Champaign. The focus of the project was to advance the hydraulic-pendulum-based CLE actuator by developing an adaptive feedback control. The complete system, using the improved controller, was limited to operating at 3000 rpm. Valve lift greater than 5 mm could not be consistently controlled [13].

More recent projects, including those at Siemens Automotive – Europe, International Truck and Engine Corporation, the University of Michigan, and the University of South Carolina represent the next phase of CLE research.

In June, 2000, Sturman Engine Systems and International Truck and Engine Corporation completed their proprietary CLE project by being the first CLE based truck to reach the summit of Pikes Peak [14]. The International engine with Sturman hydraulic valves is a diesel system. Experts familiar with the project have raised concerns about the electrohydraulic system capabilities at the higher speeds of gasoline based passenger cars [15].

Concurrent to the International Truck and Engine and Sturman Engine Systems development, Siemens Automotive has been advancing their version of the CLE. Siemens claims their Electromechanical Valve Train (EVT) is nearing production capabilities [16]. Again, the system is proprietary, but some details are available. The Siemens' CLE is operating on the anticipated 42 volt bus and comes from an electrical distributor and crankshaft-mounted starter-generator. Special algorithms provide the current signal to the engine actuators and are capable of controlling valve velocity and soft-seating. Their actuator is based on a spring/mass system. Early results show a fuel savings of 10 percent. Siemens is currently working on the next phase, which will focus on sensorless actuator control [16].

At the University of Michigan, Anna Stefanopoulou and her research team are working on improvements to an electromechanical actuator's control algorithm. These improvements are designed to better control the seating velocity of the engine valve using an electromagnetic field to decelerate the actuator. Improvements to the controller allow for the valve to seat quietly and reduce component wear [17].

Piezoelectric Piloted CLE Actuator Development

A common factor among the publicized CLE work is the need for significantly improved control algorithms. This stems from the means by which the engine valves are actuated and the need to tightly control a variety of parameters: timing, displacement, and seating velocity. For these reasons, the University of South Carolina Advanced Actuators Research Group and its sponsor chose to approach the problem by developing a more easily controlled actuator. The newly devised system uses piezoelectric stacks to pilot a hydraulic amplifier.

The USC system uses the relationship between applied voltage and piezoelectric stack displacement to precisely control a hydraulic spool valve. The orifice areas of the hydraulic spool valve offer precise control over volumetric flow rates, and in-turn, provide control over engine valve opening and closing velocity, displacement, and timing.

Through the exploitation of the piezoelectric stacks' ease of displacement control, engine valve displacement profiles are easily altered. An example of engine valve profile shaping using simple open loop, ramp voltage input is shown in Fig. 4.

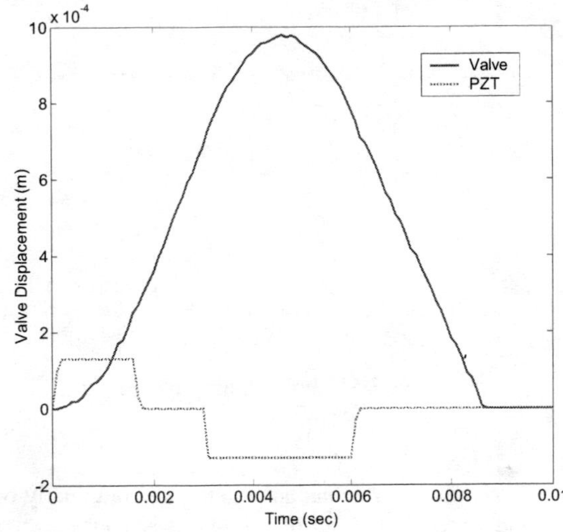

Figure 4: Valve Displacement Shaping

Similar to the previous discussion on suspension system actuators, much of the current USC CLE actuator research is focused on system modeling. Through the modeling of novel actuators, analysis and control algorithms can be completed prior to the actuators introduction onto an engine. This analysis, and especially an examination of control algorithms, is critical due to the effect an error in the controller could have on an operating internal combustion engine.

Several forms of modeling are taking place simultaneously and include Matlab/Simulink, mathematical modeling in C^{++}, and the introduction of models in the Virtual Test Bed (VTB). The VTB is a software environment developed at the University of South Carolina for simulating large-scale dynamic systems. It allows for interdisciplinary research, such as that associated with actuator development and implementation, because it is designed to simulate systems developed in a variety of discipline specific simulation packages. Through VTB, actuators, controllers, and virtual automotive systems can be combined, tested, analyzed, and altered. Furthermore, VTB is available at no charge; therefore, allowing for multiple sector collaboration.

The nuances and complexities of the piezoelectric piloted hydraulic actuators require improvements to existing models to capture the dynamics associated with the rapid actuation. Shown in Fig. 5 is the comparison among data, accepted state-space modeling, and a newly derived time step integration model. Although the time step integration model better captures the transport delay and step response, further improvements can be made.

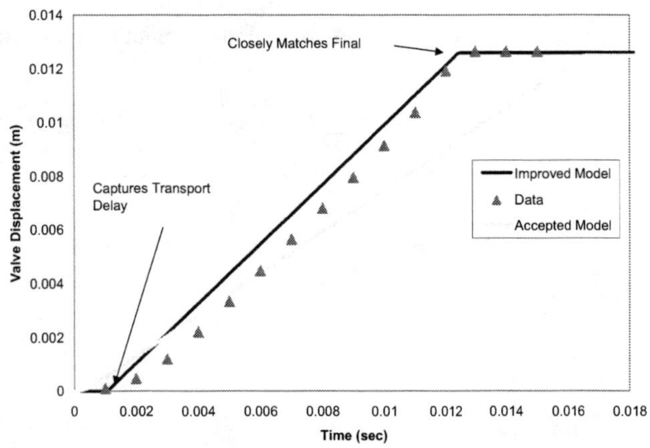

Figure 5 Modeling Comparison

NEAR FUTURE DEVELOPMENTS

Both advanced actuation systems are continually being revised and updated. Much of this work is concentrated on improvements to modeling, new simulation environments, actuator system prototypes, control algorithm development, and hardware testing. These programs are the focus of the University of South Carolina's Advanced Actuators Research Group.

The active suspension project is currently addressing comparison testing between the four corner model and empirical data. Upon verification of the model, it will be integrated into VTB and updated for user-friendly applications. This integration requires the four corner model to be expanded to its most fundamental parameters, so these parameters are accessible to the VTB user. Doing so will allow for a user with little knowledge of the model's intricacies to make changes to suspension, vehicle, and road parameters. This will allow for the broad implementation of the four corner model for improved suspension simulation.

Similar research is underway on the camless engine actuator. Modeling of asymmetric underlapped hydraulic spool valves with piezoelectric control is being developed. Models of the system components are also being created for use in VTB and will allow for complete camless valvetrain design and analysis to be performed. The system model is also being verified with empirical data from the actuator hardware. Furthermore, the actuator hardware is being integrated onto an internal combustion engine for variable inlet valve timing, displacement, and overlap control. For this integration of actuator and engine, control algorithms are being developed for controlling piezoelectric piloted hydraulic systems.

CONCLUSION

Considerable effort has been focused on the improvements to automotive components over the previous decade. Much of the early work has demonstrated both the potential and the limitations of drive-by-wire components. However, through the effort of a variety of researchers, the introduction of fully active suspension and camless engine valvetrains is becoming closer to a reality.

REFERENCES

[1] Leighton, N J; Pullen, J., 1994 "A novel active suspension system for automotive application," *Proc Instn Mech Engrs*, **208**, pp. 242-250.

[2] Gordon, Jacques. Feb, 1999, "Understanding Electronic Suspension Systems," *Motor Age*.

[3] ElBeheiry, E.M., Karnopp, D.C., 1996 "Optimal Control of Vehicle Random Vibration with Constrained Suspension Deflection," *Journal of Sound and Vibration*, **189**, pp. 547-564.

[4] Gillespie, Thomas D., 1992, Fundamentals of Vehicle Dynamics, Society of Automotive Engineers, Inc., Warrendale, PA., pp. 146-168.

[5] Wood, Giles D., 2003, "Simulating Mechanical Systems in Simulink with SimMechanics." Math Works, Inc., Natick, MA.

[6] Lyons, J.S., Morehouse, J.H., Young, E.F., 2001 "Developing a Systems Approach to Engineering Problem Solving and Design of Experiments in a Racecar-Based

Laboratory Course," *Journal of Engineering Education*, **90**, No. 1.

[7] Lamm, J., November 1980, "V-8-6-4, Cadillac Tries Multi Displacement," *Road & Track.*

[8] Anonymous, May 21, 2001, "General Motors – GM and the Environment." Date Accessed: June 13, 2001 <http://www.gm.com/company/gmability/environment/gm_and_the_env/releases/displacement_052101.htm>

[9] Mori, K., 2001, "Honda's High-Output LEV Engine" Date Accessed: June 13, 2001 <http://s2000.vtec.net/jdmhtml/high-output-lev.htm>

[10] Anonymous, 2001, "Lexus – Variable Valve Timing a First in an SUV," Autoworld. Date Accessed: June 18, 2001 <http://www.autoworld.com/RX300.htm>

[11] Gould, L; Richeson, W; and Erickson, F., 1991, "Performance Evaluation of a Camless Engine Using Valve Actuation with Programmable Timing," SAE Paper No. 910450.

[12] Schechter, M.; and Levin, M., 1998, "Camless Engine," SAE Paper No. 960581.

[13] Anderson, M; Tsao, T-C; and Levin, M., 1998, "Adaptive Lift Control for a Camless Electrohydraulic Valvetrain," SAE Paper No. 981029.

[14] Anonymous, Spring 2001, "First Camless Truck Reaches the Summit of Pikes Peak!" Power Stroke Registry.

[15] Murray, C., April 14, 2000, "Electronic Controls to Replace Camshafts in Diesel Engines," Date Accessed: July 7, 2000 <http://www.eetimes.com/story/industry/system_and_software_news/OEG20000414S0050.htm>.

[16] Ladd, D., September 13, 1999, "Camless Engine is Gaining Momentum," Siemens Automotive. Date Accessed: July 4, 2000 <http://media.siemensauto.com/mediacenter2/queries/releasefull.phtml?prjob_num=1195>.

[17] Sharke, P., 2003 "Controlled Breathing Better Air Management Improves Engine Efficiency, but at the Cost of Elaborate Controls," *Mechanical Engineering* **125** No. 5.

Proceedings of IMECE'03
2003 ASME International Mechanical Engineering Congress
Washington, D.C., November 15–21, 2003

IMECE2003-41226

AIR-ASSISTED MARINE VEHICLES FOR FUTURE USE IN TRANSOCEANIC TRANSPORTATION

Konstantin I. Matveev

California Institute of Technology, Pasadena, CA

ABSTRACT

The technical aspects of two types of advanced marine vehicles, Air Cavity Ship and Wing-In-Ground, are considered for future use in transoceanic transportation. Both concepts utilize the air medium to improve technical and economic characteristics of transportation means. The air is supplied under the bottom hull sections of the Air Cavity Ship, reducing the overall wetted surface and consequently hydrodynamic drag. A contact with water is completely eliminated in the case of the Wing-in-Ground, which moves above the water. This results in enhancing the lift-drag ratio in comparison with a flight in the open air. The principles, experience, and research opportunities on the way to make these concepts applicable for transportation are discussed.

INTRODUCTION

International transoceanic transportation is expanding fast due to a growth of the world trade and economy. Ships are the primary but slow means for carrying freight across oceans. Aircraft are used for transportation of express cargo, however at high cost. A growing demand for fast delivery of many cargo categories at low price shows an opportunity for intermediate transport means that would allow shipping at speeds higher than that of conventional marine vessels and at a cost lower than aviation can offer. Another factor influencing the development of future transportation is the energy-conservation policy. New transport means must have significantly better fuel efficiency than characteristic for existing options. Traditional ships and airplanes are already optimized to a point at which further considerable improvement of their performance is unlikely. Radical technological modifications are needed.

There is also a significant demand from military to accelerate a shipping of heavy equipment or a large amount of troops, as well as to reduce a time delay with respect to the delivery of forces brought by airlift. Besides, due to relatively easy sea access, maritime transportation is preferable.

Thus, the surface ships were, are, and will likely remain to be a main type of transportation across oceans. The increase of speed of power sea craft since their creation is significantly smaller than that of ground or air vehicles. The reason for this sad state of affairs is the media encountered by a ship. Water is much heavier and more viscous that air. Besides, ships operate in time-varying sea conditions, and additional structural requirements must be fulfilled to withstand traveling across oceans.

There are ideas to create a competitive ocean transportation making incremental improvements in hull hydrodynamics and propulsion systems; perhaps the most publicized example is FastShip Atlantic [1]. However, despite a lot of studies and activities on such projects over the years, they have not yet been realized (probably due to lack of funding and low projected economic gains), and their future is questionable.

A logical solution to improve the ship performance drastically is to reduce a contact between hull surfaces and water. Since the middle of last century, two types of craft aiming at this goal have found widespread application. Hydrofoils lift the boat hull above the water, leaving wetted the foils, struts, and propulsion mechanisms only. Hovercraft create an air cushion beneath the hull, limited by flexible skirt, or in the case of Surface Effect Ships (SES), by combination of solid and flexible hull components, which remain in contact with water. These boat types successfully found operation niches: hydrofoils are the best small- and mid-sized passenger fast ferries; hovercraft are used for military purposes and in remote areas inconvenient for other transportation means. Nevertheless, the transoceanic transportation is not suited for these craft. Efficiency of hydrofoils drops at a large scale (above a few hundred tons), which is necessary to cover large distances over rough water; and hovercraft and even SES require too much power (and respectively fuel), and therefore, are economically unfeasible for ocean routes.

Two other ship types based on the idea of reducing a contact with water and without size limitation are the subjects of this paper. An attractive option for conserving the energy spent by moving ships is to employ the air lubrication of wetted hull surfaces, which leads to a significant reduction of hydrodynamic drag. This concept utilizes the artificial cavitation, creating air cavities in the flow adjacent to ship hulls [2]. Air lubrication technology has already been tested on full-scale vessels classified as Air Cavity Ships (ACS). Significant drag reduction is achieved at a low power expense. The accompanied energy savings can be used either to reduce fuel consumption, keeping speed the same, or to increase ship speed without modifying the propulsive power.

In order to fill a niche between traditional air and ship cargo transportation, a vehicle with properties of both an airplane and a surface ship can be used. Such a craft, called Wing-In-Ground (WIG) and usually classified as a marine vessel, is a high-speed low-altitude flying vehicle that utilizes the surface effect to enhance its lifting characteristics [3]. The realistic lift-drag ratio of optimized WIG can reach 25-30. Operational speed of these vehicles is from 60 to 350 knots. Therefore, WIG provides superior efficiency at speeds between those of fast ships and airplanes. WIG's have been already developed for several decades and found practical application as large military machines and small passenger vehicles, but not yet as civil cargo transportation means.

The next sections of this paper are dedicated to the discussion of the principles, characteristics, configurations, experience, as well as research and development opportunities, to make these concepts suitable for transoceanic transportation.

AIR CAVITY SHIP

Air-cavity ships are advanced marine vehicles that use air injection at the wetted hull surfaces to improve their hydrodynamic characteristics. The concept of drag reduction by supplying gas under the ship's bottom was proposed more than a century ago. This idea looks deceptively simple, that is why there were many unsuccessful attempts to implement it in practice. To achieve positive results, physical understanding of multi-phase flows under a ship bottom is required. Based on the systematic research, several successful ACS's have been created and found practical application in recent years [4].

A schematic view of the ACS concept (high-speed variant) is shown in Fig. 1. An air cavity is generated behind a step by supplying gas using a blower or exhaust gases. The bottom geometry is designed to maximize the hull area covered by the cavity, and to minimize the amount of airflow rate delivered for supporting the cavity. The ACS performance is very sensitive to the hull lines chosen, in particularly to those that surround the cavity [5]. Drag reduction achieved on a properly designed full-scale ACS reaches 30%, while the power spent on the cavity-maintaining gas flow is always less than 3% of the total propulsive power. This performance enhancement is quite significant; it reduces fuel consumption and can save about $200K per year on a vessel with power system of 5,000 KWt. Pressure inside the cavity is higher than atmospheric, hence it carries an additional ship load.

The ACS principle may appear to resemble the SES concept; however, there are significant differences. First, there are no flexible seals on an ACS boat. The air layer is contained by solid hull parts, which not only prevent air leakage from the cavity, but also influence the air cavity characteristics. Secondly, the airflow rate needed to support the air cavity on an ACS is of order of magnitude smaller than on SES. Therefore, the ACS vehicle is a more economical means of transportation.

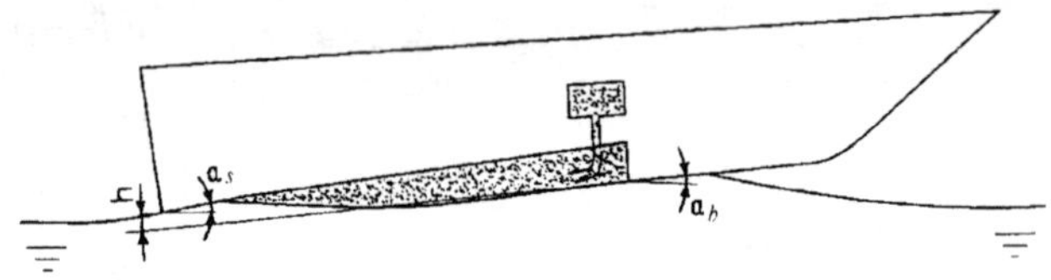

Fig. 1 Principal scheme of a high-speed ACS

Drag reduction by air lubrication can also be achieved on relatively slow vessels, such as tankers and cargo ships designed for transoceanic transportation. Because of the stability limit on the cavity dimensions, a different arrangement of the air cavity systems must be employed. If the ship length is large and its speed is not sufficiently high, the entire bottom of a vessel cannot be covered by a single cavity. This explains some of previous unsuccessful attempts to reduce drag by supplying gas through only a single nozzle in low speed regimes. Several air cavities must be created on ACS operating in the displacement mode (Fig. 2a). This scheme can be installed on already existing vessels, reducing investments needed for implementation of the ACS concept. It is also possible to create the optimized ACS vessel with a single air cavity formed in a certain speed range (Fig. 2b). This configuration has certainly better hydrodynamic properties, but requires more investment, since a new and more sophisticated ship must be designed and built.

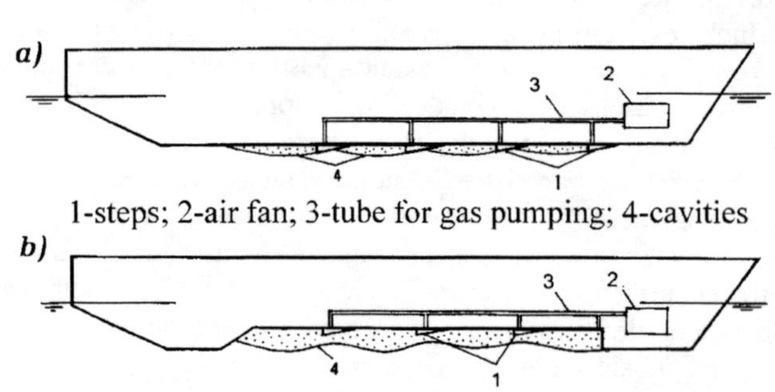

Fig. 2 Schemes of air cavity systems on displacement ships [6]

When a ship is moving in a semi-displacement regime, a significant portion of the hydrodynamic resistance is of the wave nature. In this case, the artificial cavitation is not as effective as in the case of slow or planning ships. However, the presence of a compressible air cavity smoothes the pressure gradients along the ship hull. This effect leads to the wave drag reduction and lower wash wake generated by a ship. A total increase of the efficiency of an ACS moving in a semi-displacement regime should reach 15-25% in comparison with conventional vessels [7].

There were numerous attempts to develop the ACS craft over the last hundred years, and most of them failed. The systematic and profound studies were carried out at the Krylov Shipbuilding Institute and at the Central Hydrofoil Design Bureau in Russia. These efforts resulted in creating small boats and sub-mid-size fast vessels using the artificial cavitation to reduce a hydrodynamic drag. Such ships are produced in series nowadays. Also, several successful full-scale cargo vessels operating in inland waters were equipped with air cavity systems (as depicted in Fig. 2).

Due to specific features of the ACS system, the design methodology is somewhat different from development of traditional ships [4]. In the first stage of design, the following steps are accomplished:

- Pressure distribution over the ship hull is determined either experimentally (towing model testing) or numerically (using tools of Computational Fluid Dynamics).
- Approximate calculation of the cavity system having parameters close to optimal (maximum cavity area at minimum gas consumption) is accomplished.
- Model testing (towing) with further modifications of the cavity system aimed to optimizing ship performance at various conditions (cruising regime; full-load and ballast mode; maneuvering; rough seas) is carried out.
- Hull-propulsor interaction is estimated. The selection and optimization of a propulsor is followed by a self-propelled testing.
- The results (resistance, trim, draft, cavity parameters, air consumption) are scaled to the full-scale ship configuration.

Knowledge and experience available from previous studies on ACS is a solid foundation for further developing the vessels of this type for transoceanic transportation. Research opportunities in this area include the following aspects:

- Laboratory study of artificial cavitation under a horizontal wall in various conditions (unsteady and non-uniform flow; various shape of cavitators; multi-cavity systems; etc.).
- Modeling and CFD study of artificially cavitating flows characteristic for ACS.
- Testing of towed models with cavities, and determining favorable configurations for different missions.
- Modeling of the ACS dynamics in rough seas and in maneuvering.
- Study of the hull-propulsor interaction.
- Study of the systems controlling cavity characteristic on the same boat for different regimes of operation.

One of the obvious economic advantages (fuel consumption per ton-mile of payload) is demonstrated in Fig. 3 for existing and projected fast sea-going ACS.

The ACS technology is proven on a full scale, its economic advantages over existing transoceanic vessels are significant, and this is an environmentally friendly technology (less amount of fuel is spent). Therefore, there is a high probability that funding, either government or private, will come to this field in the nearest future to realize a creation of ACS's for oceans. There are already several organizations in the USA, Europe, Australia, and Korea that started to work in this field. The SBIR and ONR solicitations have recently appeared that target various kinds of vehicle hydrodynamics in the presence of air at the wetted hull surfaces.

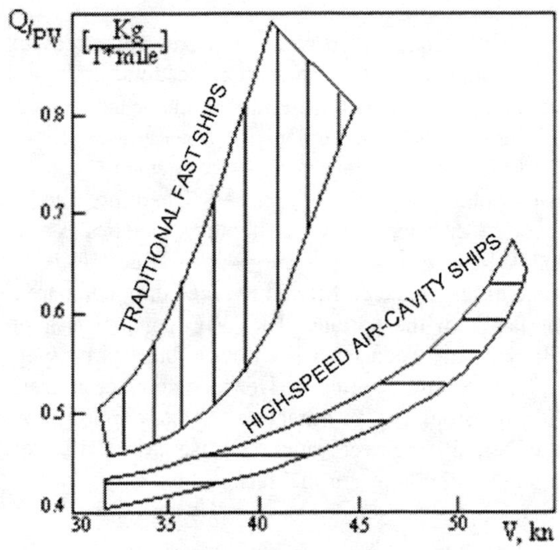

Fig. 3 Fuel consumption per ton-mile of payload for traditional fast ships and projected high-speed ACS's

WING-IN-GROUND

At the present time, there is a huge difference between two available transoceanic transportation means, aircraft and cargo ships, with regard to the cost and speed of delivery. If goods are sent via the ship option, it typically takes nearly 30 days for door-to-door service and costs about $0.03 per ton-mile of cargo transportation. Aircraft shipping provides 3 days delivery at the price of $0.30 per ton-mile. Due to this large gap in conditions of available transport service, there is an opportunity for intermediate solution that would seize some shares of both the air and sea shipping options, providing a delivery time close to the air shipping (7 days) but at a mean cost of two current services ($0.15 per ton-mile). Such a niche can be filled by the Wing-In-Ground (WIG), which combines advantageous characteristics of both aircraft and marine vessels (Fig. 4).

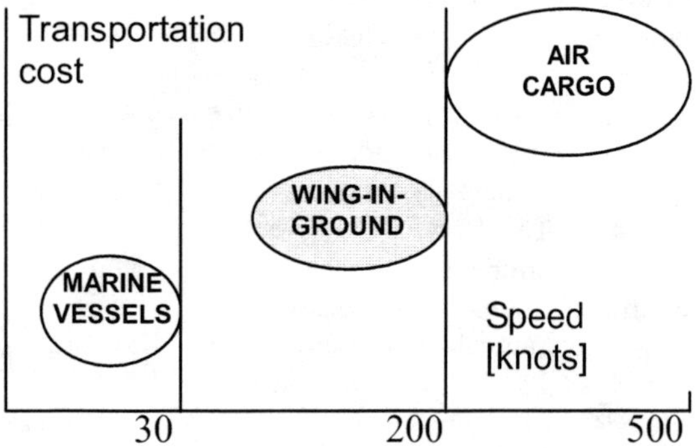

Fig. 4 Cost of transoceanic transportation by different means

The WIG concept, originated from advanced marine vehicle developments, is technically closer to aircraft, since in the cruising regime the WIG flies above the water surface (Fig. 5). However, this vehicle can land on water and can move as a surface ship; hence, it must satisfy requirements imposed on vessels capable of sustaining rough seas. Therefore, the WIG is considered by engineers and classification societies as a marine vehicle. When flying in the ground effect, WIG takes advantage of an enhanced lift and reduced drag compared with a flight far from the surface. The WIG concept is a proven technology; it has been used in military, but has not yet been applied for cargo transportation. Hence, a commercialization of the Wing-In-Ground is an attractive business opportunity, and can also benefit engineers and scientists who will work on adapting this technology for the transport industry.

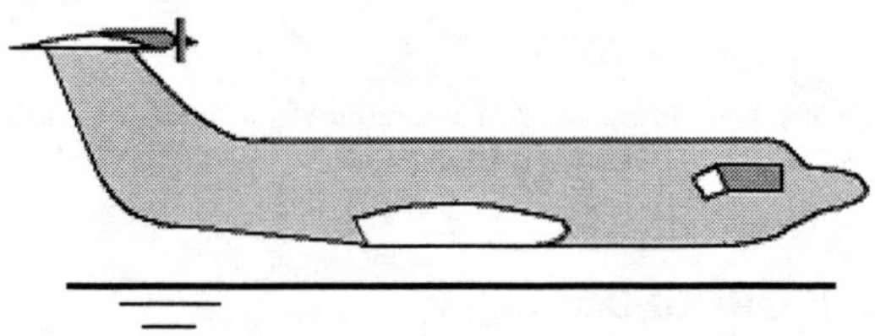

Fig. 5 Typical scheme of the second generation WIG

A dependence of the lift coefficient C_y on the reduced altitude (ratio of the flight height to the wing chord) is shown in Fig. 6 for the WIG Lun. When the flight height is less than a half of the wing chord, the lift coefficient increases significantly. Since the drag is usually decreased with an approach to a surface (due to smaller induced drag component), the lift-drag ratio becomes greatly enhanced. The total improvement in the efficiency of the WIG compared to aircraft can reach 50%; higher increments are more difficult to achieve given aero-hydrodynamic and structural constraints of WIG. It should be emphasized that the altitude-to-chord ratio is a major parameter characterizing the ground effect rather than the altitude-to-span ratio as considered in some studies, e.g. in [8]. This fact is due to dominant importance of the lift coefficient increase compared with the drag coefficient decrease.

Attempts to create the WIG craft have been undertaken by engineers since the 1930's. The most significant contribution to the progress of the WIG concept was made in Russia by the Central Hydrofoil Design Bureau under the guidance of R.E. Alekseev, who developed a number of unique test craft. The famous Caspian Monster KM with mass exceeding 500 tons still holds a record as the heaviest vehicle ever flown. This research work resulted in creation of large military craft of Orlyonok and Lun types. However, due to a low commercial potential of these craft designed for the Navy and a high cost of the development, large ekranoplans did not find practical application in the new economic and political situation in Russia in the last decade. At the present time, a development of manually controlled small craft is carried out by several companies over the world. The recent success of the FS8 vehicle, designed by the German firm AFD [9], raises a hope that WIG craft will find a niche in the transportation system in the near future.

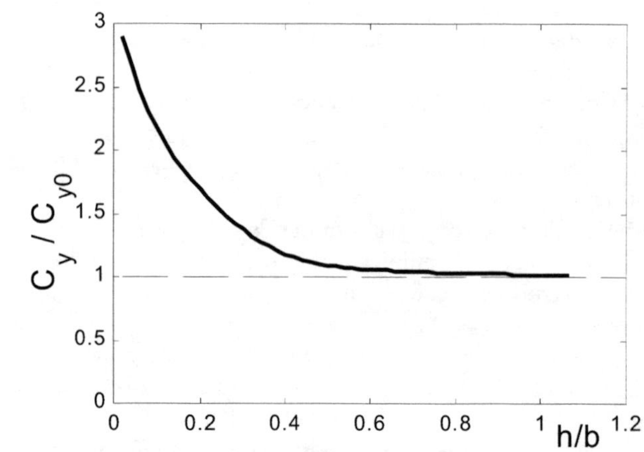

Fig. 6 Dependence of the lift coefficient of WIG Lun on the reduced altitude (C_{y0} – lift coefficient far from the ground)

Unique characteristics of the WIG concept make it a convenient platform for various tasks. Amphibious properties of WIG are suited for all-season operations in northern areas. The use of WIG craft has been studied for rescue operations in the ocean and as the first stage in the sea-based launch of reusable aerospace planes [10]. Boeing Phantom Works has recently announced a project to develop a high-capacity cargo plane, using the ground effect, for military and commercial purposes [8].

Commercialization of the WIG concept, in particular for transoceanic transportation, depends on the success in marketing, classification issues, and a creation of prototypes that will demonstrate the economic advantages of the WIG concept. We will discuss only the last concern here. The effective design of the WIG craft operating in the vicinity of the air-sea interface remains a challenging problem. Distrust of customers and investors with regard to WIG is caused to the great extent by periodic crashes occurring with these craft. Most accidents involve newcomers, who do not have sufficient skills to accomplish successful full-scale flights. However, even experienced designers had a series of accidents, which reasons are not clear. Due to complex physics of the flight in the ground effect (especially transient), it is difficult to understand the WIG behavior in certain flight regimes.

Basic theory of WIG is developed quite well. Analytical approaches based on conformal mapping, τ-parameter, and matched asymptotic expansions [11] provide a useful knowledge but cannot accurately predict the unsteady motion of actual complex WIG's in real conditions (waves, wind, etc.). Numerical methods, based on the potential theory, as well as viscous solvers, can treat realistic geometries; however, most CFD techniques require unreasonably large CPU time, therefore they cannot handle most real situations.

The breakthrough in that direction is a recent development of the mathematical model [12,13] capable of accurately predicting dynamics of the real WIG vehicles at acceptable consumption of computing resources. This model simulates the WIG motion in all regimes, such as planing, take-off, transition to flight, and flight itself. Aero-hydrodynamic coefficients in the model are computed using nonlinear hydro-aerodynamics [14]. Nonlinearity is essential when dealing with flying in the ground effect. Application of the linear theory often gives wrong results. This is another reason why newcomers, using simple linear calculations, often obtain incorrect results when they try to estimate performance of their WIG projects.

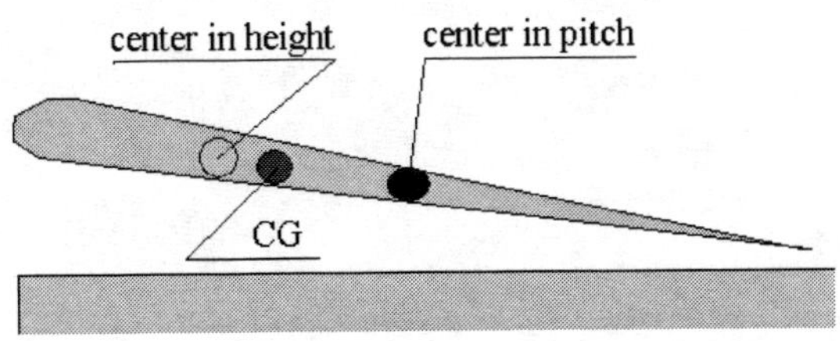

Fig. 7 Favourable positions of the center of gravity (CG), aerodynamic center of altitude (flight height) and center of pitch

The first necessary step in determining the WIG dynamics is a selection of the locations of the gravity center and aerodynamic centers of pitch and altitude. Notice that the last center is specific for the craft flying in the ground effect only. A favorable disposition of this point is demonstrated in Fig. 7. The center of altitude should be located in front of the center of pitch, and the gravity center must be between aerodynamic centers closer to the center of altitude. The admissible intervals for these points of actual craft can be computed by the model mentioned; some results are available in [13].

Since most accidents with WIG's occurred during the take-off phase, this mode is considered to be the most dangerous regime of motion. To overcome the hydrodynamic hump drag, a large thrust, exceeding that in cruising flight three times, is required. The typical drag and thrust curves are shown in Fig. 8. In the second case (Drag 2), the thrust must be abruptly reduced at a detachment moment. If this action is not accomplished, then the over-thrust appears that leads to the sudden loss of a longitudinal stability followed by the stall effect. This is a typical scenario of most crashes happened with WIG craft. The risk of the pitch-up tendency during the take-off regime can be reduced by a proper positioning of the center of gravity and by correct piloting strategy.

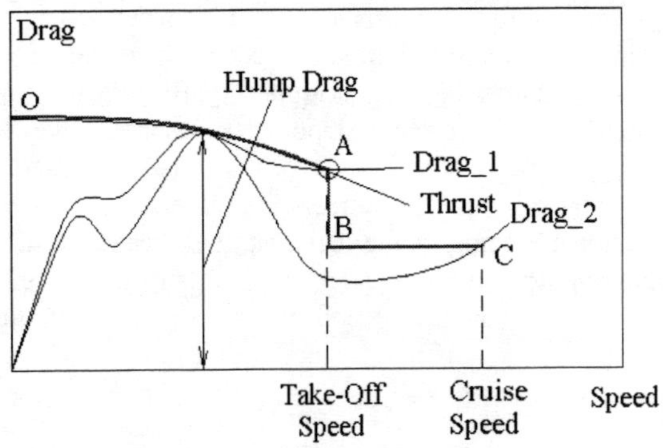

Fig. 8 Typical curves of Drag and Thrust

Two different strategies for the engine control aimed at avoiding the unstable behavior of the WIG craft during the take-off mode have been investigated for the second type of Drag-Thrust diagram. A simple recommendation is the following: a steep reduction of the thrust should be provided down to the value needed for the cruise regime immediately after the detachment moment, as indicated by curve O-A-B-C. If this action is not accomplished (characterized by appearance of excessive speed), an acceptable way to avoid a crash is to turn the engine off immediately, and turn it on again when the speed of motion becomes equal to the speed at the detachment moment.

As a future work aimed at establishing the Wing-In-Ground as a long-range transportation means, the following research areas have been identified:

- Improvement of practical numerical tools for simulating dynamics of the Wing-In-Ground and for optimizing their aerodynamic configurations.
- Investigation of the hydroelastic and structural mechanics of a craft moving in the vicinity of the sea-air boundary.
- Development of reliable control systems capable of providing the optimal flying condition and fast executing of maneuvers in emergency situations.
- Construction and testing of prototypes of the commercial WIG's demonstrating their economic advantages.

CONCLUDING REMARKS

Two types of advanced marine vehicles are considered in this paper that have a great potential to become an economic means for future transoceanic transportation. The Air Cavity Ship is a concept that is superior to traditional marine vessels. Wing-In-Ground is suited for a niche between the ship and air transport. Both concepts have been proven technologically, and serial vehicles of these types are produced for specific purposes. In order to transfer these technologies to the transportation industry, re-engineering work is needed (as well as obtaining marine society classification). That creates an attractive research opportunity for scientists and engineers.

The actual implementation of the described concepts to the great extent depends not only on R&D progress but also (and perhaps dominantly) on business and political support for these ideas, since the amount of funding needed for development of these advanced marine craft is quite significant, and various regulations in marine and shipping industries have to be modified to allow for these types of transportation. The important factors are the price of fuel and global energy conservation concerns. If the fuel cost goes steadily upward in the future and the environmental concern gains more weight in politics, then introducing of both the ACS and WIG concepts would not be so difficult, since these are truly energy-conserving technologies.

REFERENCES

[1] www.fastshipatlantic.com

[2] Matveev, K.I. (2003). "Air Cavity Ships," *Speed at Sea*, **9**(1), 13-16.

[3] Maskalik, A.I., Rozhdestvensky, K.V., and Synitsin, D.N. (1998). "A View of the Present State of Research in Aero- and Hydrodynamics of Ekranoplans," RTO AVT Symposium on Fluid Dynamics Problems of Vehicles Operating Near or in the Air-Sea Interface, Amsterdam.

[4] Butuzov, A., Sverchkov, A., Poustoshny, A., and Chalov, S. (1999). "State of Art in Investigation and Development for the Ship on the Artificial Cavity," IWSH, China.

[5] Matveev, K.I. (2003). "On the Limiting Parameters of Artificial Cavitation," *Ocean Engineering*, **30**(9), 1179-1190.

[6] Sverchkov, A.V. (2002). "Perspectives of Artificial Cavity Application Aimed on Resistance Reduction of Ocean/River Ships," ISC, 95-100.

[7] Butuzov, A., Sverchkov, A., Poustoshny, A., and Chalov, S. (1999). "High-Speed Ships on the Cavity: Scientific Base, Design Peculiarities and Perspectives for the Mediterranean Sea," HSMV, Capri, Italy.

[8] Rawdon, B.K., and Hoisington, Z.C. (2003). "Air Vehicle Design for Mass-Market Cargo Transport," 41st Aerospace Sciences Meeting & Exhibit, Reno, USA. AIAA paper No. 2003-0555.

[9] Fischer, H., and Matjasic, K. (1998). "From Airfisch to Hoverwing," Workshop WISE up to ekranoplan GEMs, The University New South Wales, 69-89.

[10] Aframeev, E.A. (1998). "Conceptual Bases of WIG Craft Building: Ideas, Reality and Outlooks," RTO AVT Symposium on Fluid Dynamics Problems of Vehicles Operating Near or in the Air-Sea Interface, Amsterdam.

[11] Rozhdestvensky, K.V. (2000). *Aerodynamics of a Lifting System in Extreme Ground Effect*, Springer-Verlag, Berlin.

[12] Benedikt, K., Kornev, N.V., Meyer, M., and Ebert, J. (2001). "Complex Mathematical Model of the WIG Motion Including the Take-off Mode," *Ocean Engineering*, **29**, 315-357.

[13] Kornev, N.V., and Matveev, K.I. (2003). "Complex Numerical Modeling of Dynamics and Crashes of Wing-in-Ground Vehicles," 41st Aerospace Sciences Meeting & Exhibit, Reno, USA, AIAA paper No. 2003-0600.

[14] Kornev, N.V., and Taranov, A.E. (1998). AutoWing 2.4, http://www.cl.spb.ru/ taranov/Index.htm

Proceedings of IMECE'03
2003 ASME International Mechanical Engineering Congress
Washington, D.C., November 15–21, 2003

IMECE2003-55233

ADVANCES IN THE RAILROAD LOCOMOTIVE CRASHWORTHINESS

Mr. Swamidas Punwani
Federal Railroad Administration
1120 Vermont Avenue, NW
Washington, DC
(202) 493-6369
john.punwani@fra.dot.gov

Gopal Samavedam
Foster Miller, Inc.
350 Second Avenue
Waltham, MA
(781) 684-4275
gsamavedam@foster-miller.com

Steve Kokkins
Foster Miller, Inc.
350 Second Avenue
Waltham, MA
(781) 684-4265
skokkins@foster-miller.com

ABSTRACT

The paper describes locomotive and fuel tank crashworthiness research being conducted by the Federal Railroad Administration for improved safety of the locomotive crew under collision scenarios. The research involves static and dynamic impact strength evaluations of locomotive structural components. These evaluations which are based on full scale tests and simulations using finite element analysis are described in this paper. Correlations between the test and simulation results are also presented in some cases.

INTRODUCTION

The Federal Railroad Administration (FRA) has been active in the Research and Development of railroad locomotive crashworthiness for improved safety of the crew in collision situations. Two major sources of threat to the crew safety are the intrusion of the colliding vehicle components into the crew compartment and the high deceleration levels experienced by the crew that could result in secondary collisions with the interior consoles, windows and walls. This paper discusses how the threats of intrusion can be mitigated.

The locomotive front end components such as the collision posts, hood, and window posts, if properly designed, can resist the intrusion of the colliding body (or the fractured components of the locomotive under consideration) into the crew compartment. Assessments of the static and dynamic impact strength of the front end components and structures are required for crashworthiness evaluations. In addition to the front end components, structural integrity of locomotive fuel tanks is important for safety in accident scenarios. This paper describes crashworthiness assessments of locomotives and fuel tanks being carried out by the Federal Railroad Administration on existing locomotives in revenue service.

Means of mitigation for high deceleration effects are also being considered by the FRA using seat restraints such as belts and airbags, de-lethalization of cab interior and novel crew seat designs. These mitigating measures will be discussed in a future publication.

1. OBJECTIVES

The main objective of this paper is to present the static and dynamic strength and crashworthiness evaluations of typical front end components of a freight locomotive structure, and correlate the test data with finite element simulations. Another objective is to provide assessments and discuss issues related to crashworthiness of locomotive fuel tanks.

2. LOCOMOTIVE STATIC STRENGTH EVALUATIONS

A measure of crashworthiness of structural components is the minimum load level that each locomotive front end structural component must withstand without significant permanent deformation. For example, the collision post is expected to withstand a longitudinal load of 500 kips when applied at a height of 30 in above the sill. Therefore, test evaluation of static strength is important.

2.1 Locomotive Structural Component Test Fixture

In order to perform these static strength evaluations, a facility was designed and constructed that can accommodate the load levels and orientations. The locomotive test fixture is located in Fitchburg, MA and operated by Foster-Miller. It can accommodate a wide range of loading configurations, including axial, lateral and angled loads up to 1 million lbs. The test fixture is shown in Figure 1.

The facility is made up of three elements:

- The Test Fixture
- The Hydraulic System
- The Instrumentation and Data Acquisition System.

The fixture is 40 ft long x 16 ft high x 26 ft wide, and has a total weight of about 100,000 lbs. It is constructed of welded steel beams, and is designed to accommodate load levels far above those proposed in this program. All of the components of the structure are bolted together, allowing for partial disassembly of the structure when test articles are installed or removed. The design of the frame completely contains all forces within the structure, eliminating the need for external reaction points.

Figure 1. FMI/FRA Locomotive Component Structural Test Fixture

The forces are applied to the test article by the hydraulic system which consists of a hydraulic cylinder, a pump, and a control system. To achieve the high loads that are required for these tests and minimize the size and complexity of the hydraulic components, a 10,000-psi system was used. The hydraulic pump and reservoir tank were sized for these cylinders. A pressure transducer in line with the input to the cylinder provides feedback to the data acquisition system on the pressure, which is converted to force by the data acquisition program. A cooling system helps to maintain the temperature of the hydraulic fluid.

The instrumentation and data collection system are used to collect and record all of the data during the test. Strain gauges and linear variable differential transducers are used to measure strain and displacement, respectively. The data collection is controlled by a LabView software program written specifically for this test facility. The data is converted to engineering units and stored as a tab-delimited spreadsheet file to ease post-test processing.

Table 1 describes the types of tests being performed on the locomotive using the fixture and the load levels for the tests.

Details of the test and results are provided in Reference [1]. As an example, the test on the windshield center post is shown in Figure 2. The load carrying capability in this particular case is found to be 16,000 lbs.

Table 1. Test Matrix

Test	Load Levels
1. Axial loading of single collision post	500,000 lb applied 30 inches above the sill
2. Axial distributed loading of short hood	Distributed load over 12 in wide x 30 in high top corner of hood
3. Axial loading of windshield corner post	45,000 lb at any location on the post
4. Axial loading of windshield center post	65,000 lb at any location on the post
5. Lateral loading of windshield corner post	45,000 lb at any location on the post
6. Axial loading of both collision posts	1,000,000 lb at 30 in above the sill
7. Angle loading of single collision post	500,000 lb at 30 in above the sill
8. Upward loading of anticlimber	200,000 lb

Figure 2. Windshield Center Post After Loading to Failure

3. LOCOMOTIVE DYNAMIC STRENGTH EVALUATION

Dynamic strength and crashworthiness evaluations were performed by full scale collision tests between a locomotive consist and other vehicles, which were 1) a hopper car consist, 2) a trailer truck carrying logs, and 3) a trailer truck carrying heavy steel coils. The dynamic tests were performed at the Transportation Technology Center (TTC) under funding by the FRA with instrumentation and test conduct by Transportation Technology Center, Inc. personnel. The test requirements definition and simulations were performed at the Foster Miller facility. These tests and results are described in Ref [2].

3.1 Test 1: Locomotive Consist Colliding with Stationary Hopper Consist

3.1.1 TEST DESCRIPTION

A moving consist including a modern locomotive trailing three loaded hopper cars, was used as the striking car with a velocity of 32.1 mph. The stationary consist on the same track had a total of 35 loaded hopper cars. Figures 3 and 4 show the vehicles before impact.

After the collision, the locomotive came to a stop with its lead truck overriding and resting on the first stationary hopper car. The final position of the locomotive, with its front end elevated, was at a vertical angle of about 7° relative to the tracks. Figure 5 shows the test locomotive after impact. Figure 6 illustrates the crushed hopper car after the collision and after the removal of the locomotive from atop the hopper car. Note the front bogie from the locomotive left inside the hopper car.

Figure 3. Locomotive and Moving Consist Before Impact

Figure 4. Stationary Hopper Cars Before Impact

Figure 5. Locomotive and Hopper Cars After Collision

Figure 6. Hopper Car After Test Collision

3.1.2 SIMULATION MODEL

A three dimensional finite element model has been developed using Hypermesh™ software. Basic structural and mechanical components (including the locomotive, trailing cars, separate bogies and suspension, and draft gear) were modeled using shell, plate, beam and solid finite elements. Nonlinear material properties were assigned for all deformable structures with elastic, elasto-plastic, and fully ductile (if applicable) behavior up to fracture (ultimate strength). The dynamic simulations were solved with the finite element analysis software LS-DYNA. Figure 7 shows the Finite Element model for the striking locomotive.

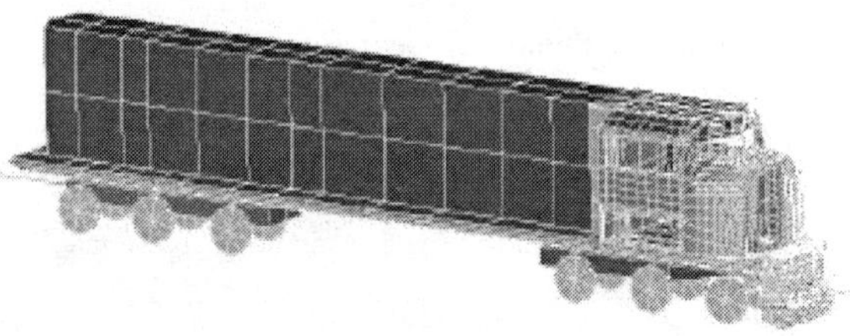

Figure 7. Finite Element Model of Striking Locomotive

The three follow-on hopper cars are modeled as shown in Figure 8. The first three hopper cars in the stationary consist of 35 were modeled in detail. The fourth and remaining cars are combined into a single mass/stiffness system equivalent to the remaining 31 cars. The entire locomotive consist has about 50,000 degrees of freedom.

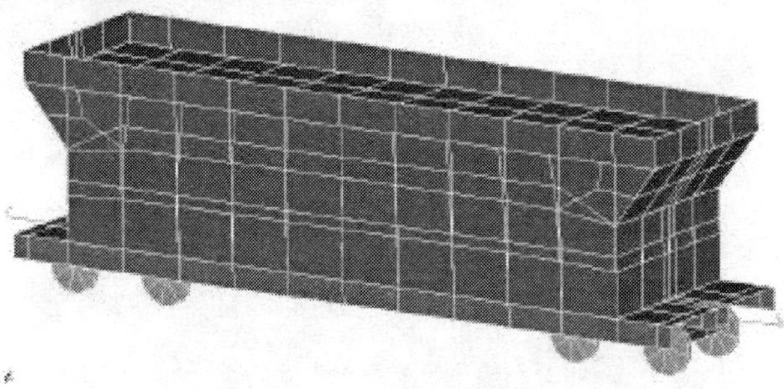

Figure 8. Following Hopper Car of Striking Consist

3.1.3 CORRELATIONS

Damage

Figures 9 through 11 illustrate the damage to both the locomotive and hopper car from the simulated collision.

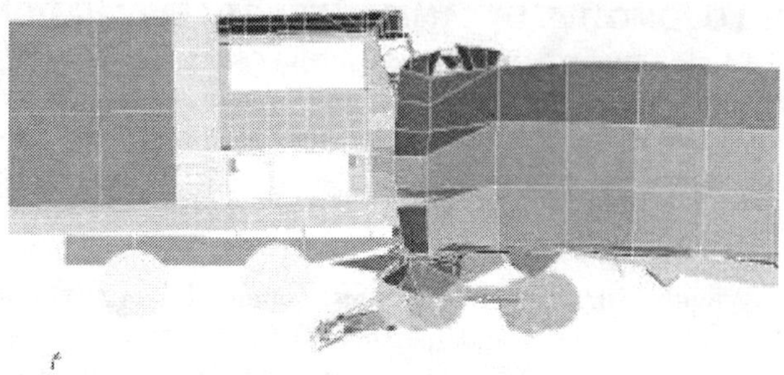

Figure 9. Locomotive Overriding Hopper Car

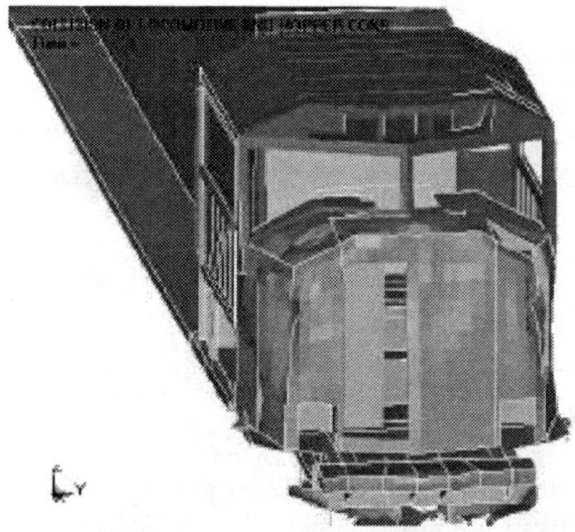

Figure 10. Damage to Locomotive in Simulation

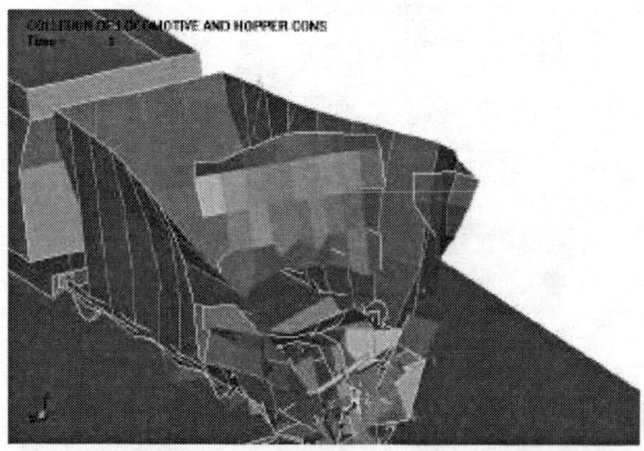

Figure 11. Hopper Car Damage in Simulation

Dynamic sequences were also compared. The agreement between the actual dynamic sequences (from high speed video) and the Finite Element Analysis sequences (computer generated) was reasonable. The damage seen in the simulations is also in reasonable agreement. Locomotive override of the hopper car occurs in both the test and simulation. Also, there is no intrusion into the operator's cab and a minimum amount of damage to the cab in both the simulated and test scenarios.

Acceleration

Node accelerations in the model were compared with those measured from the test. Figure 12 shows an example of an acceleration correlation between test vehicles and results from the computer simulation, for a location on the locomotive cab near the crew seat.

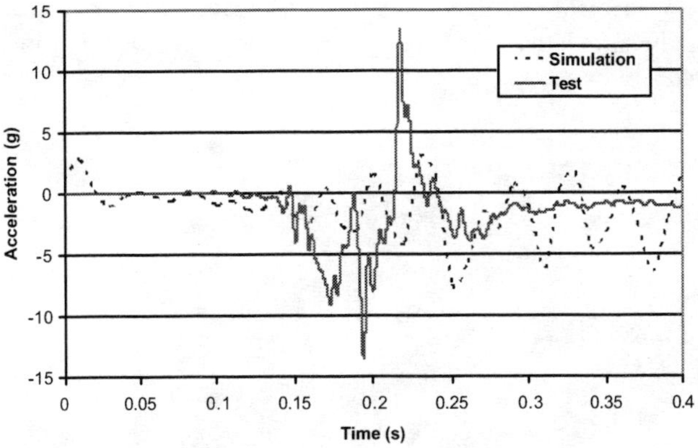

Figure 12. Locomotive Cab Floor X Acceleration

Strain

Locomotive collision posts were instrumented with strain gauges and the data was compared with the simulated strain from the computer analysis. Figure 13 shows the strain correlation between the values measured in the test and generated from the computer analysis.

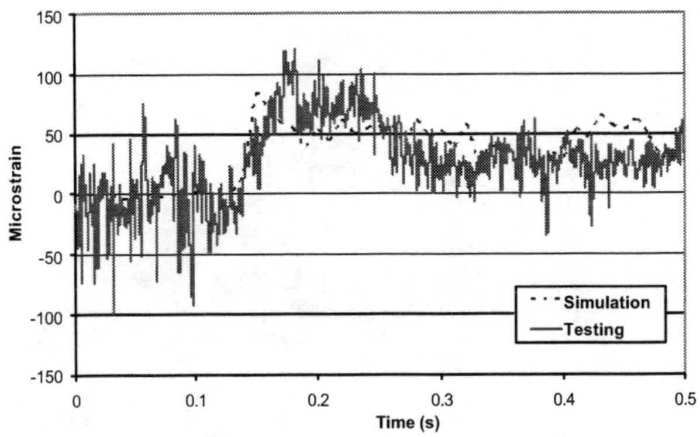

Figure 13. Correlation of Strain on Collision Post

3.2 Test 2: Locomotive Consist Colliding with Log Truck at Grade Crossing

3.2.1 TEST DESCRIPTION

A moving consist similar to that used in Test 1 trailing three loaded hopper cars was used as the striking car with a velocity of 52.4 mph. A loaded log truck was used for the stationary target and is comprised of an 18 wheeled tractor trailer combination with a ten wheeled tractor. The trailer was fully loaded with logs fastened by two cables to the trailer. The stationary truck and trailer were located at a grade crossing perpendicular to tracks with the moving consist. Figure 14 and Figure 15 show the vehicles before impact.

Figure 14. Moving Test Consist Prior to Collision.

Figure 15. Loaded Log Truck and Trailer Prior to Collision

During the collision, the locomotive plowed through the stationary truck and trailer. Most of the logs cabled to the trailer were snapped into pieces while the trailer and truck were pushed to the other side of the track. The grade crossing after the collision, including the damaged truck, trailer and log debris is shown in Figure 16. The right side of the tracks in the figure shows the rear end of the test trailer and log debris. Left of the track is the forward part of the trailer, the truck and more log debris.

Figure 16. Stationary Vehicles at Grade Crossing After Test Collision.

After the collision, the locomotive proceeds beyond the grade crossing and is stopped by a ballast obstruction located a few hundred yards down the track. Figure 17 shows the damage to the front of the locomotive after the collision. As seen in the image, the locomotive is derailed by the ballast obstacle, the three trailing hopper cars are shown in the background.

Figure 17. Damage to Locomotive After Test Collision.

3.2.2 SIMULATION MODEL

The truck, trailer and logs were modeled using shell, plate and beam elements. Non-linear material properties were also assigned for deformable structures. Initial positioning of the modeled log truck was perpendicular to the track direction and situated at a grade crossing. Logs were tied down to the trailer using two cables with an initial positive offset length representing an initial tensile force. The cables were also modeled using "cable elements".

3.2.3 CORRELATIONS

Damage

Figure 18 shows the damage to the locomotive at the end of the simulation. Figure 19 is a top view of the locomotive and log trailer after the collision. The logs are omitted for clarity. The modeling of the logs for the collision was difficult due to the inconsistent nature of the material. During the testing , all of the logs were broken into pieces while the top layers of logs in the simulation did not break. After the initial impact with the collision posts, the unbroken logs caused additional damage to the windshield and upper cab area. However, neither the test nor the simulated scenarios showed substantial intrusion into the operator cab space.

Figure 18. Simulated Damage to Locomotive

Figure 19. Top View of Simulated Collision

Acceleration and Strain

Figures 20 and 21 show examples of an acceleration and a strain correlation respectively between test vehicles and results from the computer simulation.

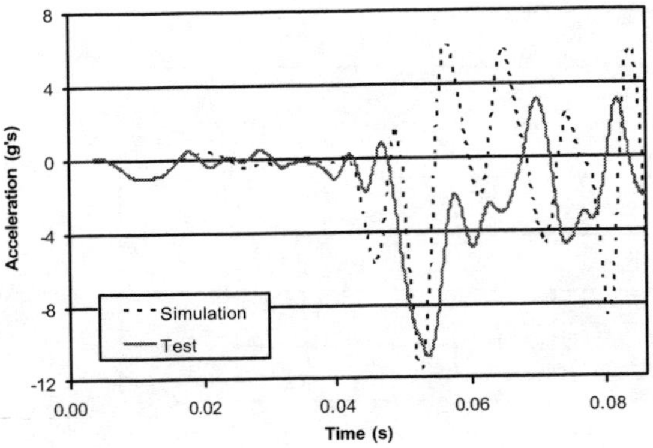

Figure 20. Locomotive Cab Floor X Acceleration

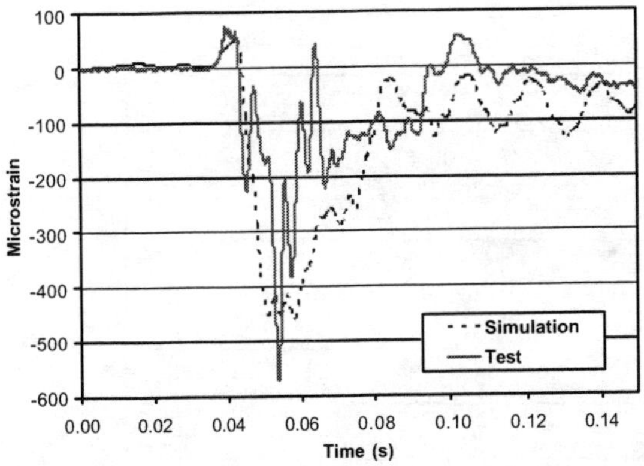

Figure 21. Locomotive Collision Post Strain

3.3 Test 3: Locomotive Consist Colliding with Steel Coil Truck at Grade Crossing

3.3.1 TEST DESCRIPTION

A moving consist similar to that used in Tests 1 and 2 trailing three loaded hopper cars was used as the striking car with a velocity of 62.4 mph. A truck and trailer loaded with two steel coils was used for the stationary target. The stationary target is comprised of an 18 wheeled tractor trailer combination with a ten wheeled tractor. Two coils of steel were loaded onto the trailer. The coil located on the rear of the trailer was aligned with the right collision post of the moving consist. The stationary truck and trailer were located at a grade crossing perpendicular to tracks of the moving consist. Figure 22 shows the stationary target vehicle before impact.

Figure 22. Truck Loaded with Steel Coils Prior to Collision

The rear coil is embedded into the locomotive during impact while the truck, trailer and forward coil were pushed to the side of the tracks.

After the collision, the locomotive proceeded beyond the grade crossing and was stopped by a ballast obstruction located a few hundred yards down the track. Prior to stopping, the steel coil lodged in the locomotive fell to the ground. Figure 23 shows damage to the front of the locomotive after the collision.

Figure 23. Damage to Locomotive After Collision

3.3.2 SIMULATION MODEL

The truck, trailer and coils were modeled using appropriate shell, plate, and solid elements. Non-linear material properties were also assigned for deformable structures.

3.3.3 CORRELATIONS

Damage

The simulated damage to the locomotive is shown in Figures 24 and 25. Correlation between the test and simulated scenarios are consistent. Both cases experience major damage to the cab including complete shear failure of the impacted collision post. Cab intrusion also occurs after the collision with the steel coil.

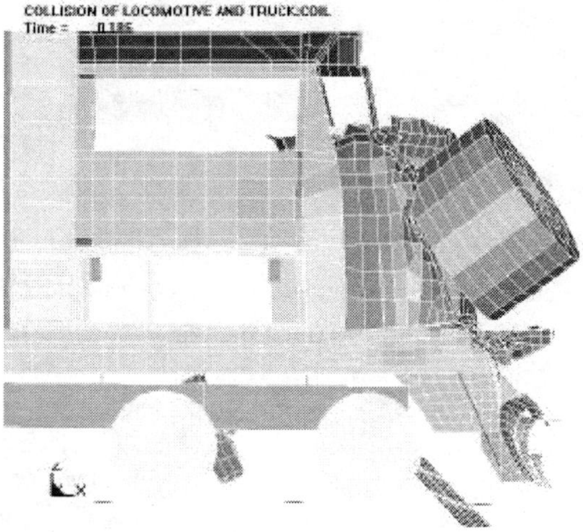

Figure 24. Locomotive Cab Damage due to Steel Coil

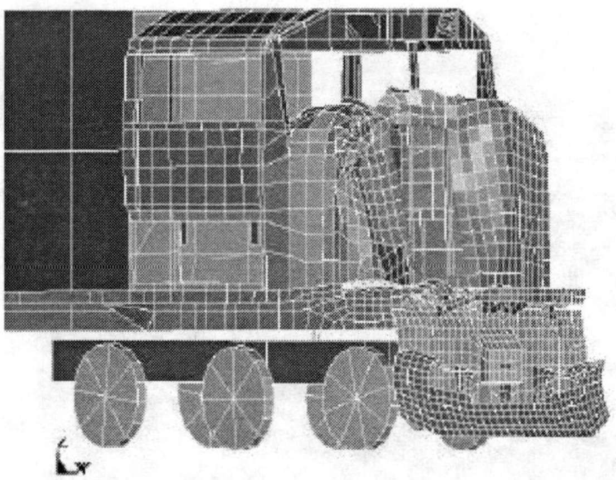

Figure 25. Simulated Damage to Locomotive

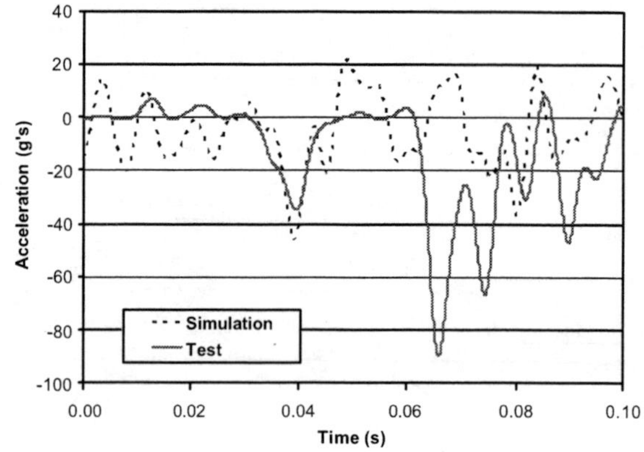

Figure 26. Locomotive Cab Floor X Acceleration

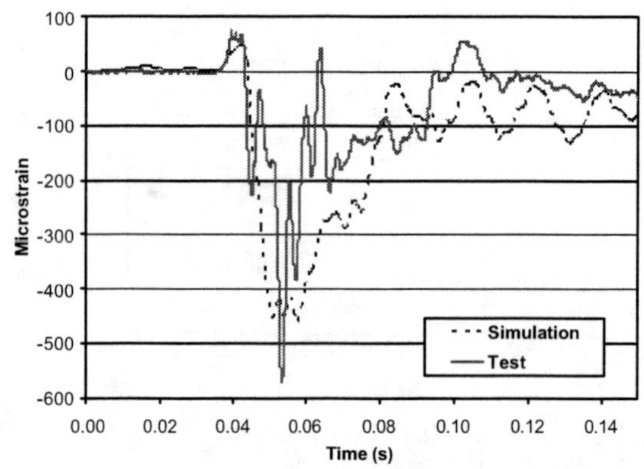

Figure 27. Locomotive Collision Post Strain

Acceleration and Strain

Figures 26 and 27 show examples of an acceleration and a strain correlation respectively between test vehicles and results from the computer simulation.

4. FUEL TANK SAFETY EVALUATION

4.1 Introduction

Insuring the structural integrity of locomotive fuel tanks in train accidents is a vital element in improving safety for train crews and the public. Damage to fuel tanks and their associated systems can lead to fuel leaks and fire, one of the most dangerous situations for the train crews and passengers. Such accidents also have the potential for serious environmental pollution. The Federal Railroad Administration is conducting research on optimization of fuel tank structural parameters for improved safety and cost savings in manufacturing. Improved safety and structurally efficient designs can be achieved using advanced Finite Element solutions and considerations of more inclusive accident scenarios than what have been considered in the literature to date. Advanced finite-element methods can provide realistic assessments of structural behavior, including effects of large deflections, yielding and material failure.

4.2 Development of simulation techniques and design issues

"Generic" fuel tank designs as shown in Figure 28 were developed for large, conventional road locomotives up to the 400,000 lb range, using representative materials, thicknesses, baffle layout, etc. The tank behavior was simulated using finite-element modeling techniques capable of representing large deflections, yielding and material failure of the structure.

Safety requirements at present use static load scenarios intended to represent in-line and jackknife derailments, and truck side impacts, in simplified form. This has been successfully simulated with results used both to develop improvements and to understand the relative effects of key parameters such as wall thickness, material strength, and weld sizes.

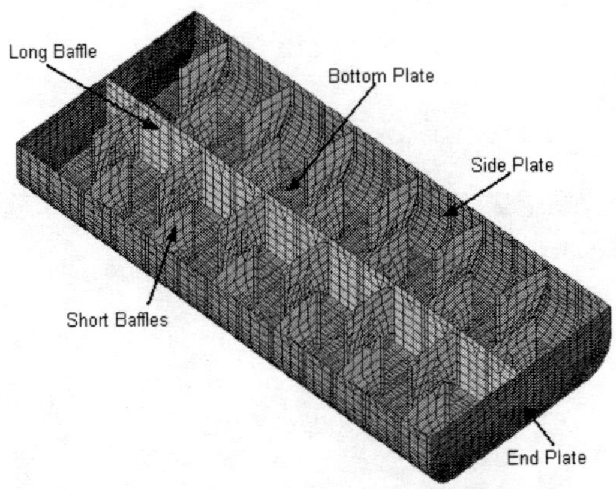

Figure 28. Generic Fuel Tank

In a "minor derailment" scenario, the locomotive is assumed to derail but stay closely aligned with the rails. The end wall bears the rail load, with locally concentrated high stresses as shown in Figure 29.

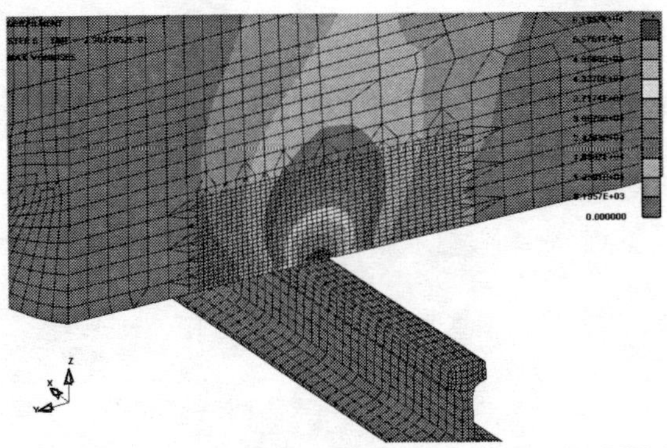

Figure 29. Concentrated Stress on Fuel Tank End Wall

Parametric analyses were used to gauge the benefit of several strengthening strategies, where the load capacity vs. deflection behavior is shown for original and strengthened versions (See Figure 30).

In a jackknife derailment scenario, the locomotive is assumed to derail, and then land perpendicular to the rails. Large deflections and distributed stresses, viewed down through the tank top are shown in Figure 31.

Studies like the above showed that a tradeoff between thickness and material strength for the bottom plate could be employed for cost efficiency: a thickened, but non-high-strength steel could preserve or increase the loading capacity of the fuel tank. Slight thickening of the interior baffles also increased their stability and load-bearing.

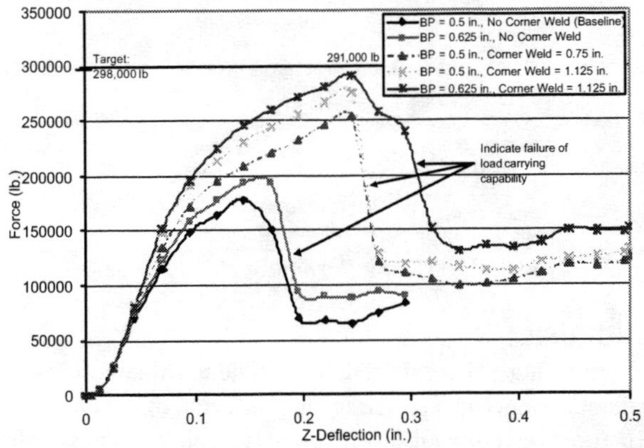

Figure 30. Load Capacity versus Deflection

Figure 31. Stress in Fuel Tank Perpendicular to Track

4.2 Dynamic penetration conditions

Dynamic penetrations can occur in severe accidents. Tank behavior can be assessed using two hypothetical cases: a rail section moving in-line, impacting the end plate of the fuel tank, and same rail section impacting the side wall at 45° angle (Figure 32).

Finite element modeling techniques can provide assessments of structural behavior of fuel tanks under both static and dynamic loading conditions. The tools can be used for insight into tank behavior in accident scenarios, and also to develop design improvements. Dynamic impact scenarios, such as end wall penetration and angled side wall penetration can be simulated to understand and improve tank performance and designs. The simulations will be validated by testing in the future.

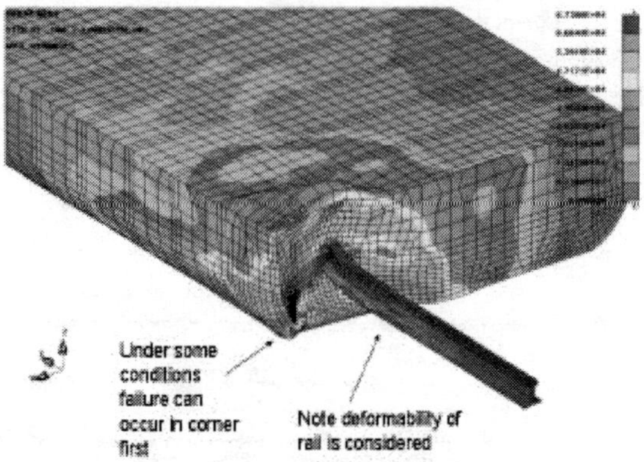

Figure 32. Stress in Fuel Tank Perpendicular to Track

5. CONCLUSIONS

1. Dynamic finite element models provide a valuable tool in understanding and predicting the behavior of locomotive structures in collisions with other vehicles. Reasonable correlations were obtained between the analysis predictions and the results of three full scale dynamic impact tests in the areas of locomotive front end damage, acceleration levels at the cab floor and the strains in the collision post.
2. The strength of the locomotive front end components can be evaluated using the test fixture at Fitchburg, MA. The fixture can apply large loads (up to 1 million pounds) in any direction on the locomotive structures.
3. The simulation tool can be used for improved crashworthiness design of locomotives and fuel tanks. The improved design can be validated by full scale static and dynamic testing as illustrated in the paper.

ACKNOWLEDGMENTS

The authors wish to thank Mr. Mark Yachmetz, Associate Administrator for Railroad Development and Ms. Jo Strang, Deputy Associate Administrator for Railroad Development, who oversee the R&D Programs, for their support in the publication of this paper. The authors also thank Ms. Claire Orth, Chief of Equipment and Operating Practices, of the Office of Research and Development of the Federal Railroad Administration for her support of the locomotive crashworthiness research program.

REFERENCES

Kristoff, S, "Locomotive Crashworthiness: Static Component Evaluations", DOT/FRA/ORD-03/XX (In Preparation).

Samavedam, G., and Kidd, J., "Full Scale Locomotive Dynamic Crash Testing and Correlations", DOT/FRA/ORD-03/XX (In Preparation).

Proceedings of IMECE'03
2003 ASME International Mechanical Engineering Congress
Washington, D.C., November 15–21, 2003

IMECE2003-42833

VIRTUAL DETECTION OF A RADIAL-PLY TIRE'S THREE-DIMENSIONAL FREE VIBRATION MODES TRANSMISSIBILITY

Y. P. Chang, Ph.D.
Mechanical Engineering Department
Oakland University
Rochester, MI 48309-4478

ABSTRACT

A full nonlinear finite element P205/70R14 passenger car radial-ply tire model was developed and run on a 1.7-meter-diameter spinning test drum model at a constant speed of 50 km/h in order to investigate the tire transient response characteristics, i.e. the tire three-dimensional free vibration modes transmissibility in X, Y and Z directions. The tire model was constructed in extreme detail with three-dimensional solid, layered membrane, and beam elements. The reaction forces of the tire axle in longitudinal (X axis), lateral (Y axis) and vertical (Z axis) directions were recorded when the tire rolled over a cleat on the test drum, and then the FFT algorithm was applied to examine the transient response information in the frequency domain. The result showed that this tire has clear peaks of 45, 40, and 84 Hz transmissibility in the longitudinal, lateral and vertical directions. This result was validated against previous studies by analytical, experimental, and FEA model approaches and showed excellent agreement. The parameters adopted in this FEA tire model were also compared with experimental work and the extraordinary agreement was also confirmed. The tire three-dimensional free vibration modes transmissibility was successfully detected virtually.

1. INTRODUCTION

Tires, from the standpoint of ride dynamics, behave primarily as springs that absorb the irregularities and roughness features on the roads. The tire is also a dynamic system with resonant vibrations, which will be associated with the transmission of the contact forces from the road to the vehicle and may interact with the vehicle's own resonances, affecting ride comfort and handling stability [1 and 2].

Hammer impacts, bumpy roads, oscillating platforms, or spinning drums produce hub oscillations via tire tread and sidewall vibrations distributed along the tire circumference. The question of whether a correlation exists between these distributed vibrations/inputs and the complex transfer phenomena of hammer/road/platform/drum input motion to hub motion is the major concern and focus of this study.

The literature records many studies of tire free vibration modes transmissibility phenomena, carried out primarily by analytical derivation [3, 4, 9, 10, 12 ~ 15] and four different empirical methods (the modal analysis hammer impact test [3 and 4], the real road test with a running vehicle [5], the lab test with a tire on a vibration platform [3, 6 ~ 9], and the lab test with a tire running on a spinning drum [4, 7, 10, 11]). Generally similar results were obtained in every case, whether using an analytical or empirical approach, i.e. a vibration peak will appear between 30 and 50 Hz in the longitudinal (fore/aft) X direction, and a peak between 35 and 40 Hz will appear in the lateral Y direction, and a peak between 75 and 100 Hz in the vertical Z direction. A virtual tire/drum testing system was established in the present study, the tire spindle (hub) reaction forces were recorded, and then the FFT algorithm was applied to determine the transient response information in the frequency domain

2. SIMULATION APPROACH

Tire free vibration modes play a key role in vehicle ride comfort and handling stability, for applications ranging from passenger cars to earthmoving equipments. In this study, a full nonlinear FEA tire model will be constructed; the tire free vibration modes transmissibility will be successfully detected, using a virtual tire/drum rotating test machine created with PAM-SHOCK. The results of the simulation will show excellent agreement, quantitatively and qualitatively, with previous research.

Many parameters can affect the resonant frequencies of tire free vibration modes: tire size, construction, inflation pressure, and operating conditions such as load and temperature. In order to fully concentrate on the major issue of tire free vibration modes transmissibility detection, while still maintaining the numerical computations under a manageable level, several practical assumptions were proposed in this research:

- Any effects on or from the suspension system were neglected.
- No wind or temperature effects were considered.
- The rim/wheel assembly was assumed to rotate at the same angular velocity as that of the tire, so no slipping or velocity difference between the tire and the rim/wheel assembly will be considered.
- Further, whenever the subject of tire vibration is mentioned, the effects caused by either built-in tire non-uniformities or any out-of-balance conditions of the tire-rim/wheel combinations come immediately to mind. These effects will not be discussed in this research. The tire model is assumed to be perfectly symmetric.

Inputs or stimuli to the tire from such things as impacts, rough road surfaces, tire non-uniformities, and tread patterns can potentially excite tire free vibration modes. Many studies about tire free vibration modes transmissibility have been carried out, primarily by four empirical methods as discussed before. When road irregularities, such as a single impact from a bump or a pothole, or a series of small impacts from rough road surfaces, are enveloped by the tire, these impacts excite the tire free vibration modes in much the same way as a hammer impact does during a modal analysis test [11]. It has also been shown in previous research that the transmissibility of the force into the spindle of a rolling automobile wheel is very similar to that of a tire on a stationary vibration platform [5 and 8]. It should also be noted that regarding the transmission of vibrations, qualitative similarity exists between results of the tests using the vibrating platform and the spinning drum [7]. These important research concluded that these four empirical methods, the modal analysis hammer impact test, the real road test with a running vehicle, the lab test with a tire on a vibration platform, and the lab test with a tire running on a spinning drum, will all give the similar results. Therefore the tire industry standard lab test with a tire running on a spinning drum was adopted in this virtual simulation study.

A passenger car tire P205/70/R14 operating under typical conditions will be investigated in this research by applying a nominal vertical load of 4500 N with a recommended inflation pressure of 0.207 MPa. Tire free vibration modes and transmissibility, tire/drum FEA models, simulation procedures and the results and discussion will be presented in sections 3 to 6; the tire parameters verification will be demonstrated in section 7, and the conclusions will be addressed in section 8.

3. TIRE FREE VIBRATION MODES AND TRANSMISSIBILITY

A relatively large portion of the tire mass is concentrated in the tread, which is connected to the rim/wheel by the compliant sidewalls. This combination of mass and compliance presents a typical vibration system as shown in Figure 1, which permits the tread to resonate when excited by road inputs, such as bumps and potholes [1].

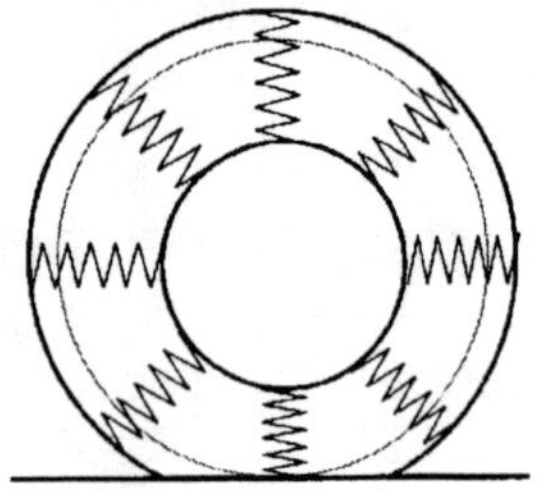

Figure 1 Tire Radial Spring Model [1]

Figure 2 shows examples of the first five vertical free vibration modes of a tire. The first mode, which will occur somewhere near 85 to 87 Hz in a passenger car tire, involves a simple vertical motion of the entire tread band without distortion. This mode can be easily excited by vertical inputs from the contact patch. Since the entire tread band moves up and down in unison, the force associated with this resonance is transmitted to the rim/wheel, the tire axle, and therefore the chassis and the vehicle [1].

The second mode contrasts with the first one, in that the tread band is oscillating in an elliptical fashion, always remaining symmetrical about the vertical and horizontal axes. The top and bottom of the tread are always moving out of phase so that a net vertical force of zero is imposed on the tire spindle. Although this resonance can be excited by vertical inputs from the contact patch, the tire very effectively absorbs the inputs without transmitting forces to the axle because the phenomenon is completely symmetric. The third- and the higher-order modes also behave in a pattern similar to the second mode [2].

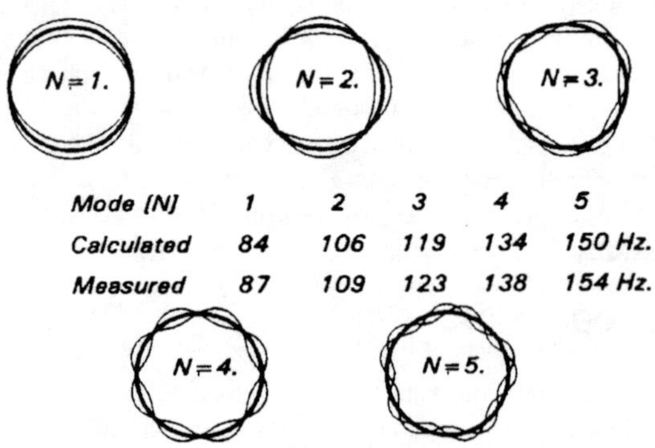

Figure 2 First Five Vertical Free Vibration Modes and Shapes of a Tire [13 and 14]

From this simple picture of a tire as a dynamic system, it is possible to begin building an understanding of the tire's dynamic behavior in transmitting road irregularities, how the tire vibrates the chassis of a vehicle, how the tire as a dynamic system with its own free vibration modes affects the transmission of vibrations to the vehicle, how the tire interacts with the vehicle's own resonances, and therefore how and why the tire's dynamic characteristics play a critical role in vehicle ride comfort and handling stability.

As mentioned above, only the first free vibration resonance in the vertical direction can be transmitted to the vehicle; all of the second- and higher-order modes' vibrations will be absorbed without transmitting forces to the tire axle because they are completely symmetric phenomena. This transmission behavior can be demonstrated by measuring the accelerations at the spindle of a tire as the tire envelops such an impact. The vibration transmissibility of the tire is defined as:

$$\text{transmissibilty} = \frac{\text{tire axle force (acceleration) amplitude}}{\text{ground contact displacement (force) amplitude}}$$

This quotient must be calculated for each input frequency in order to completely describe the tire's dynamic characteristics.

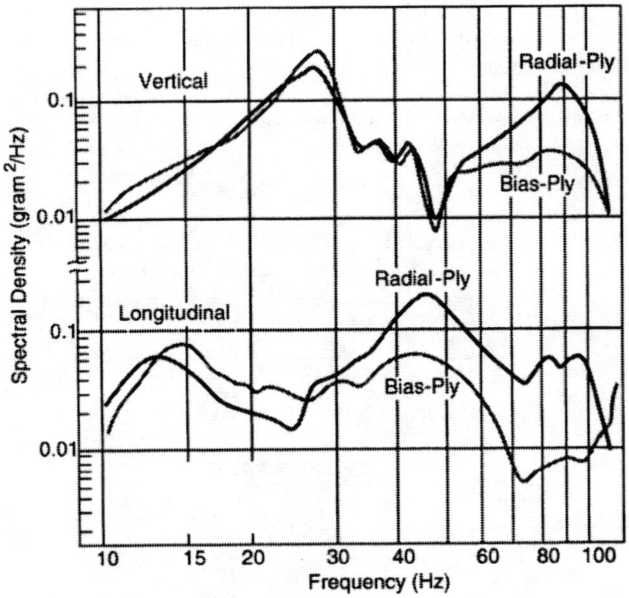

Figure 3 Spectra of Forces Measured When a Tire Encounters an Obstacle [1 and 5]

Barson et al. [7] reported this tire first vertical free vibration mode transmissibility phenomenon. Figure 3 shows the typical spectra of forces measured when a passenger car tire encounters a small obstacle at a speed of 50 km/h (30 ft/min). Data were shown for both radial-ply and bias-ply tires for forces in the vertical and longitudinal directions. The radial-ply tire shows increasing amplitude of the vertical force during the ranges of 20 to 30 and 80 to 100 Hz and reaches its peaks at frequencies of approximately 27 and 86 Hz. Since these data were collected from a road test of a running vehicle, the first peak at 27 Hz represents the axle hop (suspension system) resonance, therefore of greater interest is the second, higher-frequency peak at around 86 Hz. Clearly, from the previous discussion and Figure 2, it would be expected that this behavior is the result of the high transmissibility of the tire's first vertical free vibration mode.

The force in the longitudinal direction behaves similarly to the vertical curve. The amplitude of the longitudinal force increases in the ranges of 10 to 20 Hz, 40 to 60 Hz, and 80 to 100 Hz, and reaches its peaks at frequencies of approximately 14, 46 and 83 Hz. The first peak at 14 Hz will be due to the natural frequency of the system, just as in the vertical direction it is mainly owing to the longitudinal compliance of the suspension system. The third peak at 80 to 100 Hz corresponds to the mode found and matches with the first vertical free vibration mode. The second peak at 46 Hz represents the first rotational free vibration mode, i.e. the rotational (fore/aft) natural vibration resonance; this mode exhibits purely uniform rotational displacements of the tread band. These phenomena imply that the tire appears as a very stiff, rather than compliant, element with regard to road inputs at these specific frequencies. The force in the lateral direction behaves also similarly to the vertical and longitudinal curves.

The spinning test drum with a cleat on it will serve as an impact input in order to excite the tire three-dimensional free vibration modes. Since this simulation only modeled and focused on the tire itself without the axle or the vehicle suspension system, only the first longitudinal/lateral/vertical free vibration modes transmissibility detections were expected.

4. TIRE/DRUM FEA MODELS

A tire consists primarily of reinforced rubber composites and rubber materials, therefore the tire has been modeled as an assembly of three-dimensional Mooney-Rivlin hyperelastic solid finite elements (bricks) for rubber material, fiber-reinforced layered membrane finite elements for reinforced rubber composites, and beam elements for two beads as shown in Figure 4.

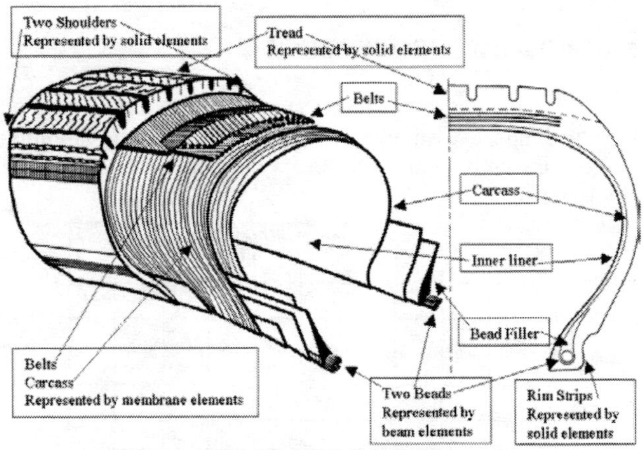

Figure 4 The FEA Tire Model Elements

This FEA tire model construction is based on a specific passenger car tire, i.e. a P205/70/R14 tire. The cross section is symmetric against the centerline of the tire and consists of three types of elements built under the PAM-SHOCK environment as shown in Figure 5. Rotating this cross section one full revolution around the tire axle axis and then cutting it into 60 slices established the complete three-dimensional tire model.

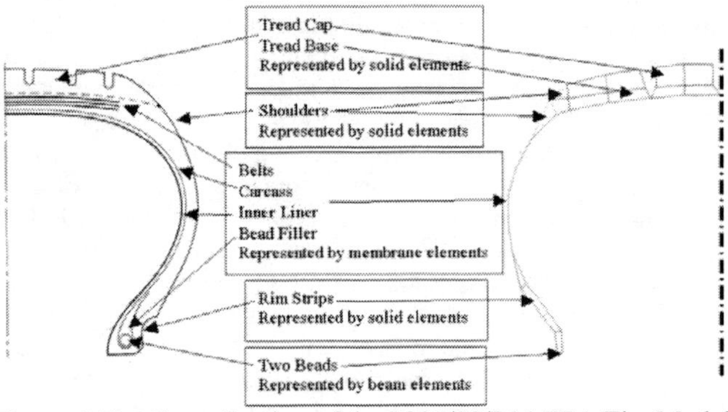

Figure 5 The Cross-Section of the P205/70/R14 FEA Tire Model

The virtual finite element radial-ply passenger car tire model was built successfully as shown in Figure 6 (a), includes over 18,000 nodes and 24 different material definitions. The half-cleat, as shown in Figure 6 (b), was built on the test drum in order to excite a tire's free vibration modes in all three directions.

(a) Tire/Drum Assembly (b) Half-Cleat Detail
Figure 6 The Complete Tire/Drum/Cleat FEA Model

5. SIMULATION PROCEDURES

The main purpose of this simulation is use a sophisticated FEA tire model to simulate and detect the tire free vibration modes transmissibility in the computer virtual testing environment.

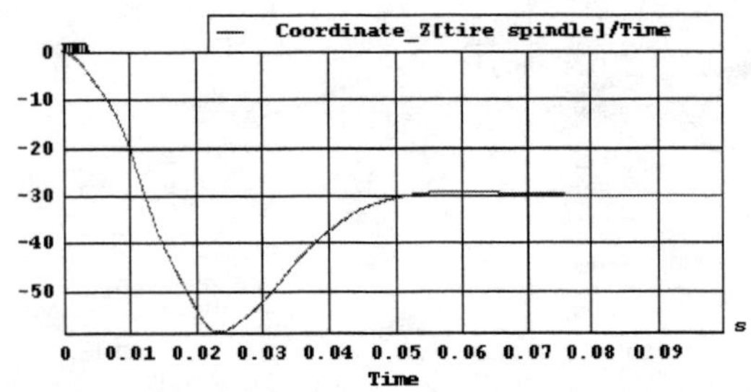

Figure 7 Tire Spindle Vertical Deflection Time History

Simulation phase I is only about preparing the tire to be rotated on the test drum. This stage includes three major goals: inflating to 0.207 MPa, loading with 4500 N, and totally resting the tire on the drum statically. The whole simulation time is from 0 to 0.1 second for phase I.

Figure 7 illustrates the tire spindle vertical deflection time history; the tire spindle deflection history showed that the tire was resting on the test drum stably with a static deflection 29.98 mm.

With the drum and the tire ready to be rotated, phases II are the major simulation stages. The primary goal of phase II is to accelerate the tire and drum to the speed of 50 km/h linear velocity at the contact surface. The friction force will be generated between the contact surface of the tire and drum by applying an appropriate and sufficient friction coefficient, so that the tire will be driven to the same speed as the drum at all times during the whole simulation period. The drum and the tire then were kept rotating at this constant speed until the tire encountered the cleat on the drum without any penetrations or numerical errors.

6. RESULTS AND DISCUSSIONS

The tire three-dimensional free vibration modes transmission phenomena were simulated and virtually detected successfully in this section.

6.1 Tire Spindle Vertical and Longitudinal Reaction Force Curves

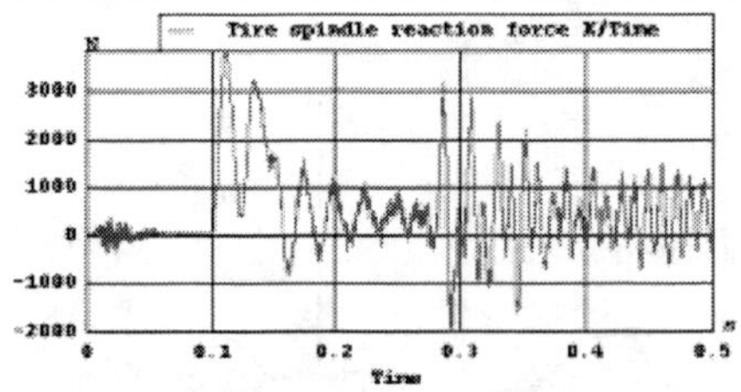

(a) Longitudinal Direction (X axis)

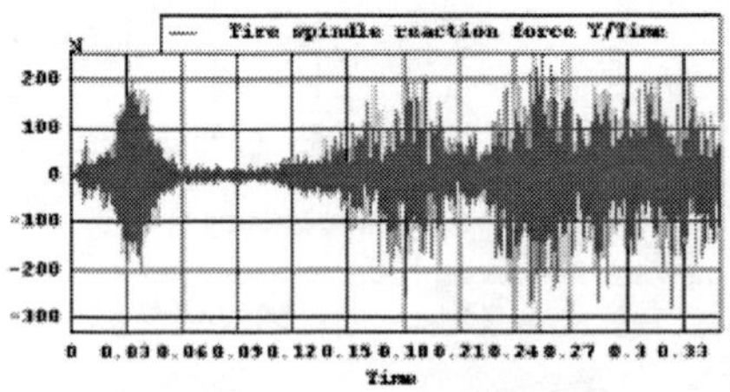

(b) Lateral Direction (Y axis)

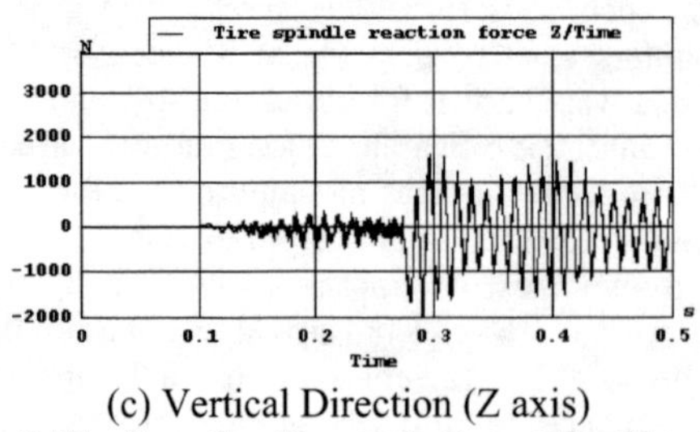

(c) Vertical Direction (Z axis)
Figure 8 The Reaction Forces Acting at the Tire Axle

As shown in Figure 8, the tire reaction forces acting at the tire axle in all three X, Y, and Z directions begin vibrating wildly at around 0.275 second, which is when the tire starts meeting the cleat on the drum (as shown in Figure 9).

It is extremely important to make sure that there were no penetrations when the tire met the cleat, and throughout the whole simulation. The simulation was carefully examined, there were no penetrations found.

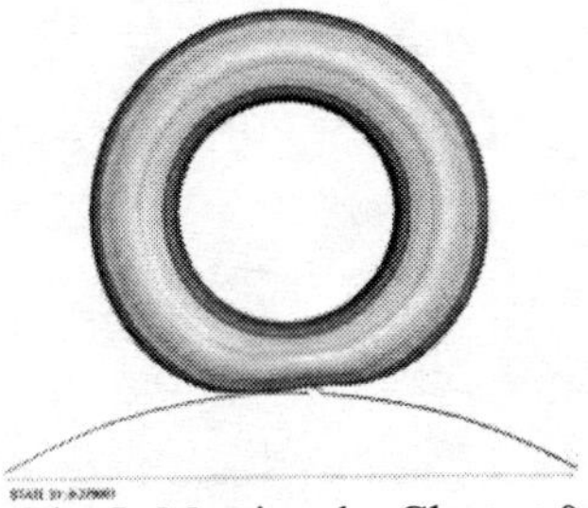

Figure 9 The Tire Is Meeting the Cleat at 0.275 Second

6.2 FFT Results of the Tire Spindle In-Plane Reaction Force Curves

Since the cleat on the test drum provides a three-dimensional input to the tire, tire three-dimensional free vibration mode excitations were expected. The FFT algorithm was applied to the tire reaction forces' curves in Figure 8 to extract information in the frequency domain; the results are shown in Figure 10. In order to compare the simulation results of this research to the typical previous research, Figure 3 is presented here again with Figure 10.

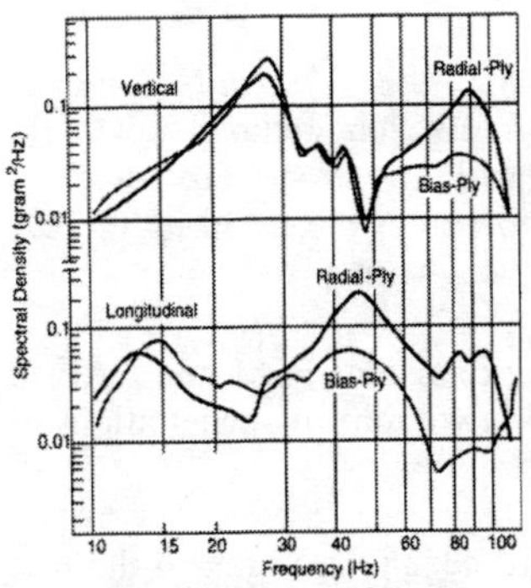

Figure 3 Spectra of Force Measured When a Tire Encounters an Obstacle

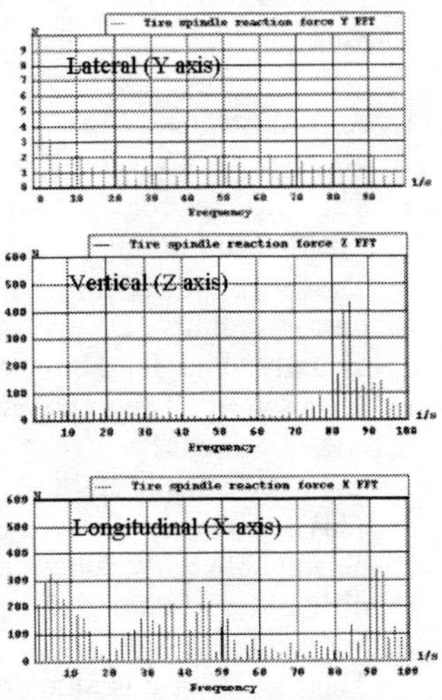

Figure 10 FFT Results of Tire Reaction Forces in X, Y and Z Axes

By comparing Figures 10 and 3, the simulation results in Figure 10 are seen to match perfectly with the free vibration modes transmission phenomena of a real vehicle test in Figure 3. The FFT results in Figure 10 show the amplitude increasing in the vertical direction and reaching its peak at a frequency around 84 Hz. As mentioned before, there was no suspension system effect modeled and considered in this simulation; hence this is the only peak that would be expected (there could be no axle hop), and this behavior is the result of the high transmissibility of the tire first vertical free vibration mode. In the longitudinal direction, the amplitude is increasing and reaches its peaks at frequencies of approximately 1 to 4, 45 and 90 Hz. Again, without considering the suspension system's longitudinal compliance effect in this simulation, the first peak between 1 to 4 Hz is the impact input frequency from the cleat on the drum. (The drum's angular velocity is 15 rad/sec, i.e., around 2.5 Hz.) The third peak at around 90 Hz corresponds to the first vertical free vibration mode. The second peak at 45 Hz represents the first rotational/longitudinal free vibration mode.

6.3 Comparisons and Discussions

The conclusion can be drawn from this simulation that this tire has its first longitudinal, lateral and vertical free vibration modes at 45, 40, and 84 Hz respectively, and transmitted these vibrations to the vehicle tire axle in the X, Y and Z directions. The simulation and the detection were complete and successful.

In order to verify this simulation result quantitatively and qualitatively, the validations against previous studies by analytical, experimental and FEA model approaches are summarized in Table 1 and show excellent agreement.

Table 1 Tire Free Vibration Resonances/Transmission

Authors (reference) (Publish year)	Tire Type (Tire Model)	Load	Pressure	Methodology	Longitudinal (X) Resonance	Lateral (Y) Resonance	Vertical (Z) Resonance
Barson & Dodd (5) (1971)	Radial-ply	N/A	N/A	Vehicle road tests at 50 km/h and meets a cleat	46 Hz	N/A	86 Hz
Pacejka (9) (1981)	Cylindrical beam on elastic foundation	N/A	N/A	Analytical approach	45.5 Hz	39 or 40 Hz	83.7 or 87 Hz
Ellis (13 and 14) (1989 & 1994)	Cylinder with tension on elastic foundation	N/A	N/A	Analytical approach Measured	N/A	N/A	84 Hz 87 Hz
Huang (15) (1992)	Ring on elastic foundation	N/A	206.7 kPa	Analytical at 0 km/h Analytical at 70 km/h	N/A	N/A	93 Hz 80 Hz
Scavuzzo et al. (11) (1993)	P205/70R14	4890 N	207 kPa	Computational FEA model	46 Hz	35 Hz	84 Hz
Y. P. Chang (2003)	P205/70R14	4500 N	207 kPa	FEA model running on a cleated-drum	45 Hz	40 Hz	84 Hz

7. TIRE PARAMETERS VERIFICATION

It is important to verify this FEA tire model's parameters against real experimental results. Two different verifications will be presented and discussed in detail, and the FEA tire model's simulation results will be seen to agree excellently with experimental measurements and the previous research data.

7.1 Static Vertical Stiffness versus Pressure Curve

The tire static vertical stiffness versus inflation pressure curve can be made from the different vertical static deflections due to different tire inflation pressures with a fixed load 4500 N from phase I simulation which will be discussed and shown in Figure 11. The deflections under a nominal load 4500 N and different inflation pressures from the simulation results can be summarized in Table 2 (a).

Experiments were set up to verify the FEA simulation results. A P205/70R14 tire was used to measure the static deflections according to different inflation pressures. The load on the tire is 4050 N, i.e. 413 kg; the experimental deflections are summarized in the Table 2 (b).

Table 2 will give the vertical static stiffness versus inflation pressure curves in Figure 11; the simulation results match the experimental measurements very well.

Table 2 The Deflections Under Fixed Load and Different Inflation Pressures

(a) Fixed load 4500 N by simulations

Tire Pressure	Static Deflection
0.138 MPa	40.01 mm
0.207 MPa	30.51 mm
0.276 MPa	24.46 mm

(b) Fixed load 4050 N by experiments

Tire Pressure	Static Deflection
0.138 MPa	36 mm
0.207 MPa	27 mm
0.276 MPa	22 mm

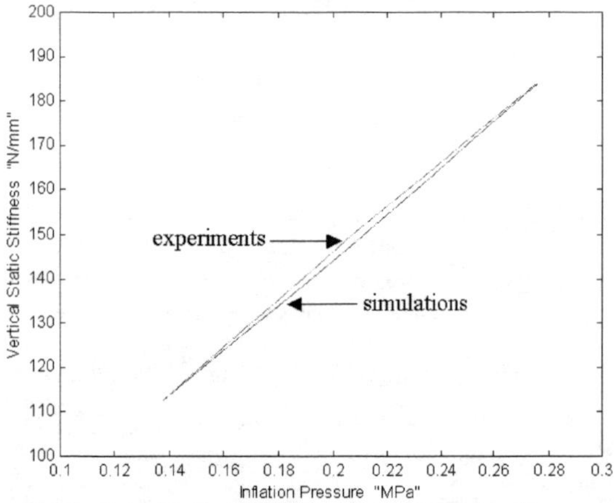

Figure 11 Vertical Static Stiffness versus Inflation Pressure Curves

This curve is well known as a linear relationship. Both the simulation results and the experimental measurements agree with the previous researches excellently [16].

7.2 Vertical Load versus Static Deflection Curve

The vertical tire load versus static deflection curve can be made from the different vertical static deflections due to different loading conditions with a fixed tire inflation pressure 0.207 MPa from phase I simulation which will be discussed later. The deflections under a fixed tire inflation pressure 0.207 MPa and different loading conditions from the simulation results are summarized in Table 3 (a).

Table 3 The Deflections Under Fixed Inflation Pressure 0.207 MPa and Different Loading Conditions

(a) Simulation results

Tire Load	Static Deflection
4500 N	29.98 mm
3000 N	20.09 mm
1500 N	10.01 mm

(b) experiment results

Tire Load	Static Deflection
4050 N (413 kg)	27 mm
2650 N (270 kg)	18 mm
1305 N (133 kg)	9 mm

Experiments were set up to verify the FEA simulation results. A P205/70R14 tire was used to measure the static deflections according to different loading conditions. The inflation pressure of the tire is 0.207 MPa; the experimental deflections are summarized in Table 3 (b). These data give the vertical load versus static deflection curves in Figure 12. Again, the simulation results match the experimental measurements very well.

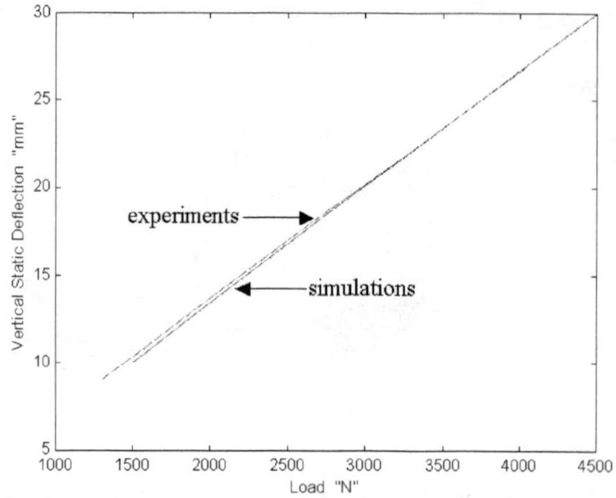

Figure 12 Vertical Load versus Static Deflection Curves

It has been found that for a given inflation pressure, the vertical load versus static deflection characteristics for both radial-ply and bias-ply tires are more or less linear, thus both the simulation results and the experimental measurements agree with the previous researches excellently [2, 13, 14, 16].

8. Conclusions

Conclusions of this research can be summarized as follows:

- A sophisticated FEA model of a general passenger car radial-ply tire was built.
- The artificial rotating tire/drum/cleat testing system was established thoroughly.
- The tire three-dimensional free vibration modes transmissibility at and 45 Hz longitudinally, 40 Hz laterally, and 84 Hz vertically were virtually detected successfully. The simulation results are listed in Table 1 and show excellent agreement with previous analytical and experimental studies.
- This simulation shows the promise of FEM and CAE to be valuable tools, functioning without any penetrations or difficulties of numerical instability.

The results of this study are beginning to reveal the true nature of what happens when a tire meets irregular road inputs. The challenge facing vehicle manufacturers and tire suppliers is to apply the knowledge of tire vibration resonances and transmissibility in cascading a series of small changes in vibration performance, in a series of test designs, into an overall optimal tire and to the task of defining and optimizing the tire-vehicle system's performance.

ACKNOWLEDGEMENT

The research presented in this paper is an extension of the Ph.D. thesis of the author, which was completed at the Department of Mechanical and Nuclear Engineering, The Pennsylvania State University under the supervision of Dr. M. El-Gindy, director of the Vehicle Simulation Research Center of Pennsylvania Transportation Institute (PTI) and graduate faculty of the Department of Mechanical and Nuclear Engineering.

The author would like to express his gratitude to Dr. M. El-Gindy for his patient guidance and assistance to complete the preparation work during his Ph.D. study at Penn State, which eventually made this research happen.

REFERENCES

1. Gillespie, T. D., (1992), Fundamentals of Vehicle Dynamics, published by Society of Automotive Engineers, Inc.
2. Clark, S. K., (1981), Mechanics of Pneumatic Tires, published by U.S. Department of Transportation.
3. Negrus, E., Anghelache, G., and Stanesch, A., (1997), "Finite Element Analysis and Experimental Analysis of Natural Frequencies and Mode Shapes for a Non-Rotating Tire," *Vehicle System Dynamics Supplement 27*, pp. 221-224.
4. Negrus, E., Anghelache, G., and Sorohan, S., (1998), "Tire Radial Vibrations at High Speed of Rolling," *SAE Paper 980260.*
5. Barson, C. W., and Dodd, A. M., (1971), "Vibrational Characteristics of Tires," *Institute of Mechanical Engineers,* paper C94/71, 12p.
6. Chiesa, A., Obert, L., and Tamburini, L., (1964), "Transmission of Tire Vibrations," *Automobile Engineer*, Vol. 54, Dec., pp. 520-530.
7. Barson, C. W., James, D. H., and Mocrombe, A. W., (1967 ~ 68), "Some Aspects of Tire and Vehicle Vibration Testing," *Proceeding of Institute of Mechanical Engineers,* Vol. 182, No. 3B, 32p.
8. Potts, G. R., and Csora, T. T., (1975), "Tire Vibration Studies: The State of the Art," *Tire Science and Technology*, Vol. 3, No. 3, pp. 196-210.
9. Pacejka, H., (1981), Chapter 9 "Analysis of Tire Properties" of Mechanics of Pneumatic Tires, edited by Clark, published by U.S. Department of Transportation.
10. Takayama, M., and Yamagishi, K., (1984), "Simulation Model of Tire Vibration," *Tire Science and Technology*, Vol. 1, No. 1, pp. 38-49.
11. Scavuzzo, R. W., Richards, T. R., and Charek, L. T., (1993), "Tire Vibration Modes and Effects on Vehicle Ride Quality," *Tire Science and Technology*, Vol. 21, No. 1, pp. 23-39.
12. Kung, L. E., Soedel, W., and Yang, T. Y., (1986), "Free Vibration of a Pneumatic Tire-Wheel Unit Using a Ring on an Elastic Foundation and a Finite Element Model," *Journal of Sound and Vibration*, Vol. 107, No. 2, pp. 181-194.
13. Ellis, John R., (1989), Road Vehicle Dynamics, published by author.
14. Ellis, J. R., (1994), Vehicle Handling Dynamics, published by Mechanical Engineering Publications Limited, London.
15. Huang, S. C., (1992), "The Vibration of Rolling Tires in Ground Contact," *International Journal of Vehicle Design*, Vol. 13, No. 1, pp. 78-95.
16. Wong, J. Y., (1993), Theory of Ground Vehicles, 2nd edition, published by John Wiley & Sons, Inc.

Proceedings of IMECE'03
2003 ASME International Mechanical Engineering Congress
Washington, D.C., November 15–21, 2003

IMECE2003-55324

CHARATERIZATION OF HIGHWAY TRAFFIC NOISE GENERATED BY RIGID PAVEMENT CONTRACTION JOINTS

Lawrin Ellis/University of Florida, Dept. of Mechanical and Aerospace Engineering, Gainesville, FL

Christopher Niezrecki//University of Florida, Dept. of Mechanical and Aerospace Engineering, Gainesville, FL

David Bloomquist//University of Florida, Dept. of Civil and Coastal Engineering, Gainesville, FL

ABSTRACT

Contraction joints in rigid (concrete) pavements are required to permit expansion of each monolithic section of roadway. At higher speeds, the major source of highway noise is attributed to vehicle tire/roadway interaction. Current concerns about limiting the impact of highway traffic noise, has forced transportation agencies to consider strategies to control noise generated by tire/roadway interaction. Within this work the difference in noise generated by 1/4 versus 3/8-inch joint widths is conducted. The study focuses on passenger vehicles including a sedan and a light duty van/truck. Both vehicle in-cabin and roadside noise levels are measured for vehicle speeds of 50, 60, and 70 miles per hour. For the sedan, the minimum and maximum observed in-cabin differences were determined to be 1.08 and 1.82 dB(A), respectively. Minimum and maximum observed roadside differences are 1.19 and 2.58 dB(A), respectively. Van tests resulted in minimum and maximum observed in-cabin differences of 0.60 and 1.09 dB(A) and minimum and maximum observed roadside differences of 1.05 and 3.18 dB(A), respectively. This paper contains details of reference standards, test methods, and the results obtained.

INTRODUCTION

Highway traffic noise impacts more people than virtually any other source of environmental noise [1]. In the United States, the total number of vehicle miles traveled is projected to increase by 65% during the 20-year period from 1989 to 2009 [2]. As more residential areas develop near highways, traffic noise impacts to communities escalate.

Traditionally, transportation noise control has been approached by considering three aspects: the noise source, the path, and the receiver. The increase in traffic noise impacts was addressed in a report sponsored by the International Organization for Economic and Cooperative Development. The report recommended placing an emphasis over the next 10 years on noise control at the source. Controlling noise at the source reduces costly methods of controlling the path and the receiver. Path and receiver control methods may involve continual efforts to mitigate traffic noise due to increasing pressure to locate residential developments near existing roadways. On the other hand, source control strategies can produce benefits that may be realized at all times wherever vehicles operate [2].

The factors affecting tire/pavement noise are primarily vehicle speed, or, more precisely, the rotational speed of the tire, and the texture of the road surface. The tire pattern and construction of the tire can have a significant but smaller effect. For example, changing the tire pattern from a smooth or bald tire to a deeply patterned tread will generally only increase the noise levels by 2 to 3 dB(A) on most surfaces. However, changing the road surface texture from a smooth polished surface to a deeply textured surface can increase tire/pavement noise by 8 to 10 dB(A). For these reasons, changing road surface textures, rather than redesigning tires, offers greater opportunities to control tire/pavement interaction nose.

Florida Portland Concrete Cement (PCC) highway pavements are typically constructed using contraction joints with a minimum width of 3/8-inch (~9.53 mm). A potential method of lowering tire/pavement noise is to minimize the noise generated as tires contact contraction joints by decreasing the minimum width to 1/4-inch (~6.35 mm). Proponents' justification for considering this as a means of minimizing highway traffic noise is both environmental and economical. Decreasing the noise levels emitted results in less environmental noise pollution and therefore less annoyance to persons and other living beings. Furthermore, reduced noise emission levels may result in less funding required for methods of abatement and/or potentially reduced property setback requirements to meet local noise ordinances. These economic impacts can lead to savings on highway construction costs as well as increasing the amount of property for development adjacent to highways. Opponents of the narrower 1/4-inch joint are concerned about the integrity of the joint seal. This concern centers around the potential difficulty of sandblasting each face of the lime-rock aggregate clean enough to ensure long term adhesion of the silicon sealants. The seal is designed to eliminate water intrusion, which, in time, can compromise the structural integrity of the pavement at the joint resulting in pavement damage and costly repairs. Moreover, technical guides published by manufacturers of the sealant and many concrete experts recommend the use of the 3/8-inch joint. Additional considerations are the difficulty of installing the backer rod and sealant and the increased stresses exerted on the sealant as the slabs are subjected to thermal expansion and contraction.

The Florida Department of Transportation (FDOT) constructed a transversely tined PCC roadway test section using 1/4-inch contraction joints to compare the effects of using narrower joints. The objective of this study is to quantify the difference in noise emitted by 1/4-inch versus 3/8-inch contraction joints. The study includes measuring and quantifying the in-cabin and roadside noise emitted as vehicles cross the contraction joint sections. Since the majority of vehicles on the nation's highways are passenger vehicles, two types of common passenger vehicles are chosen for the study. Both in-cabin and roadside noise measurements are made using a Buick sedan (automobile) and a Ford van (light truck/passenger vehicle). Measurements are made as each vehicle travels at highway speeds of 50, 60, and 70 mph (~80, 97, and 113 km/h, respectively). Sound measurements of both roadside and in-cabin noise are conducted during the summer and winter months to compare the potential effects of contraction in the slab. Additionally, average joint width measurements are made to quantify width differences observed between warm and cold weather testing. A quantitative analysis of the results will yield a relative difference in noise emitted by the difference in joint widths. Further qualitative subjective listener experiments using the resultant data will explore whether or not these differences, if any, are perceptible.

This paper presents the experimental methodologies and test set-ups employed, followed by a summary of results for noise and joint width measurement data collected during both summer and winter conditions. The measured differences are then compared to the results of subjective listener experiments designed to quantify whether or not the average person can perceive the change in noise levels. The paper concludes with a discussion of additional means for quantifying and improving tire/pavement noise interaction.

TEST SITE LOCATION

Measurements are conducted on the northbound side of Florida State Road 9A just southeast of Jacksonville, Florida. The 1528 ft, 1/4-inch (~466 m, ~6.35 mm) joint test section begins approximately 1.5 miles (~2.41 km) north of the intersection of SR9A and Phillips Highway. Signs mark the beginning and ending points of the test section. A 1-inch (25.4 mm) contraction and the north end of the section clearly locates the transition from 1/4-inch joints back to 3/8-inch joints. Contraction joints are spaced approximately 16 ft (~4.88 m) on center. Measurements for the 3/8-inch joints are conducted on the north side of the 1-inch contraction joint. The shoulder along roadside joint measurement sections is essentially free of acoustically reflective objects.

ROADSIDE EXPERIMENTAL METHODOLOGY

Experimental Setup

The design of the roadside experimental setup is based on and conducted largely in accordance with Federal Highway Administration (FHWA) Standard Number USDOT/FHWA-PD-96-046 [3] and International Organization for Standardization (ISO) Standard Number ISO 11819-1 [4]. These standards are primarily used to measure and predict statistical data for highway traffic noise generated by all vehicle types at various highway speeds. Data, such as vehicle type, speed, and climatic conditions, is collected continuously, sometimes over the course of days, as vehicles pass specified observer and microphone locations. A recent roadway surface noise comparison study conducted in California used the USDOT/FHWA and ISO standards as measurement procedure guidelines [5].

Unlike the broad focus of the standards, this study concentrates on a single comparison for a specific pavement condition change as observed for two test vehicles. The roadside test setup borrows from and conforms as closely as possible to the intention of the standards. Minor modifications are made based on specific measurement objectives and site conditions.

Instrumentation consists of two Brüel & Kjær (B&K) 1/2-inch (12.7 mm) free-field microphones, Type 4165 and 4190, each using a B&K Type 2669 pre-amplifier, connected to channels one and two respectively, of a B&K Type 2690 Nexus signal conditioning amplifier. A foam windscreen protects each microphone. The Nexus is programmed to output a 100mV/Pa signal from each microphone input signal using 200V as supply voltage. Output voltage signals are sent to a Spectral Dynamics DSPT SigLab Model 20-42 digital signal analyzer. The frequency range of the SigLab analyzer is programmed to measure from 0 - 20,000 Hz at a differential frequency of 12.5 Hz. The analyzer record length is set to 4096 data points. The differential frequency and record length correspond to a sampling rate of 51,200 Hz, which is sufficient to prevent aliasing. SigLab operation is controlled within MATLAB® using the Virtual Noise Analyzer (VNA) and Long Capture (VCAP) modules. Recorded data is saved and converted into a MATLAB® file for processing. An instrumentation schematic is shown in Fig. 1.

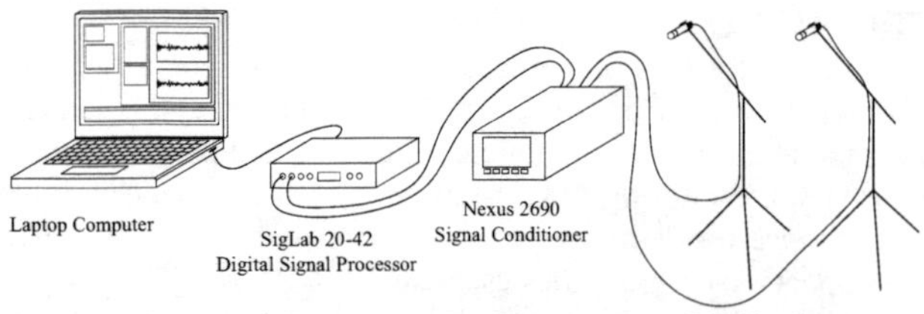

Figure 1. Roadside instrumentation setup

In order to record maximum and average L_P, each microphone is located in line with a contraction joint and positioned 5 ft (~1.5 m) above the road surface at a distance of 33 ft (~10.0 m) from the centerline of the right-hand traffic lane as shown in Fig. 2. The distance is a compromise between that specified by the USDOT/FHWA and the ISO standards' microphone placement distance from the centerline of the traffic lane being measured, 50 ft (~15.0 m) and 25 ft (~7.5 m) respectively. Microphones are positioned using microphone boom stands with tripod bases for stability. Each microphone is mechanically isolated by using a foam wrap (approximately 1-inch thick) around the pre-amplifier section and held in place with a spring loaded clip-type holder.

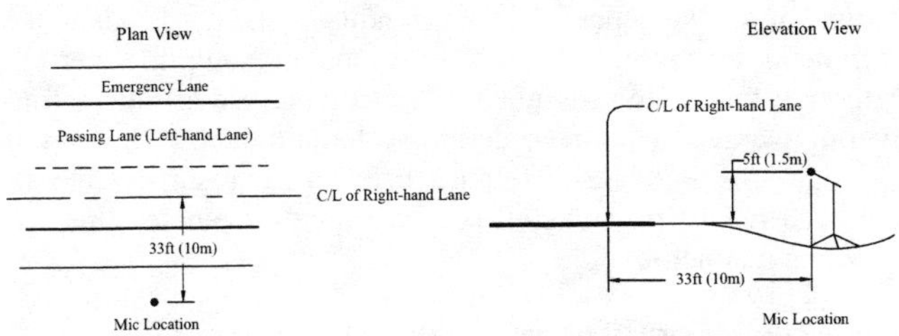

Figure 2. Roadside microphone positioning schematic

Although grazing incidence is typically used for the measurement of line source noise, such as highway traffic, to capture the maximum L_P (assumed to occur the instant the front tires contact the contraction joint), the microphone head is positioned for normal incidence. Normal incidence is best suited for point-source measurements in which the sound being measured propagates from a single, known direction [4].

Testing Procedure

The maximum and average L_P levels are measured at four different joint locations within each set of contraction joints. Each test is conducted using three personnel: the driver of the test vehicle, a person to signal as the test vehicle passes a specified location, and an operator to initialize data collection. Hand-held two-way radios are used for communication between all personnel.

Noise data for each test section are recorded as the vehicle travels at highway speeds of 50, 60, and 70 mph, ± 2 mph (~80, 97, and 113 km/h, respectively, ± 3.2 km/h). The vehicle speed tolerance of ± 2 mph is assumed throughout testing. First, the vehicle test speed is confirmed and the signal person is positioned at a specified distance from the microphone location closest to the approach of test vehicle approach. Signal positions are roughly 300 ft (~91 m), 350 ft (~107 m), and 400 ft (~122 m) for vehicle speeds of 50, 60, and 70 mph, respectively. Next, the test vehicle accelerates to and maintains a constant test speed. As the vehicle passes the signal position, the operator is signaled to begin capturing data. A total of 8 seconds of data is recorded for each vehicle pass. After the data is recorded, the operator observes the data to determine if the run is valid. Tests are considered invalid and rerun if another vehicle, opposing traffic, or background noise influences the results. The procedure is repeated for each speed.

Because only two microphones are used simultaneously, tests are conducted with microphones initially positioned at locations 1 and 2. After successful completion of all three speeds, the microphones are repositioned at locations 3 and 4 and each speed is repeated (Fig. 3 and 4). The location of the equipment table remains constant serving as a reference position as microphones are moved from locations 1 and 2 to locations 3 and 4. The contraction joints at each microphone location are permanently marked and numbered for future testing. Full measurements are first performed within the 1/4-inch joint section and then repeated within the 3/8-inch joint section.

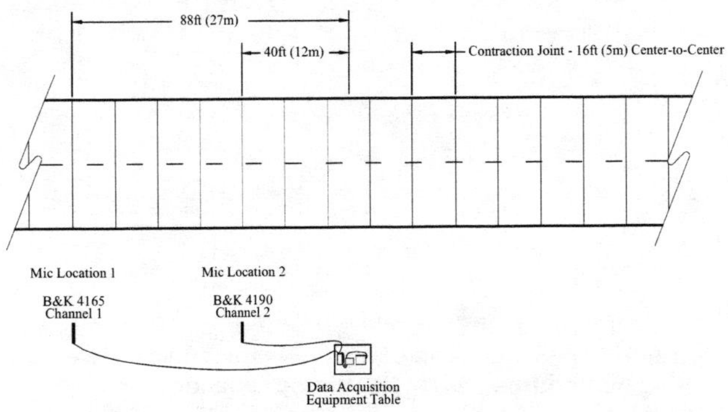

Figure 3. Roadside schematic for microphone locations 1 and 2

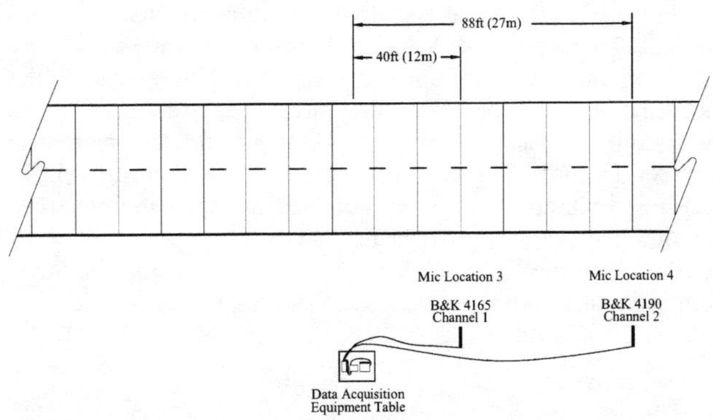

Figure 4. Roadside schematic for microphone locations 3 and 4

Calculation of Sound Pressure Levels of Roadside Noise

Sound propagation close to the ground for horizontal distances of less than 300 feet (~100 meters) is essentially independent of atmospheric conditions such as ambient temperature, relative humidity and vertical wind gradients. For this case, the atmosphere may be considered homogeneous and sound ray paths as straight lines [1]. If no reflecting surfaces other than the ground are present, the only environmental component to adversely affect sound propagation is ground attenuation. Harris provides tables containing values of ground attenuation in decibels for short-range propagation in discrete frequencies over varying surface hardness and source height/receiver distances.

The roadside measurements made in this study models the passing vehicle as a single point monopole source. Sound radiates as spherical waves from the point of closest approach as shown in Fig. 5 [6]. Although wind and atmospheric conditions are periodically recorded, the effects of temperature and humidity are neglected. Ground level wind effects are considered negligible for wind speeds of less than 10 mph (~16 km/h). Since the roadside measurements are conducted over the same type of terrain and the results are compared as relative from location to location, ground attenuation is not considered in the final analysis. Finally, since differences between the sets of contraction joints are relative to one another, attenuation due to microphone windscreens is neglected.

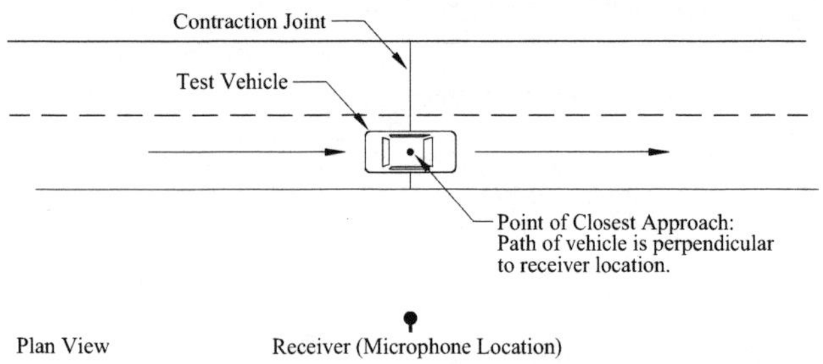

Figure 5. Approximation of a single moving vehicle as a point source passing a microphone location

The maximum sound pressure level, L_{Pmax}, and average sound pressure level, L_{Pavg}, are calculated from continuous A-weighted noise measurements made as the vehicle passes each microphone. Recorded voltage data is converted to pressure by dividing by microphone sensitivity resulting in instantaneous pressure versus time. A MATLAB® program is used to locate the maximum observed pressure as the vehicle passes the microphone. The maximum pressure, P_{max}, is assumed to be the impulse noise created as the front tire contacts the contraction joint. The maximum sound pressure, in dB(A), for each microphone location is calculated by using the observed maximum value as shown in Eq. (1), where P_{ref} is defined as 20μPa. This method of calculating the L_{Pmax}, using the strict definition of maximum pressure, assumes the existence of a tone with amplitude P_{max}.

$$L_{P\,max}, dB(A) = 20 \log_{10} \left[\frac{\left(\frac{P_{max}}{\sqrt{2}} \right)}{P_{ref}} \right] \quad (1)$$

After calculating L_{Pmax}, the program defines an event length based on the time it takes one vehicle wheelbase to pass a given contraction joint at the given test speed. The program further defines a pass-by time using the event length and maximum pressure point determined earlier. Total pass-by time is three vehicle wheelbase lengths in time. The pass-by time begins when the front tires of the vehicle are one wheelbase from the contraction joint, prior to contacting the joint. Time concludes when the rear tires of the vehicle have crossed the joint and are one vehicle wheelbase beyond the joint. The wheelbase for the sedan is 109" (~2.77 m). Sedan pass-by times are approximately 0.372, 0.310, and 0.265 seconds, respectively, for speeds of 50, 60, and 70 mph. The van wheelbase is 120.7" (~3.06 m) and pass-by times are approximately 0.411, 0.343, and 0.293 seconds, respectively, for speeds of 50, 60, and 70 mph. RMS values are calculated using the defined pass-by time. The average sound pressure level, in dB(A), for each microphone location is calculated using Eq. (2).

$$L_{Pavg}, dB(A) = 20 \log_{10} \left(\frac{P_{RMS}}{P_{ref}} \right) \quad (2)$$

Finally, maximum and average pressure levels are determined as the arithmetic average over all microphone locations for a given set of contraction joints at each test speed. This method of calculating an overall average is done as a matter of convenience. The referenced standards also use arithmetic averaging of each independent sampling period. Although this method is not as strict or precise as calculating overall averages using recorded time-domain data, it should not introduce a significant difference in the final results. Overall averaged results for in-cabin pressure levels are determined in the same manner.

IN-CABIN EXPERIMENTAL METHODOLOGY

Experimental Setup

Instrumentation consists of two Brüel & Kjær (B&K) 1/2-inch pressure-field microphones, both Type 4192, each using a B&K Type 2669 pre-amplifier, connected to channels 1 and 2 respectively, of a B&K Type 2690 Nexus signal conditioning amplifier. The Nexus is programmed to output a 100mV/Pa signal from each microphone input signal using 200V as supply voltage. Microphone signals are sent to the SigLab analyzer model 20-42 (channels 1 and 2). Although not required by the standard, vibration measurements are recorded using a PCB Piezotronics model 338B04 accelerometer. Accelerometer output voltage signal is sent to SigLab channel 3 via a PCB model 480C021CP sensor signal conditioner. Finally, a 1.5V DC manual trigger signal, sent to SigLab channel 4, is used to clearly identify the point in time where the roadway transitions from the 1/4-inch to 3/8-inch contraction joint sections (location of 1-inch contraction joint). An instrumentation schematic is shown in Fig. 6.

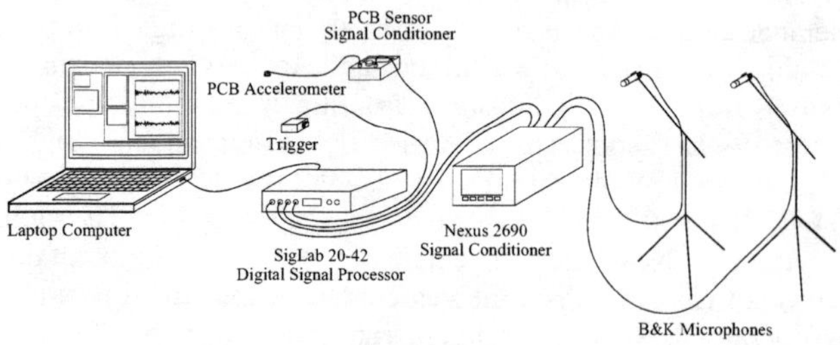

Figure 6. In-Cabin instrumentation setup

Microphones are located with respect to the driver and passenger seats as shown in Fig. 7. Each microphone is positioned horizontally pointing in the direction of travel. Microphones are stabilized using microphone boom stands with tripod bases. Each microphone is mechanically isolated by using a foam wrap (approximately 1-inch thick) around the pre-amplifier section and held in place with a spring loaded clip-type holder. When possible, the accelerometer is mounted directly to the body of the test vehicle using standard accelerometer wax as an adhesive. However, due to mounting restrictions in the sedan during winter testing, the accelerometer is mounted on the driver's seat frame where it is bolted to the vehicle body.

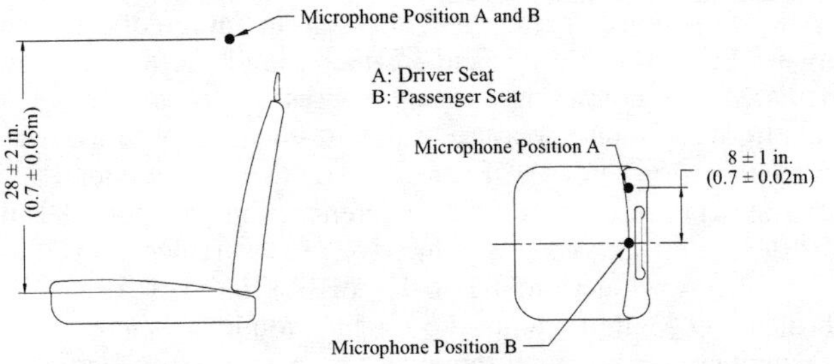

Figure 7. In-cabin microphone positioning with respect to driver and passenger seats

Calculation of Sound Pressure Levels of In-cabin Noise

In rooms of fairly regular shape with smooth walls, such as the passenger compartment of transportation vehicles, sound is not statistically diffuse and depends upon the acoustical response of the enclosure [6]. The actual distribution depends upon the mode shape of the enclosure. While acoustical finite-element modeling methods to approximate the actual pressure distribution exist, noise measurements made within the vehicle cabin assume that a steady state pressure field is present after the vehicle has reached and maintained a constant speed for an extended period of time. Vehicle speeds are held constant during in-cabin noise testing; therefore, aerodynamic and engine noises are assumed constant for each section. Also, since FDOT specifies the design of tining rakes [7], noise generated by the road surface is further assumed to be relatively constant for each section and lane. Observed differences between the final in-cabin sound pressure levels are considered to be due to the impulse noise contribution of the contraction joints. The number of joint impulses varies with vehicle speed and is equal to the vehicle speed in feet/second (ft/s). The total number of joint impulses for each section is approximately 73, 88, and 102 for speeds of 50, 60, and 70 mph, respectively.

The in-cabin average sound pressure level for each microphone is calculated using a time-offset length that is analogous to the event length used for roadside noise calculations. A-weighted data is recorded continuously for 24 seconds. After approximately 12-seconds have elapsed, the road section containing 1/4-inch joints transitions to 3/8-inch joints. A trigger signal is used to clearly identify the transition point (approximate location of the 1-inch contraction joint). Recorded voltage data is converted to pressure data by dividing by microphone sensitivity resulting in instantaneous pressure versus time. Two time vectors, one for 1/4-inch joint section and one for 3/8-inch joint section, each 8 seconds in length, are created using the trigger point. The vector for the 1/4-inch joint section begins exactly 9 seconds prior to the trigger point, ending 1 second prior to reaching the joint transition location (resulting in 8 seconds of continuous data). Similarly, the vector for the 3/8-inch joint section begins exactly 1 second after the trigger point and continues for 8 seconds. RMS values for each section are calculated from the time vectors and the average sound pressure level for each microphone is calculated by substituting P_{RMS} in-cabin values into Eq. 2. Finally, overall average in-cabin levels for each section and a given speed are determined as the arithmetic average of each microphone observed average sound pressure levels in both the right-hand and left-hand lanes.

ROADSIDE NOISE RESULTS

Roadside tests were conducted during both summer and winter conditions to compare potential differences in average sound pressure levels due to increases in joint width. Summer tests were conducted on September 20, 2002 and winter tests were conducted on January 14, 2003.

Summer and Winter Testing

As described earlier, pass-by time is used to calculate the average sound pressure, L_P, levels. Sound pressure levels observed during roadside testing are presented in a tabular format for each vehicle. Data includes the averaged L_{Pavg} values and the averaged L_{Pmax} values for all four microphones. Differences are calculated by subtracting the1/4-inch joint results from 3/8-inch joint results. Tables 1 through 4 contain roadside test results for both summer and winter conditions.

Table 1. Buick sedan summer roadside results

	L_{Pavg}, dB(A)			L_{Pmax}, dB(A)		
Speed,	Joint Width		Difference	Joint Width		Difference
mph	1/4"	3/8"	dB(A)	1/4"	3/8"	dB(A)
50	72.7	75.2	2.5	101.2	105.2	4.0
60	75.7	76.9	1.2	104.7	106.0	1.3
70	78.5	80.1	1.6	107.4	108.8	1.4

Table 2. Ford van summer roadside results

	L_{Pavg}, dB(A)			L_{Pmax}, dB(A)		
Speed,	Joint Width		Difference	Joint Width		Difference
mph	1/4"	3/8"	dB(A)	1/4"	3/8"	dB(A)
50	74.3	77.5	3.2	103.8	106.5	2.7
60	78.6	79.6	1.0	107.0	108.9	1.9
70	80.1	82.7	2.6	108.6	111.5	2.9

Table 3. Buick sedan winter roadside results

	L_{Pavg}, dB(A)			L_{Pmax}, dB(A)		
Speed,	Joint Width		Difference	Joint Width		Difference
mph	1/4"	3/8"	dB(A)	1/4"	3/8"	dB(A)
50	74.3	75.9	1.6	104.6	105.8	1.2
60	76.3	77.9	1.6	105.2	107.2	2.0
70	79.5	80.1	0.6	108.9	110.2	1.3

Table 4. Ford van winter roadside results

	L_{Pavg}, dB(A)			L_{Pmax}, dB(A)		
Speed,	Joint Width		Difference	Joint Width		Difference
mph	1/4"	3/8"	dB(A)	1/4"	3/8"	dB(A)
50	76.3	77.7	1.4	104.9	109.4	4.5
60	79.7	82.0	2.3	108.6	111.2	2.6
70	81.0	83.6	2.6	110.5	112.8	2.3

Summary of Roadside Results

Roadside measurements clearly result in measurable differences between the 1/4-inch joints and 3/8-inch joints with the larger joints consistently emitting higher average and maximum noise levels. Differences are calculated by subtracting the sound pressure of the 1/4-inch joints from that of the 3/8-inch joints. Summer L_{Pavg} measurements result in differences ranging from 1.0 to 3.2 dB(A) while summer L_{Pmax} measurements result in differences ranging from 1.3 to 4.0 dB(A). Winter L_{Pavg} measurements result in differences ranging from 0.6 to 2.6 dB(A) while L_{Pmax} measurements result in differences ranging from 1.2 to 4.5 dB(A). Comparisons between summer and winter measurement results are further discussed later.

IN-CABIN NOISE RESULTS

Summer and Winter Testing

As with roadside testing, in-cabin tests were conducted during both summer and winter conditions to compare potential differences in average sound pressure levels due to increases in joint widths. For the van, summer tests were conducted on August 8, 2002 and winter tests were conducted on January 14, 2003. Sedan summer and winter tests were performed on September 20, 2002 and January 14, 2003, respectively.

Also as discussed earlier, differences observed for in-cabin noise measurements are assumed to arise due to the impulse noise contribution of the contraction joints. Sound pressure levels observed during in-cabin testing are presented in tabular format for each vehicle. Data includes the averaged L_{Pavg} values for both microphones and the averaged g_{RMS} values for the accelerometer. Differences are calculated by subtracting the1/4-inch joint results from 3/8-inch joint results. Tables 5 through 8 contain in-cabin test results for both summer and winter conditions.

Table 5. Buick sedan summer in-cabin results

	L_{Pavg}, dB(A)			Accelerometer, g_{RMS}		
Speed,	Joint Width		Difference	Joint Width		Difference
mph	1/4"	3/8"	dB(A)	1/4"	3/8"	g_{RMS}
50	64.0	65.8	1.8	0.0936	0.1018	0.0083
60	65.9	67.4	1.5	0.1251	0.1339	0.0088
70	67.8	68.9	1.1	0.1555	0.1725	0.0170

Table 6. Ford van summer in-cabin results

	L_{Pavg}, dB(A)			Accelerometer, g_{RMS}		
Speed,	Joint Width		Difference	Joint Width		Difference
mph	1/4"	3/8"	dB(A)	1/4"	3/8"	g_{RMS}
50	68.3	68.9	0.6	0.1228	0.1314	0.0087
60	70.0	70.7	0.7	0.1531	0.1603	0.0072
70	72.0	72.6	0.6	0.1783	0.2088	0.0305

Table 7. Buick sedan winter in-cabin results

	L_{Pavg}, dB(A)			Accelerometer, g_{RMS}		
Speed,	Joint Width		Difference	Joint Width		Difference
mph	1/4"	3/8"	dB(A)	1/4"	3/8"	g_{RMS}
50	63.5	62.7	-0.8	0.0323	0.0347	0.0025
60	65.3	64.9	-0.4	0.0388	0.0398	0.0009
70	67.7	67.4	-0.4	0.0471	0.0468	-0.0003

Table 8. Ford van winter in-cabin results

	L_{Pavg}, dB(A)			Accelerometer, g_{RMS}		
Speed,	Joint Width		Difference	Joint Width		Difference
mph	1/4"	3/8"	dB(A)	1/4"	3/8"	g_{RMS}
50	67.0	67.6	0.6	0.1021	0.1109	0.0088
60	68.7	69.4	0.7	0.1287	0.1411	0.0124
70	70.9	71.2	0.3	0.1651	0.1749	0.0098

Summary of In-cabin Results

In-cabin noise measurements result in minor differences between the 1/4-inch joints and 3/8-inch joints. Differences are calculated by subtracting the sound pressure of the 1/4-inch joints from the sound pressure of the 3/8-inch joints. However, the results vary. For the Ford van, summer and winter L_{Pavg} measurements result in similar differences ranging from 0.3 to 0.7 dB(A). For the sedan, summer L_{Pavg} measurements result in differences ranging from 1.1 to 1.8 dB(A), indicating that the 3/8-inch (~9.53 mm) joints are slightly louder. Winter L_{Pavg} measurements result in differences ranging from -0.8 to -0.4 dB(A), indicating that the 1/4-inch joints are slightly louder.

JOINT WIDTH MEASUREMENT RESULTS

Although the main function of rigid pavement contraction joints is to control cracking, a secondary benefit is space for contraction or expansion. Joint gap widths are expected to vary as the temperature of each pavement changes – i.e., widths decrease as temperatures increase; widths increase as temperatures decrease. The average joint widths reported in Table 9 are calculated from 10 separate measurements made within the joint at each microphone location. Joint gap widths are measured during both summer and winter testing to further compare noise differences measured between summer and winter. These differences include those observed between the 1/4-inch joints and the 3/8-inch joints as well as measurable differences observed within the same section of joints.

Table 9. Difference in average joint width between summer and winter conditions

	Average Joint Expansion [inches]	
Mic Location	1/4-inch Section	3/8-inch Section
1	0.019	0.041
2	0.030	0.038
3	0.039	0.041
4	0.014	0.044
Avg. Expansion	0.026	0.041

As shown in Table 9, the 1/4-inch joint widths increased by approximately 9.45% while the 3/8-inch joint width increased by approximately 9.05%.

DIFFERENCE-LIMEN TESTING

The range of sound pressure levels and frequencies over which the human ear receives useful information is remarkable. The frequency range is generally considered to be from 20 Hz to 20,000 Hz. The range of pressure from the threshold of perception to the threshold of pain is 20 μPa (~2.901 x 10-9 psi) to 200 Pa (~2.901 x 10-2 psi). The ear primarily perceives small changes in sound pressure levels as differences in loudness. Table 10 contains loudness changes in dB(A) and typical human response as published by the Federal Highway Administration and most states' Department of Transportation. A change of 3 dB(A) is considered just noticeable while the average listener can readily distinguish a loudness change of 5 db(A).

Table 10. Decibel changes, relative loudness and acoustic energy loss

Sound Level Change, dB(A)	Relative Loudness	Acoustic Energy Loss, %
0	Reference	0.00
-3	Barely Perceptible	50.0
-5	Readily Perceptible	66.7
-10	Half as Loud	90.0
-20	1/4 as Loud	99.0

More recent data suggests that as sound levels increase, so does the ability to differentiate loudness. For sound pressure levels of 60 dB or more above the threshold of hearing, listeners may detect a loudness change of 0.5 dB or less as shown in Fig. 8. In addition to sound pressure level and frequency, the just-noticeable difference in sound pressure level depends on signal duration for short time intervals. As duration increases up to a critical period of 0.1 seconds, the just-noticeable difference decreases [1].

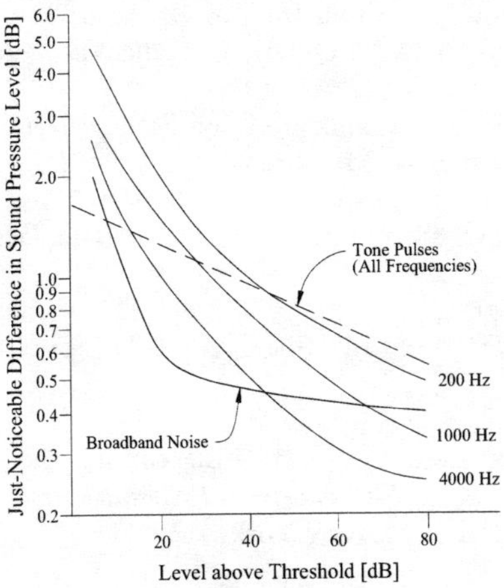

Figure 8. Just-noticeable differences in sound pressure level for tone pulses, pure tones, and broadband noise

In order to qualify the disagreement between the data in Table 10 and Fig. 8, a subjective listening experiment was designed using vehicle in-cabin noise measurement data. A discussion of just-noticeable difference testing and rationalization for experimental method follows.

Two general stimulus paradigms are typically used for loudness discrimination experiments: loudness memory and loudness modulation. Loudness memory experiments require the listener to judge which is the more intense of two alternately presented stimuli. The loudness difference-limen (DL) is usually defined as the intensity difference that produces 75% correct responses. Loudness modulation experimental methods typically introduce periodic changes in intensity into a continuous signal. The loudness DL corresponds to the smallest detectable change in intensity [8].

Most loudness DL testing involves the use of pure tones of single frequencies. Research conducted in 1949 used a loudness DL test that introduced abrupt intensity changes into a continuous pure tone. A loudness memory DL experiment conducted in 1950 used two tones, each 1.7 seconds in length, presented in succession separated by 0.33 seconds of silence. These tests were conducted at base levels of 4, 24, and 44 dB across a wide frequency range of normal hearing [9]. Variations of these tests have been conducted since the 1950s.

Unlike the sinusoidal tones often used in the assessment of hearing, sounds encountered in everyday life are generally complex containing multiple frequency components. One of the most important characteristics of the auditory system is that it acts as a limited-resolution frequency analyzer: complex sounds are broken down into individual sinusoidal frequency components [10]. Due to the subjectivity of listeners and difficulty qualifying the results, limited research on loudness DLs for broadband noise has been conducted. A method to attempt to quantify the Just-Noticeable Difference (JND) of broadband roadway noise is needed to settle the existing discrepancies previously mentioned.

Threshold of audibility tests present a tone of 1-second duration to both ears. For tones shorter than 0.3 seconds, apparent loudness increases with increasing tone duration and for tones longer than 3.0 seconds fatigue sets in and the apparent loudness diminishes with time [11]. In order to produce reasonably reliable results in a simple subjective listener experiment, the 0.3 to 3.0 second range limits duration of exposure time to the signal. The loudness memory method described above is used as the basis for the experimental procedure.

Loudness DL Experiment Using Vehicle In-Cabin Broadband Noise

To create the experiment, actual data recorded during the in-cabin tests for the Buick Century is used. The pressure data is converted into a wave file format using MATLAB®. Next, the converted wave file is played back into a set of Sony Model MDR-CD180 stereo headphones. Sound pressure levels within the headphone are measured using a B&K Model 4939 1/4-inch (~6.35 mm) microphone. A multiplier applied to the output is used to adjust the playback level until it reaches the actual recorded level. An adjusted wave file is then saved as a base for creating comparison noise. A total of four base files are created from data recorded in both the 1/4-inch and 3/8-inch joint sections. Next, a base file is opened for playback using Nero Wave Editor. Playback levels are again measured to ensure that they are comparable to the level recorded in the vehicle. Approximately two seconds of data are then cut from the original file. Playback levels are again monitored for agreement. Following monitoring, the playback level is varied using volume control within Nero and measured in the headsets. The noise level is increased from 0 to 3 dB in increments of 0.5 dB. Finally, comparative listening test files are created. In order to ensure that the tonal character of the comparative noises is the same, files are paired only with files created from the same base file. Comparative listening files are paired using the known pressure level to create comparative differences from 0 to 3 dB in increments of 0.5 dB. Each noise is separated by 2 seconds of silence. The 28 comparative listening sound samples each contain a noise duration of 2 seconds, followed by 2 seconds of silence, followed by an additional 2 seconds of noise. A second set of 28 comparative tests is created using 1/2 second of noise separated by 2 seconds of silence, followed by an additional ½ second of noise. A final total of 56 comparative listening tests of known pressure level and pressure differences are created. Pressure levels ranged from 72.8 to 80.2 dB as a representative range of values measured during roadside testing.

A total of 21 male and female test subjects ranging in age from 21 to 66 years participated in subjective listening experiments. Tests are administered by playing the calibrated comparative listening files into the Sony stereo headphones. Subjects are asked to listen to each set of noise data and identify whether the first noise or second noise seemed loudest. If the subject cannot distinguish between the two sounds or considers the sounds to be of equal loudness, they indicate no noticeable difference.

The results from the subjective listening experiments suggest that the difference-limen for the in-cabin broadband noise used for the experiments is approximately 2.5 dB. Results also suggest that listeners are able to distinguish when no difference occurs, 66.7% correct for long duration and 77.4% for short duration. Since volunteers are asked to indicate "No Difference" when they are unable to distinguish which sound is louder, the result may be misleading. There are four tests within each round of testing where no difference exists between the two noises presented. On average, respondents answered "No Difference" 41.8% of the time during each round of testing.

Comparison of Roadway Noise Results to Difference-limen Experiments

The 2.5 dB DL determined by the study is close to the 3.0 dB suggested as the just-noticeable difference in most national and state highway noise programs. Based on the DL results above, the average listener is unable to distinguish differences measured during all in-cabin testing. A total of 12 roadside L_{Pavg} values are reported. Four of the values are above the 2.5 dB threshold. There are also 12 roadside L_{Pmax} values reported. Of these, five exceed the 2.5 dB threshold. In all cases, the noise emitted by the 3/8-inch joint is louder than the 1/4-inch joint. The DL result suggests that the average listener might be able to distinguish a difference in pass-by noise 33% to 37.5% of the time.

The results of the study indicate that measurable differences between the noise emitted by 1/4-inch joints and 3/8-inch joints due to passenger vehicles traveling at highway speeds exists both within the vehicle cabin and at the highway roadside. For the highway roadside noise, the empirical difference may be enough to require abatement measures or may slightly influence property setback requirements. The difference-limen experiment results, however, suggest that the average listener will not recognize any difference in in-cabin noise and may only distinguish a difference at the highway roadside less than 38% of the time.

CONCLUSIONS

The results of the study indicate that measurable differences between the noises emitted by 1/4-inch joints and 3/8-inch joints due to passenger vehicles traveling at highway speeds exists both within the vehicle cabin and at the highway roadside. For the highway roadside noise, the empirical difference may be enough to require abatement measures or may slightly influence property setback requirements. The difference-limen experiment results, however, suggest that the average listener will not recognize any difference in in-cabin noise and may only distinguish a difference at the highway roadside less than 38% of the time.

ACKNOWLEDGMENTS

The authors would like to thank the Florida Department of Transportation (Contract # S55010032E) for sponsoring this research.

REFERENCES

[1] Harris C., editor, 1998, 3rd ed., *Handbook of Acoustical Measurements and Noise Control*, Acoustical Society of America, Melville (NY).

[2] Herman, L.A., 1997, "Analysis of Strategies to Control Traffic Noise at the Source", Transportation Research Record, Vol. 1626, pp. 41-48.

[3] United States Department of Transportation/Federal Highway Administration (USDOT/FHWA) PD-96-046, 1995, "Highway Traffic Noise Analysis and Abatement Policy and Guidance", United States Department of Transportation, Federal Highway Administration, Office of Environment and Planning, Noise and Air Quality Branch, Washington (DC).

[4] International Organization for Standards (ISO) 11819-1, 1997, "Acoustics - Measurement of the influence of road surfaces on traffic noise - Part 1: Statistical Pass-By Method", International Organization for Standardization, Geneva.

[5] Gharabegian, A., Tuttle, E., 2002, "Diamond Grinding for Roadway Noise Control", Sound and Vibration, Vol. 36 No. 12, pp. 26-29.

[6] Beranek, L.L., and Vér, I.L., editors, 1992, *Noise and Vibration Control Engineering: Principles and Applications*, John Wiley and Sons, New York.

[7] Florida Department of Transportation (FDOT), 1991, *Standard Specifications for Road and Bridge Construction*, Florida Department of Transportation, Tallahassee.

[8] Ballantyne, D., 1990, *Handbook of Audiological Techniques*, Butterworth-Heinemann, London.

[9] Jacobson, J.T., and Northern, J.L., 1991, *Diagnostic Audiology*, Pro-Ed, Austin.

[10] Crocker, M.J., editor, 1998, *Handbook of Acoustics*, John Wiley and Sons, New York.

[11] Kinsler, L. E., Frey, A.R., Coppens, A.B., and Sanders, J.V., 1982, 3rd ed. *Fundamentals of Acoustics*, John Wiley and Sons, New York.

Proceedings of IMECE'03
2003 ASME International Mechanical Engineering Congress
Washington, D.C., November 15–21, 2003

IMECE2003-55330

PROBABILISTIC ANALYSES FOR THE PERFORMANCE CHARACTERISTICS OF THE MAIN BEARINGS IN AN OPERATING ENGINE DUE TO VARIABILITY IN BEARING PROPERTIES

Jin Wang
University of Michigan

Nickolas Vlahopoulos
University of Michigan

Zissimos P. Mourelatos
Oakland University

Omidreza Ebrat
Federal-Mogul Corporation

Kumar Vaidyanathan
Federal-Mogul Corporation

ABSTRACT

This paper presents the development of surrogate models (metamodels) for evaluating the bearing performance in an internal combustion engine. The metamodels are employed for performing probabilistic analyses for the engine bearings. The metamodels are developed based on results from a simulation solver computed at a limited number of sample points, which sample the design space. An integrated system-level engine simulation model, consisting of a flexible crankshaft dynamics model and a flexible engine block model connected by a detailed hydrodynamic lubrication model, is employed in this paper for generating information necessary to construct the metamodels. An optimal symmetric Latin hypercube algorithm is utilized for identifying the sampling points based on the number and the range of the variables that are considered to vary in the design space. The development of the metamodels is validated by comparing results from the metamodels with results from the actual simulation models over a large number of evaluation points. Once the metamodels are established they are employed for performing probabilistic analyses. The initial clearance between the crankshaft and the bearing at each main bearing and the oil viscosity comprise the random variables in the probabilistic analyses. The maximum oil pressure and the percentage of time (the time ratio) within each cycle that a bearing operates with oil film thickness less than a user defined threshold value at each main bearing constitute the performance variables of the system. The availability of the metamodels allows comparing the performance of several probabilistic methods in terms of accuracy and computational efficiency. A useful insight is gained by the probabilistic analysis on how variability in the bearing characteristics affects its performance.

INTRODUCTION

In this paper metamodels are developed from an engine simulation solver and utilized in probabilistic analyses for main bearing lubrication applications for an operating engine. The Kriging method for generating metamodels is employed along with probabilistic algorithms for this purpose. Probabilistic analyses typically require a large number of evaluations of the performance parameters for different combinations of the random variables. Thus, probabilistic solutions become computationally intensive. As an alternative, metamodels are developed from the values of the performance parameters evaluated at a limited number of sample points.

A great deal of research has been done on metamodel development. Polynomials, interpolating and smoothing splines [Craven and Wabba 1978], neural network [Hajela and Berke 1993], [Rumelhart et al 1994], [Cheng and Titterington 1994] and [Ellacott et al 1997], radial basis functions [Dyn et al 1986], and wavelets [Jansen et al 1996] have been investigated in the past. [Cressie 1988], [Sacks et al 1989a] and [Sacks et al 1989b] introduced the concept of the Kriging method. The Kriging method provides an efficient predictor for a given set of data. It treats the deterministic output as the realization of a stochastic process and provides a statistical basis for efficient prediction. [Currin et al 1988] and [Sacks et al 1989a] established that the maximum likelihood estimation (MLE) is useful in estimating the optimal parameters required by Kriging model at the analysis stage.

A metamodel (or surrogate model) is comprised by equations that replace the actual simulation model. The equations that define the metamodel are evaluated from actual simulations performed at a group of sample points. Different methods can be employed for selecting the group of sample points. [Park 1994] and [Morris and Mitchell 1995] developed an optimal Latin Hypercube selection algorithm that combines good projective properties and computational efficiency. [Ye et al 2000] developed an Optimal Symmetric Latin Hypercube (OSLH) algorithm that offers a combination between computational effort and optimality. It was demonstrated that the OSLH has advantages over the regular Latin Hypercube method and the Symmetric Latin Hypercube selection algorithm with respect to criteria such as entropy and the minimum intersite distance. The OSLH is employed in this paper for selecting the group of sample points based on the range of variation of the random variables. Metamodels are developed based on the group of sample points selected by the OSLH algorithm.

An operating V6 engine represents a complicated non-linear system and in the past engine dynamics simulations have not considered how manufacturing imperfections along with operational and material variability influences the performance of the bearings. In this paper, variability is introduced in some of the parameters of an engine system and probabilistic analyses are preformed for the main bearings of the engine. The initial uniform clearance between the crankshaft and the bearing at each main bearing and the oil viscosity comprise the five random variables considered in the analysis. All random variables are considered to have a normal distribution. The maximum oil film pressure and the percentage of time (the time ratio) within each cycle that a bearing operates with film thickness lower than a user defined threshold value are defined as the performance variables. Eight performance variables are considered in this paper. An engine dynamic simulation solver developed by [Hu et al 2002], [Hu 2002], [Ebrat 2002] is employed in this paper for evaluating the performance variables.

Probabilistic analyses are conducted by several alternative probabilistic methods. Algorithms based on sampling methods and on an advanced mean value (AMV) method are employed. The AMV method was introduced by [Wu 1987], [Wu and Wirsching 1987], and [Wu et al 1990]. The development of the metamodels makes it possible to compare the performance of the various probability methods in this application.

An overview of the system-level engine simulation solver employed in this work is presented first. The theoretical background of the OSLH and of the Kriging method is presented next. The parameters of the engine application are defined. Metamodels are developed from the results of the engine simulation solver at the sample points generated through OSLH algorithm. The metamodels are validated by comparing results from the metamodels and the actual solver for a large number of evaluation points. Finally, the developed metamodels are utilized in several probabilistic analyses.

DEVELOPMENT OF A SYSTEM-LEVEL ENGINE BEARING SIMULATION MODEL

The integrated system-level engine bearing simulation model developed by [Hu et al 2002], [Hu 2002] and [Ebrat 2002] is employed in this paper. This model consists of a flexible crankshaft model and a flexible engine block model connected by a detailed hydrodynamic lubrication model. Technical information about the system-level engine simulation solver is reviewed.

First, the Craig-Bampton method is employed for reducing the physical Degrees of Freedom (DOF) for both the crankshaft and the engine block [Craig and Bampton 1968]. The dynamic equations of motion for the engine block structure, modeled by the finite element method, are

$$[M^b]\{\ddot{x}^b\}+[C^b]\{\dot{x}^b\}+[K^b]\{x^b\}=\{F^b\} \tag{1}$$

where $[M^b]$, $[C^b]$, $[K^b]$ and $\{F^b\}$ are the mass matrix, damping matrix, stiffness matrix and load vector, respectively. Superscript "b" denotes the engine block structure. The physical DOF in the vector $\{x^b\}$ constitute the unknown variables. Due to the large size of the vector $\{x^b\}$, the Craig-Bampton reduction method is employed. The vector of the initial DOF $\{x^b\}$ can be expressed in terms of a reduced basis comprised of the retained DOF and the modal participation factors of the fixed-interface modes; i.e.

$$\{x^b\}=\begin{Bmatrix} x_i^b \\ x_r^b \end{Bmatrix}=\begin{bmatrix} X^b & T^b \\ 0 & I \end{bmatrix}\begin{Bmatrix} \alpha^b \\ x_r^b \end{Bmatrix}=[\phi^b]\begin{Bmatrix} \alpha^b \\ x_r^b \end{Bmatrix} \tag{2}$$

where, $[\phi^b]=\begin{bmatrix} X^b & T^b \\ 0 & I \end{bmatrix}$ and matrix $[T^b]=-[K_{ii}^b]^{-1}[K_{ir}^b]$ accounts for the static correction.

Matrix $[\phi^b]$ constitutes the modal matrix of the reduced basis of the Craig-Bampton DOF. The final system of equations becomes:

$$[\overline{M}^b]\begin{Bmatrix} \ddot{\alpha}^b \\ \ddot{x}_r^b \end{Bmatrix}+[\overline{C}^b]\begin{Bmatrix} \dot{\alpha}^b \\ \dot{x}_r^b \end{Bmatrix}+[\overline{K}^b]\begin{Bmatrix} \alpha^b \\ x_r^b \end{Bmatrix}=\{\overline{F}^b\} \tag{3}$$

where $[\overline{M}^b]$, $[\overline{C}^b]$ and $[\overline{K}^b]$ are the reduced mass, damping and stiffness matrices, respectively, and $\{\overline{F}^b\}=[\phi^b]^T\begin{Bmatrix} F_i^b \\ F_r^b \end{Bmatrix}$ is the vector of the modal forces.

Similar to the flexible block, the reduced equations of motion for the crankshaft are

$$[\overline{M}^c]\begin{Bmatrix} \ddot{\alpha}^c \\ \ddot{x}_r^c \end{Bmatrix}+[\overline{C}^c]\begin{Bmatrix} \dot{\alpha}^c \\ \dot{x}_r^c \end{Bmatrix}+[\overline{K}^c]\begin{Bmatrix} \alpha^c \\ x_r^c \end{Bmatrix}=\{\overline{F}^c\} \tag{4}$$

The matrices $[\overline{M}^c]$, $[\overline{C}^c]$ and $[\overline{K}^c]$ in Equation (4) are the reduced mass, damping and stiffness matrices respectively, and $\{\overline{F}^c\}$ is the modal force vector. The Craig-Bampton DOF

$\begin{Bmatrix} \alpha^c \\ x_r^c \end{Bmatrix}$ contains the modal participation factors of the fixed-interface modes and the retained physical DOF included in $\{x_r^c\}$.

The connecting mechanism between the flexible crankshaft and the flexible block is represented by a set of stiffness and damping matrices at each bearing. The stiffness and damping matrices are determined through a lubrication solver for each bearing [Ebrat 2002]. In the lubrication solver the Reynolds equation is solved and the distribution of the oil pressure in the circumference around the bearing is computed. The derivatives of the oil pressure with respect to the clearance constitute the entries of the stiffness matrices at each bearing, and the derivatives of the oil pressure with respect to the time derivative of the clearance comprise the entries of the damping matrices. For computational efficiency reasons, a diagonalization procedure of the oil film stiffness and damping matrices has been implemented [Ebrat 2002]. The diagonalization is achieved by lumping the contribution of the off-diagonal terms in an "equivalent" diagonal term for each node of the fine lubrication mesh. This reduces the computational cost substantially with minimum loss of accuracy. As a result of the diagonalization process, the fully populated oil stiffness and damping matrices are represented by their equivalent diagonal forms,

$$[K] = \begin{bmatrix} K_1^e & 0 & \cdots & 0 \\ 0 & K_2^e & \cdots & 0 \\ \vdots & \vdots & \ddots & \vdots \\ 0 & 0 & \cdots & K_N^e \end{bmatrix} \qquad (5)$$

and

$$[C] = \begin{bmatrix} C_1^e & 0 & \cdots & 0 \\ 0 & C_2^e & \cdots & 0 \\ \vdots & \vdots & \ddots & \vdots \\ 0 & 0 & \cdots & C_N^e \end{bmatrix} \qquad (6)$$

The equivalent oil film stiffness and damping matrices of Equations (5) and (6) are introduced in the system level equations between the crankshaft and the block. This information is utilized for computing the interaction forces between the crankshaft and the block.

The system of dynamic equations for the combined crankshaft and block is:

$$\begin{bmatrix} M_{ff}^c & 0 \\ 0 & M_{ff}^b \end{bmatrix} \begin{Bmatrix} \ddot{x}^c \\ \ddot{x}^b \end{Bmatrix} + \begin{bmatrix} C_{ff}^c & 0 \\ 0 & C_{ff}^b \end{bmatrix} \begin{Bmatrix} \dot{x}^c \\ \dot{x}^b \end{Bmatrix} +$$

$$\begin{bmatrix} K_{ff}^c & 0 \\ 0 & K_{ff}^b \end{bmatrix} \begin{Bmatrix} x^c \\ x^b \end{Bmatrix} = \{Q_e\} + \{Q\} \qquad (7)$$

Superscripts "c" and "b" denote the crankshaft and block, respectively. $\{Q_e\}$ is the vector of external forces and $\{Q\}$ is the vector of the interaction forces between the crankshaft and the block computed by the lubrication solver [Ebrat 2002]. For computational efficiency, Equations (7) is reduced to the following modal equations

$$[\overline{M}]\{\ddot{q}\} + [\overline{C}]\{\dot{q}\} + [\overline{K}]\{q\} = \{\overline{Q}_e\} + \{\overline{Q}\} \qquad (8)$$

where $\{q\}$are the generalized coordinates. The reduced matrices $[\overline{M}]$, $[\overline{C}]$ and $[\overline{K}]$, and the reduced vector $\{\overline{Q}_e\}$ are given in [Hu et al 2002]. The generalized force vector $\{\overline{Q}\}$, is given by

$$\{\overline{Q}\} = \begin{Bmatrix} 0 \\ -T_c^T \\ T_c^T \end{Bmatrix} \{Q\} = [T]\{Q(h,\dot{h})\} \qquad (9)$$

where $[T_c]$ and $[T_b]$ are transformation matrices between the interface of the crankshaft and the block at the bearing. The nonlinear vector $\{Q(h,\dot{h})\}$ couples the crankshaft and block dynamics. The oil film stiffness and damping matrices (Equations 5 and 6) are employed in the computation of vector $\{Q(h,\dot{h})\}$ [Ebrat et al 2003].

The nonlinear system of reduced Equations (8) is solved in the time domain using a modified Newmark method for time integration, which includes a Newton-Raphson iteration within each time step. Details for the scheme selection can be found in [Hu et al 2002]. An operating V6 engine is analyzed by employing the developed system-level engine model. Figure 1 shows the finite element model of the engine block and crankshaft, including the pulley and the flywheel.

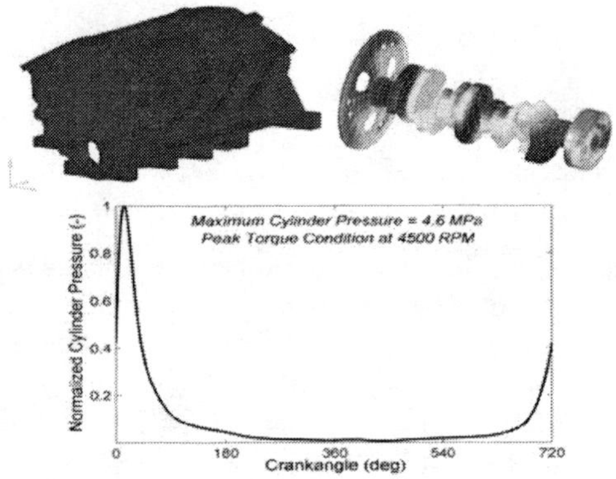

Figure 1 Crankshaft and block finite element mesh and cylinder pressure trace for an automotive V6 engine used for system level analysis

DEVELOPMENT OF A METAMODEL

Metamodels are developed for the maximum oil pressure and the percentage of time (the time ratio) within each cycle that a bearing operates with oil film thickness less than a user defined threshold value over a cycle of an operating V6 engine. The engine dynamic solver described in the previous section is employed for computing the performance variables at all the sample points used in the generation of the metamodel. An operating V6 engine represents a typical system-level application in automotive systems and comprises a highly non-linear and complicated system, which involves a great number of degree of freedom (DOF).

The maximum oil pressure in a cycle is crucial in selecting bearing materials with high enough fatigue resistance for protecting the bearing against failure under an overlay fatigue mode and is a critical indicator of the bearing capacity. The maximum oil pressure over a cycle at each main bearing constitutes four of the performance variables: $press_1$, $press_2$, $press_3$, and $press_4$. The percentage of time (the time ratio) within each cycle that a bearing operates with film thickness lower than a user defined threshold value is considered to constitute the other four of the performance variables: $ratio_1$, $ratio_2$, $ratio_3$, and $ratio_4$. This time ratio also indicates severity of the working condition of the main bearing in the investigated V6 engine system.

Eight metamodels are developed for the defined performance variables and are employed in the probabilistic analyses for the main bearing performance. The following five parameters in the bearings are considered as the random variables in this paper: the initial clearance between the crankshaft and the bearing (C1, C2, C3, C4) at each bearing and the oil viscosity (VIS). These random variables are assumed to have a normal distribution and they are considered uncorrelated.

In the operating V6 engine application, even a single run of the simulation model is computational intensive. A single run of the system-level engine simulation model requires 7 hours on a SUN workstation, or 45 minutes on a high speed SGI supercomputer. The computational time required by the operating engine dynamic simulation solver makes probabilistic analyses based on the solver impractical. The generation of the metamodels makes the probabilistic analyses feasible since it significantly reduces the computational time required for evaluating the performance variables for each set of values of the random variables. The theoretical background for developing a metamodel is presented first. Then, the specific procedure for developing a metamodel as a replacement of the operating engine dynamic simulation solver is presented and validated.

Metamodel Development

The Kriging method is employed in this paper for creating the metamodels [Cressie 1988], [Sacks et al 1989a] and [Sacks et al 1989b]. The Kriging method is based on treating $Z(\tilde{x})$, the error between the actual performance variable $y(\tilde{x})$ and an approximate value β, as a stochastic process:

$$y(\tilde{x}) = \beta + Z(\tilde{x}) \tag{10}$$

where $\tilde{x}$ is the d-dimensional vector of the random variables that defines the point where the performance variable is evaluated, and "d" is the number of random variables. $Z(\tilde{x})$ is considered as a normal process with zero mean and a covariance that can be expressed as:

$$\mathrm{cov}(Z(\tilde{x}_i), Z(\tilde{x}_j)) = \sigma^2 R(\tilde{x}_i, \tilde{x}_j) \tag{11}$$

where σ^2 is the process variance and $R(\tilde{x}_i, \tilde{x}_j)$ is the spatial correlation function (SCF). The equation used for the SCF is:

$$R(\tilde{x}_i, \tilde{x}_j) = \prod_{k=1}^{d} \exp(-\theta_k (x_{i,k} - x_{j,k})^2) \tag{12}$$

and it indicates a process with infinitely differentiable paths in the mean square sense. θ_k is the correlation parameter that corresponds to the kth component of the d-dimensional vector of the random variables $\tilde{x}$, i.e. k=1,2,...,d. For a set $\tilde{x}_s$ comprised by "n" number of sampling points $\tilde{x}_{si}$, i=1,2,...,n

$$\tilde{x}_s^T = \{\tilde{x}_{s1}, \tilde{x}_{s2}, \ldots, \tilde{x}_{sn}\} \tag{13}$$

the corresponding performance variable $\tilde{y}_s$ is considered known and its values are defined as:

$$\tilde{y}_s^T = \{y(\tilde{x}_{s1}), y(\tilde{x}_{s2}), \ldots, y(\tilde{x}_{sn})\} \tag{14}$$

The vector of correlations between the sample points $\tilde{x}_s$ and the evaluation point $\tilde{x}$ can be expressed as:

$$\tilde{r}^T(\tilde{x}) = \{R(\tilde{x}, \tilde{x}_{s1}), R(\tilde{x}, \tilde{x}_{s2}), \ldots, R(\tilde{x}, \tilde{x}_{sn})\} \quad (15)$$

The correlation matrix [R] is also defined among all the sample points:

$$[R]=[R(\tilde{x}_{si}, \tilde{x}_{sj})]_{nxn} \quad (16)$$

The spatial correlation function in Equations (15) and (16) has been defined by Equation (12). In the Kriging method the value of the performance function evaluated by the metamodel at the evaluation point $\tilde{x}$ is treated as a random variable. The computation of β and Z ($\tilde{x}$) in Equation (10) is based on minimizing the mean square error (MSE) in the response

$$MSE[\hat{y}(\tilde{x})] = E[\hat{y}(\tilde{x}) - y(\tilde{x})]^2 \quad (17)$$

subjected to the unbiasedness constraint:

$$E[\hat{y}(\tilde{x})] = E[y(\tilde{x})] \quad (18)$$

Thus, a set of parameters θ_k (k=1,2,...,d) are computed as the solution of minimizing the product $[(\det(R))^{1/n} \cdot \hat{\sigma}^2]$,

$$\text{where } \hat{\sigma}^2 = \frac{1}{n}(\tilde{y}_s - \tilde{F}\hat{\beta})^T R^{-1}(\tilde{y}_s - \tilde{F}\hat{\beta}) \quad (19)$$

$\tilde{F}$ is a vector of n-dimension with all unit entries and $\hat{\beta}$ defined as:

$$\hat{\beta} = (\tilde{F}^T R^{-1} \tilde{F})^{-1} \tilde{F}^T R^{-1} \tilde{y}_s \quad (20)$$

Once the set of optimal parameters θ_k has been computed, then the performance function at any point $\tilde{x}$ can be evaluated as:

$$y(\tilde{x}) = \hat{\beta} + \tilde{r}^T(\tilde{x}) R^{-1}(\tilde{y}_s - \tilde{F}\hat{\beta}) \quad (21)$$

where $\tilde{r}^T(\tilde{x})$ is defined in Equation (15). Equation (21) comprises the metamodel developed by the Kriging method.

Metamodels for an Operating V6 Engine

The mean values and the standard deviations corresponding to each one of the five random variables considered in the V6 engine application are summarized in Table 1. The ranges for the five random variables are representative of values encountered in industrial applications.

	mean (μ)	minimum	maximum	standard deviation (σ)	COV
C1 (micro)	30	15	45	5	16.67%
C2 (micro)	30	15	45	5	16.67%
C3 (micro)	30	15	45	5	16.67%
C4 (micro)	30	15	45	5	16.67%
VIS (Pa.Sec)	0.01	0.0058	0.0142	0.0014	14%

Table 1 Definition of random variables in the V6 engine application

Based on the definition of the random variables, an Optimal Symmetric Latin Hypercube (OSLH) algorithm developed by [Ye et al 2000] is employed for defining the best combination of sample data used for generating the metamodels. A data set consisting of 200 OSLH sample data points is employed in this paper. The engine solver is used for computing the values for the performance variables at all 200 sample points. Collecting data for the 200 sample points takes about 130 CPU hours of simulation on an SGI supercomputer. In order to validate the developed metamodels, 800 evaluation data that are different from the 200 OSLH sample data are utilized. The 800 evaluation points are identified through a random number generator from a Monte Carlo algorithm. The computations of the system-level engine solver for the 800 evaluation points require approximately 24 days of simulation on an SGI supercomputer. Results generated by the metamodel and results generated by the actual solver for the 800 evaluation points are presented in Figures 2 to 5 for the performance variables $press_2$, $press_3$, $ratio_2$ and $ratio_3$. Based on the results from the solver the values for the performance variables of bearings 2 and 3 are considerably higher than the performance variables associated with the other two bearings. Thus, in the remaining of this paper only the four performance variables $press_2$, $press_3$, $ratio_2$ and $ratio_3$ are considered.

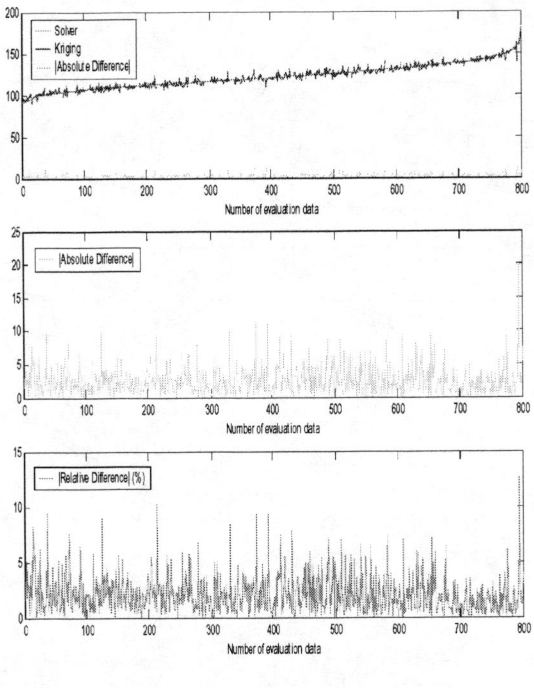

Figure 2 Validation of metamodel for performance variable $press_2$

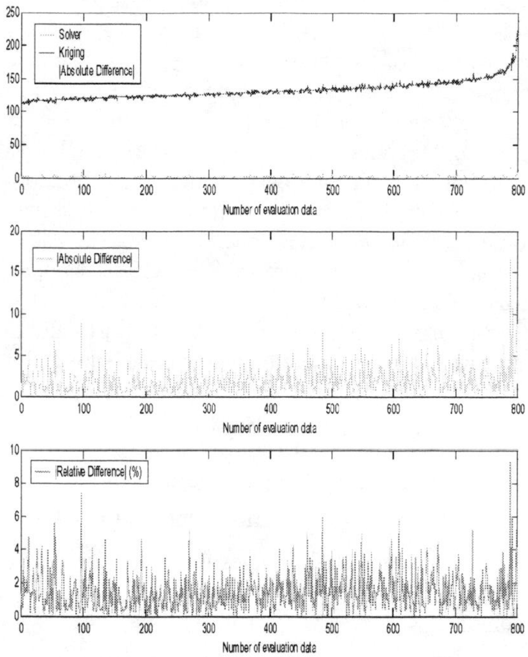

Figure 3 Validation of metamodel for performance variable press $_3$

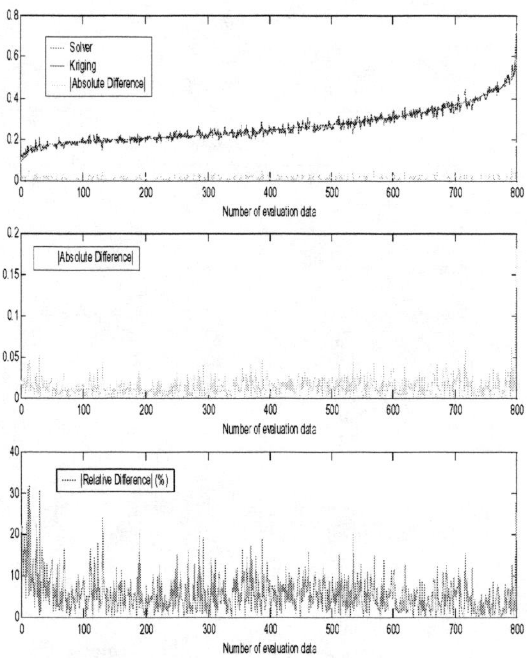

Figure 4 Validation of metamodel for performance variable ratio $_2$

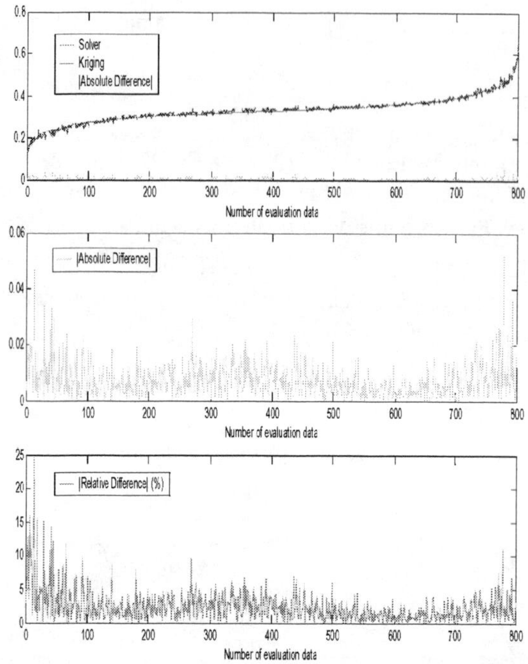

Figure 5 Validation of metamodel for performance variable ratio $_3$

The results are plotted with respect to an identification number representing each evaluation point. The numbers assigned to each evaluation point are based on an ascending order for the performance variable as evaluated by the engine simulation solver. In each figure, there are three plots. The top plot presents the comparison between the solver and the metamodel solution, along with the magnitude of the difference between the solver and the metamodel results. The middle plot presents the comparison of the magnitude of the differences between the solver and the metamodel only. The bottom plot depicts the absolute relative difference as a percentage between the two computations. Table 2 summarizes the differences encountered between the metamodel and the solver for the 800 evaluation data.

	Minimum difference	Maximum difference	Mean difference	Median difference
press $_2$	1.20E-05	12.67%	2.09%	1.67%
press $_3$	1.50E-05	9.32%	1.46%	1.20%
ratio $_2$	9.60E-05	32.08%	5.46%	4.47%
ratio $_3$	3.92E-06	24.54%	2.39%	1.85%

Table 2 Comparison of relative differences in metamodels for all performances in an operating V6 engine

It can be observed the relative mean differences and the relative median differences are very small. Large values of the maximum differences are observed only at a small number of evaluation points, thus, the differences are not expected to affect the performance of the metamodels in probabilistic analyses. Overall, the performance variables evaluated by the metamodels are in good agreement with results computed by the solver. The computational time required by the engine simulation solver during the validation analyses demonstrates that probabilistic analyses based on a sampling method is only feasible due to the development of the metamodels.

PROBABILISTIC ANALYSES

In the past, the effect of variation induced by manufacturing imperfections on the dynamic performance of an engine has not been considered. In this section the probabilistic analyses for the V6 engine are presented. The developed metamodels are extensively utilized in the probabilistic analyses.

NESSUS is a state-of-the-art code for reliability analysis and it is utilized in the work presented in this paper. Two types of reliability analysis methods are available in NESSUS. One group is based on the limit state method and the other is based on sampling techniques. Usually, sampling based methods produce more accurate results, provided that a large number of samples are used. However, sampling methods require a large number of samples and some times prohibiting computational effort. Among all the methods that are available in NESSUS the Advanced Mean Value (AMV) and the Monte Carlo methods are utilized in this work.

First, Monte Carlo simulations (MCS) with different numbers of samples are conducted for the performance variables $press_2$ and results are presented in Figure 6. The CDF appears to converge in the probability range from 1.0% to 99.0% for sample data number N greater than 1,000 and the CDF curve becomes smoother as N increases. MCS with N=20,000 is considered as the reference for all the comparisons between alternative probabilistic analyses. The CDF generated by the AMV method is compared with the CDF generated from the MCS.

Figures 7 to 10 present the comparison between the CDF curves for the four performance variables. It can be observed that the AMV results correlate well with the MCS results only for performance variable $press_3$. Since the results generated by the AMV, which does not require a large number of evaluations of the performance variables, are not correct for this application, the necessity for generating metamodels is even further amplified.

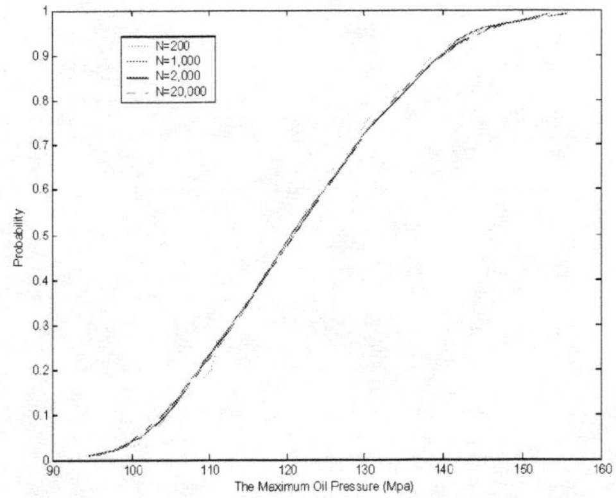

Figure 6 CDF generated by MCS with the different sample number for performance variable $press_2$

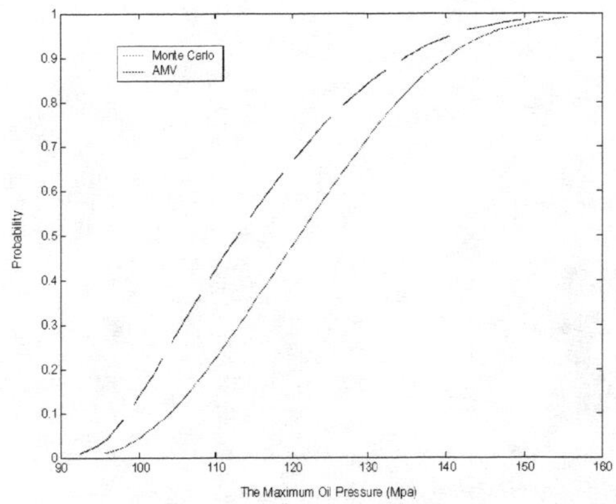

Figure 7 Comparison of CDF between AMV and MCS (N=20,000) for performance variable $press_2$

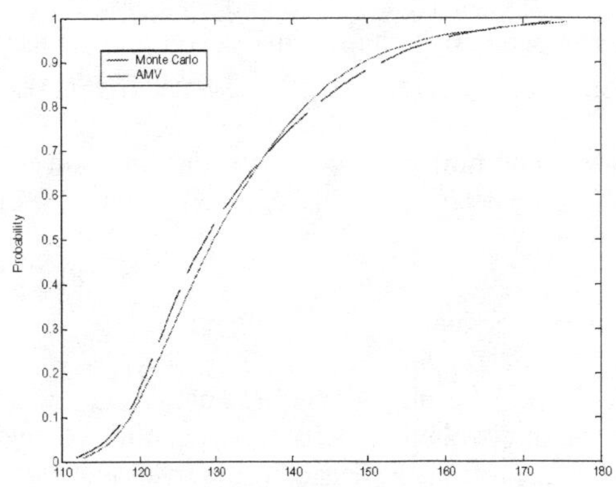

Figure 8 Comparison of CDF between AMV and MCS (N=20,000) for performance variable $press_3$

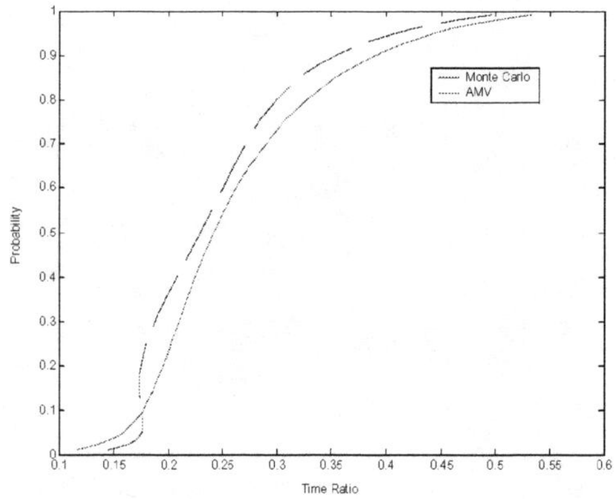

Figure 9 Comparison of CDF between AMV and MCS (N=20,000) for performance variable ratio $_2$

	mean (μ)	minimum	maximum	standard deviation (σ)	COV
press$_2$ (Mpa)	121.75	88.24	182.54	13.85	11.38%
press$_3$ (Mpa)	132.53	109.22	223.36	13.20	9.96%
ratio $_2$	0.2641	0.0646	0.7299	0.0878	33.24%
ratio $_3$	0.3348	0.0792	0.6881	0.0665	19.86%

Table 3 The statistical parameters for performance variables press $_2$, press $_3$, ratio $_2$ and ratio $_3$

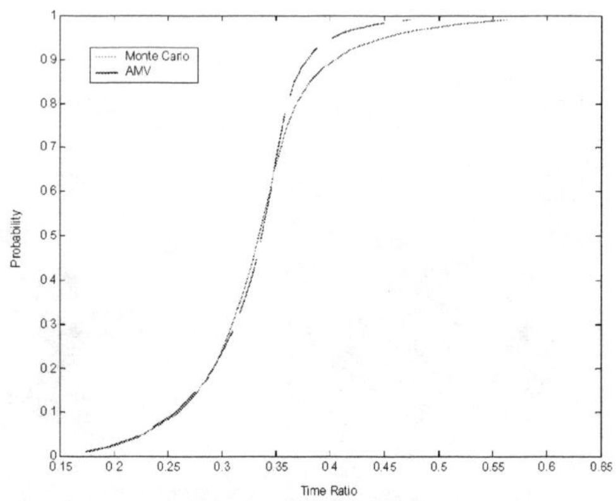

Figure 10 Comparison of CDF between AMV and MCS (N=20,000) for performance variable ratio $_3$

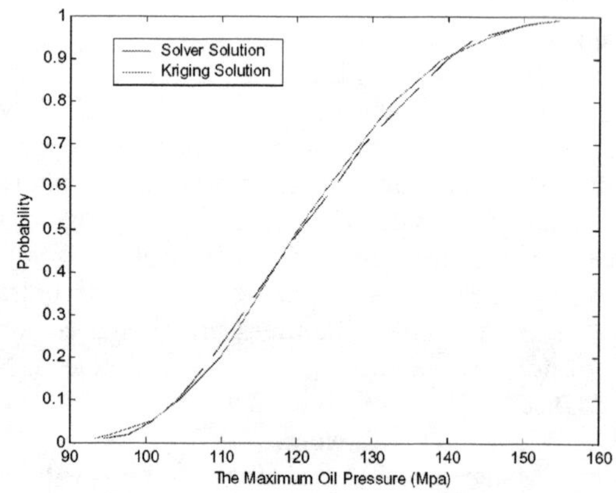

Figure 11 Comparison of CDF between the solver and the metamodel for press $_2$ when MCS (N=1,000) is employed

In order to further validate the metamodels, a comparison on the probabilistic results produced by a MCS with 1,000 samples computed by both the actual solver and the metamodel for performance variable press $_2$ is presented in Figure 11. This good agreement further substantiates that the solver can be replaced by the metamodel in the probabilistic analyses without loss of accuracy.

Finally, Figures 12 and 15 present the PDF for the performance variables press $_2$, press $_3$, ratio $_2$, and ratio $_3$, based on the MCS results. All performances demonstrate a non-normal type of distribution. Since this operating V6 engine is a highly non-linear system, it can be observed that the random distribution of the performance variable may not be a normal distribution, although the input random variables were considered to have normal distributions and to be independent of each other. Table 3 presents the statistical parameters of the performance variables.

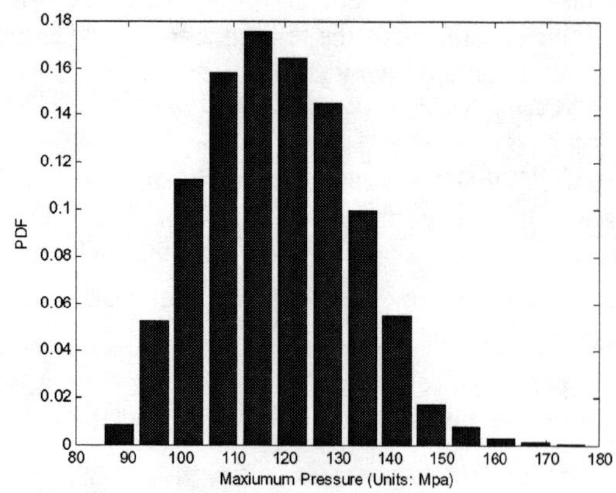

Figure 12 PDF of performance variable press $_2$

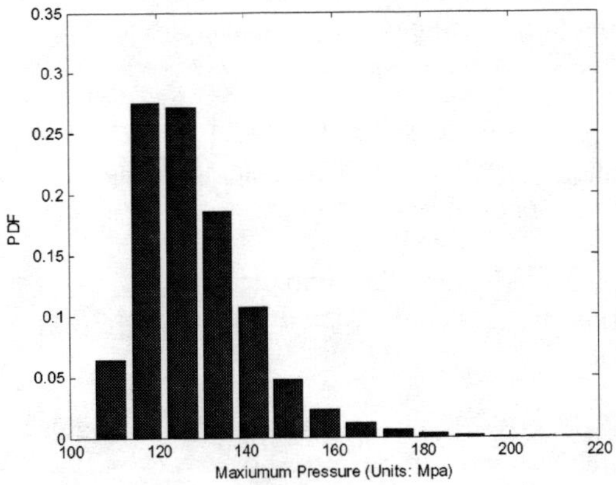

Figure 13 PDF of performance variable press $_3$

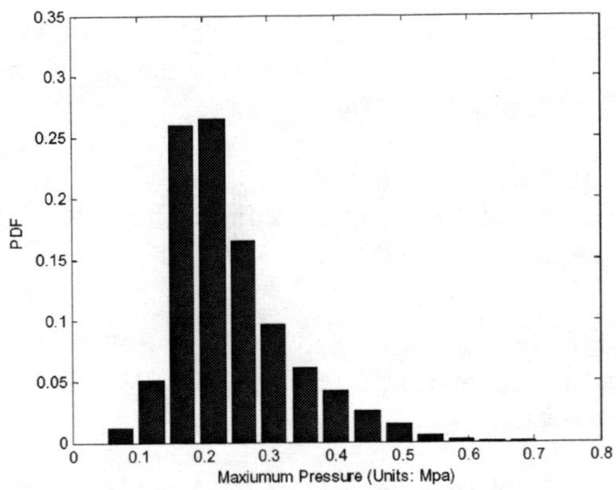

Figure 14 PDF of performance variable ratio $_2$

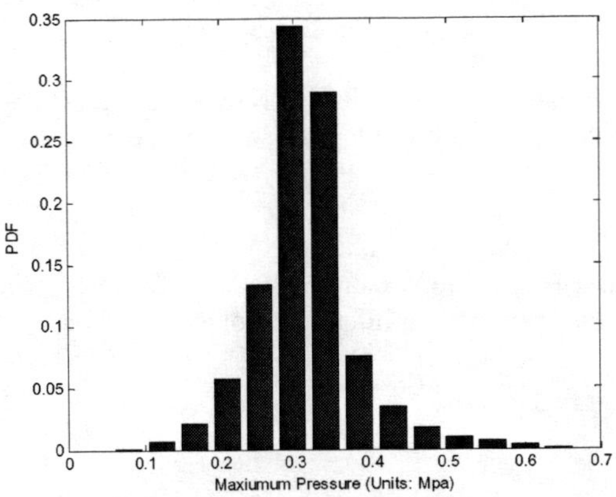

Figure 15 PDF of performance variable ratio $_3$

CONCLUSIONS

This paper presents an efficient approach for performing probabilistic analyses for the main bearings of an operating V6 engine under structural dynamic considerations. The probabilistic analyses are based on developing metamodels for the performance variables and using the metamodels along with available probabilistic algorithms. The overall approach is generic and can be applied to many engineering applications where probabilistic analyses are meaningful.

Metamodels are developed and used for computing the performance variables as a substitute to the actual simulation solvers. The metamodels are validated by comparing results computed by the metamodels with results computed by the actual simulation solver for a large number of evaluation points, which do not participate in defining the metamodels. The utilization of the metamodels provides significant computational savings, and makes the probabilistic analyses possible. In situations where only the Monte Carlo simulation can provide reliable results, the development of metamodels is critical for performing probabilistic analyses.

In the engine dynamic system simulation a comprehensive solver is employed. It captures the flexible dynamics of the crankshaft, the flexible dynamics of the engine block, and it connects the two components through a bearing lubrication solver at the main bearings of the engine. The maximum oil film pressure and the percentage of time (the time ratio) within each cycle that a bearing operates with film thickness lower than a user defined threshold are used as the performance variables. The initial uniform clearance between the crankshaft and the bearing at each main bearing and the oil viscosity of the oil comprise the random variables.

All random variables in this research are assumed to have a normal distribution and their statistical parameters are defined based on industry practices. Probabilistic analyses are performed for the performance variables. The PDF and the statistical parameters, i.e. the mean value and the standard deviation, of the performance variables are obtained. Taking advantage of the computationally efficient metamodels, CDFs are constructed by various alternative probabilistic methods for the performance variables and compared with the CDFs constructed by the Monte Carlo method. It is demonstrated that none of the alternative probabilistic methods correlate well with the Monte Carlo results in the operating V6 engine simulation.

ACKNOWLEDGMENTS

This research was supported by Federal-Mogul Corporation. The financial support is gratefully acknowledged.

REFERENCES

Cheng B. and Titterington D. M., "Neural Networks: A Review from a Statistical Perspective", Statistical Science, Vol.9, n1, pp. 2-54, (1994).

Craven, P., Wahba, G., "Smoothing Noisy Data with Spline Functions: Estimating the Correct Degree of Smoothing by the Methods of Generating Cross-Validation," Numerical Mathematics, Vol.31, pp. 377-403, (1978).

Craig, R.R. and Bampton, M.C.C., "Coupling of Substructures for Dynamic Analysis," AIAA Journal, Vol.6, n7, pp.1313-1319, (1968).

Cressie, N., "Spatial Prediction and Ordinary Kriging," Mathematical Geology, Vol.20, n4, pp. 405-421, (1988).

Currin, C., Mitchell, T., Morris, M., and Ylvisaker, D., "A Bayesian Approach to the Design and Analysis of Computer Experiments," ORNL Technical Report 6498, available from the National Technical Information Service, Springfield, VA, 22161, (1988).

Draper, N. R. and Smith, Jr. H., "Applied Regression Analysis", 2nd Edition, Wiley, New York, (1981).

Dyn, N., Levin, D., and Rippa, S., "Numerical Procedures for Surface Fitting of Scattered Data by Radial Functions," SIAM Journal of Scientific and Statistical Computing, Vol.7, n2, pp. 639-659, (1986).

Ebrat, O., "Detailed Main Bearing Hydrodynamic Characteristics for Crankshaft-Block Dynamic Interaction in Internal Combustion Engines", Ph.D. Thesis, The University of Michigan, (2002).

Ebrat, O., Mourelatos, Z.P., Hu, K., Vlahopoulos, N., and Vaidyanathan, K., "Structural Vibration of an Engine Block and a Rotating Crankshaft Coupled Through Elastohydrodynamics Bearings," SAE, 03NVC-145, (2003).

Ellacott, S.W., Mason, J.C., and Anderson, I.J., "Mathematics of Neural Networks: Models, Algorithms, and Applications," Kluver Academic Publishers, Boston, MA, (1997).

Hajela, P. and Berke, L., "Neural Networks in Structural Analysis and Design: An Overview", 4th AIAA/USAF/NASA/OAI Symposium on Multidisciplinary Analysis and Optimization, Cleveland, OH, AIAA 2, pp 901-914, AIAA-92-4805-CP, (1993).

Hu, K., "A Finite Element Formulation for Coupled Rigid and Flexible Dynamic Analysis of an Internal Combustion Engine Crankshaft System", Ph.D. Thesis, The University of Michigan, (2002).

Hu, K., Mourelatos, Z., and Vlahopoulos, N., "A Finite Element Formulation for the Coupling Rigid and Flexible Body Dynamics of Rotating Beams", Journal of Sound and Vibration, Vol.253, n3, pp. 603-630, (2002).

Jansen, M., Maifait, M., and Bultheel, A., "Generalized Cross Validation for Wavelet Thresholding," Signal Processing, Elsevier, n56, pp. 33-44, (1996).

Morris, M., Mitchell, T., "Exploratory Design for Computer Experiments," J. Statist. Plann. Inference, Vol.43, pp. 381-402, (1995).

"NESSUS /FPI Theoretical Manual", version 2.4, Southwest Research Institute, (1998).

Park, J.-S., "Optimal Latin-Hypercube Designs for Computer Experiments," J. Statist. Plann. Inference, Vol.39, pp. 95-111, (1994).

Rumelhart, D. E., Widrow, B. and Lehr, M. A., "The Basic Ideas in Neural Networks", Communications of the ACM, Vol.37, n3, pp. 87-92, (1994).

Sacks, J., Welch, W.J., Mitchell, T.J., and Wynn, H.P., "Design and Analysis of Computer Experiments," Statistical Science, Vol.4, n4, pp.409-435,(1989)(a).

Sacks, J., Welch, W.J., and Schiller, S.B., "Designs for Computer Experiments," Technometrics, Vol.31, n1, pp.41-47,(1989)(b).

Wu, Y.-T., "Demonstration of a New, Fast Probability Integration Method for Reliability Analysis," Transaction of ASME, Vol.109, pp. 24-28, (1987).

Wu, Y.-T., Millwater, H.R., and Cruse, T.A., "An Advanced Probabilistic Structural Analysis Method for Implicit Performance Functions," AIAA Journal, Vol.28, n9, pp. 1663-1669, (1990).

Wu, Y.-T., and Wirsching, P. H., "New Algorithm for Structural Reliability Estimation," Journal of Engineering Mechanics, Vol.113, n9, pp. 1319-36, (1987).

Ye, K.Q., Li, W., and Sudjianto, A., "Algorithm Construction of Optimal Symmetric Latin Hypercube Designs", Journal of Statistical Planning and Inference, Vol.90, pp.145-159, (2000).

Proceedings of IMECE'03
2003 ASME International Mechanical Engineering Congress
Washington, D.C., November 15–21, 2003

IMECE2003-55334

AN ENERGY BOUNDARY ELEMENT FORMULATION FOR PREDICTING THE ACOUSTIC FIELD AROUND A VEHICLE AT HIGH FREQUENCIES DUE TO AIRBORNE NOISE SOURCES

Aimin Wang
University of Michigan

Nickolas Vlahopoulos
University of Michigan

Jason Zhu
General Motors Corporation

Mike Qian
General Motors Corporation

ABSTRACT

An Energy Boundary Element Analysis (EBEA) formulation is presented for calculating sound radiation from a source with arbitrary shape at high frequency. The basic integral equation for the EBEA is derived including a half-space boundary condition. The time and frequency averaged acoustic energy density and acoustic intensity constitutes the primary variables of the new formulation, and the corresponding Green's functions are derived. The governing equations for the EBEA are established and the numerical formulae for the coefficients of the system matrix, the acoustic energy density, and the acoustic intensity are derived using a Gaussian quadrature. The EBEA formulation and the corresponding numerical implementation are validated by comparing EBEA results to test data for the acoustic field around a vehicle that originates from an airborne noise source. Good correlation is demonstrated between numerical predictions and test data.

INTRODUCTION

An Energy Boundary Element Analysis (EBEA) formulation has been established for the acoustic radiation analysis at high frequency in an infinite medium [1]. The frequency averaged acoustic energy density and intensity constitutes the primary variables for the EBEA. The corresponding integral formulation is derived based on the governing integral equation of the conventional Boundary Element Method (BEM) [2, 3]. The basic assumption in deriving the EBEA formulation is that the infinitesimal energy sources on each element are incoherent among different elements. This assumption is in-line with the existing high frequency Statistical Energy Analysis (SEA) [12, 16 - 19] and Energy Finite Element Analysis (EFEA) [8, 11] formulations, and it is also necessary in order to potentially utilize the incoherent acoustic power computed by the EFEA analysis on the outer part of a structure as a boundary condition for predicting the radiated acoustic field.

In this paper the EBEA is extended in order to account for a half-space boundary condition. The corresponding integral formulation is derived based on the governing integral equation of the conventional half-space BEM [15]. The half-space EBEA numerical formulation is developed by placing a continuous distribution of incoherent infinitesimal energy sources on the surface of the radiator. Finally, the validity of the half-space EBEA formulation and its numerical implementation is demonstrated by comparing the half-space EBEA results to the analytical solutions for a spherical radiator in half space and to the test data for the acoustic field around a vehicle due to an airborne noise source.

The EBEA constitutes a major improvement over the traditional SEA for propagating the acoustic energy around a vehicle in a half space. The SEA requires the artificial definition of acoustic reverberant subsystems and the definition of artificial coupling loss factors between them in order to represent the exterior acoustic space in the vicinity of the vehicle. Thus, the SEA introduces many approximations in the solution. The modeling of the acoustic propagation around a vehicle is important in airborne noise applications where information about the noise levels in the exterior of the passenger cabin due to a noise source is necessary in order to predict how much noise is transmitted in the interior of the vehicle. The EBEA can provide such information about the exterior acoustic field without modeling the unbounded acoustic medium as reverberant.

THEORETICAL FORMULATION

The half-space EBEA formulation is developed from the integral formulas for the acoustic pressure or velocity of the conventional half-space boundary element method. First, the time averaged over a period acoustic energy density and intensity at a field point Y is expressed in terms of the acoustic velocity and the acoustic pressure as [14]:

$$\langle e_{\mathrm{Y}}\rangle = \frac{1}{4}\left[\rho \hat{\mathbf{v}}_{\mathrm{Y}} \cdot \hat{\mathbf{v}}_{\mathrm{Y}}^{*} + \frac{1}{\rho c^{2}} \hat{p}_{\mathrm{Y}} \hat{p}_{\mathrm{Y}}^{*}\right], \tag{1}$$

$$\langle \mathbf{I}_{\mathrm{Y}}\rangle = \frac{1}{2}\mathrm{Re}\left(\hat{p}_{\mathrm{Y}} \hat{\mathbf{v}}_{\mathrm{Y}}^{*}\right), \tag{2}$$

where ρ is the density of the acoustic medium, subscript "Y" indicates a quantity associated with a field point Y, c is the speed of the sound in the medium, $\hat{\mathbf{v}}$ and $\hat{p}$ are the acoustic velocity and the acoustic pressure, respectively, $\langle\ \rangle$ indicates time averaging over a period (i.e., $\langle A\rangle = \left(\frac{1}{\mathrm{T}}\right)\int^{+\mathrm{T}} A(\tau)\mathrm{d}\tau$), and symbol ^ indicates complex quantities. In order to develop the primary variables of the half-space EBEA, the ensemble averaging operator $\mathrm{E}[\cdot]$ is applied on the equations for $\langle e_{\mathrm{Y}}\rangle$ and $\langle \mathbf{I}_{\mathrm{Y}}\rangle$ resulting into:

$$\widetilde{e}_{\mathrm{Y}} = \mathrm{E}[\langle e_{\mathrm{Y}}\rangle] = \frac{1}{4}\left[\rho \mathrm{E}\left[\hat{\mathbf{v}}_{\mathrm{Y}} \cdot \hat{\mathbf{v}}_{\mathrm{Y}}^{*}\right] + \frac{1}{\rho c^{2}} \mathrm{E}\left[\hat{p}_{\mathrm{Y}} \hat{p}_{\mathrm{Y}}^{*}\right]\right], \tag{3}$$

$$\widetilde{\mathbf{I}}_{\mathrm{Y}} = \mathrm{E}[\langle \mathbf{I}_{\mathrm{Y}}\rangle] = \frac{1}{2}\mathrm{Re}\left(\mathrm{E}\left[\hat{p}_{\mathrm{Y}} \hat{\mathbf{v}}_{\mathrm{Y}}^{*}\right]\right), \tag{4}$$

since the ensemble averaging in the high frequency methods is considered as equivalent to frequency averaging [11, 12, 18, 19], $\widetilde{e}_{\mathrm{Y}}$ and $\widetilde{\mathbf{I}}_{\mathrm{Y}}$ are equivalent with the time averaged over a period and frequency averaged over 1/3 octave band energy density and intensity at a field point Y.

In the conventional indirect half-space boundary integral method, the acoustic pressure at any field point Y exterior to the structure is expressed as [2, 3, 15]

$$\hat{p}_{\mathrm{Y}} = \int_S A(\mathrm{P}) g(\mathrm{P},\mathrm{Y}) \mathrm{d}S, \tag{5}$$

where S is the surface of the structure (Figure 1), P denotes the point located on the surface S, $A(\mathrm{P})$ is the complex source strength amplitude at P, In Figure 1, S_H denotes the surface of the infinite plane, S_1 is the mirror image of S, P_1 is the mirror image of P, and $g(\mathrm{P},\mathrm{Y}) = \frac{e^{-\mathrm{i}kr}}{4\pi r} + \frac{e^{-\mathrm{i}kr_1}}{4\pi r_1}$ is the half-space Green's function (with a perfectly rigid infinite plane) [15], r is the distance between points P and Y, r_1 is the distance between points P_1 and Y, as shown in Figure 1, and k is the wavenumber. The acoustical velocity vector can be obtained from the acoustic pressure

$$\hat{\mathbf{v}}_{\mathrm{Y}} = -\frac{1}{\mathrm{i}\omega\rho}\nabla \hat{p}_{\mathrm{Y}} = -\frac{1}{\mathrm{i}\omega\rho}\int_S A \nabla g(\mathrm{P},\mathrm{Y}) \mathrm{d}S, \tag{6}$$

where

$$\nabla g(\mathrm{P},\mathrm{Y}) = -\frac{1}{4\pi r^{2}}(1+\mathrm{i}kr)e^{-\mathrm{i}kr}\mathbf{E}_{r} - \frac{1}{4\pi r_1^{2}}(1+\mathrm{i}kr_1)e^{-\mathrm{i}kr_1}\mathbf{E}_{r_1} \tag{7}$$

$\mathbf{E}_r$ denotes the unit vector from P to Y, $\mathbf{E}_{r_1}$ denotes the unit vector from P_1 to Y.

Equations (5) and (6) are employed for developing expressions for the ensemble averaged quantities $\mathrm{E}\left[\hat{p}_{\mathrm{Y}} \hat{p}_{\mathrm{Y}}^{*}\right]$, $\mathrm{E}\left[\hat{\mathbf{v}}_{\mathrm{Y}} \cdot \hat{\mathbf{v}}_{\mathrm{Y}}^{*}\right]$ and $\mathrm{E}\left[\hat{p}_{\mathrm{Y}} \hat{\mathbf{v}}_{\mathrm{Y}}^{*}\right]$. The latter expressions are introduced in Equations (3) and (4) in order to develop the equations for the primary variables of the half-space EBEA formulation. The acoustic pressure square is calculated as:

$$\begin{aligned}\hat{p}_{\mathrm{Y}} \hat{p}_{\mathrm{Y}}^{*} &= \int_S A(\mathrm{P}') g(\mathrm{P}',\mathrm{Y}) \mathrm{d}S \cdot \int_S A^{*}(\mathrm{P}'') g^{*}(\mathrm{P}'',\mathrm{Y}) \mathrm{d}S \\ &= \int_S \int_S A(\mathrm{P}') A^{*}(\mathrm{P}'') g(\mathrm{P}',\mathrm{Y}) g^{*}(\mathrm{P}'',\mathrm{Y}) \mathrm{d}S' \mathrm{d}S''\end{aligned} \tag{8}$$

For incoherent sources, $\mathrm{E}\left[A(\mathrm{P}')A^{*}(\mathrm{P}'')\right]$ is non-zero only for $\mathrm{P}' = \mathrm{P}''$ [16, 17]. Thus, by applying the operator $\mathrm{E}[\cdot]$ to Equation (8) results in:

$$\mathrm{E}\left[\hat{p}_{\mathrm{Y}} \hat{p}_{\mathrm{Y}}^{*}\right] = \int_S \mu(\mathrm{P}) |A(\mathrm{P})|^{2} |g(\mathrm{P},\mathrm{Y})|^{2} \mathrm{d}S, \tag{9}$$

In Equation (9) the double integral is reduced to a single integral due to the zero cross terms in $\mathrm{E}\left[A(\mathrm{P}')A^{*}(\mathrm{P}'')\right]$. The variable $\mu(\mathrm{P})$ appears in the equation in order to retain the proper units during the reduction from a double to a single integral, and $\mu(\mathrm{P})$ is equal with the differential area associated with the location P of the integration.

Following a similar process for $\mathrm{E}\left[\hat{\mathbf{v}}_{\mathrm{Y}} \cdot \hat{\mathbf{v}}_{\mathrm{Y}}^{*}\right]$ and $\mathrm{E}\left[\hat{p}_{\mathrm{Y}} \hat{\mathbf{v}}_{\mathrm{Y}}^{*}\right]$ results in:

$$\mathrm{E}\left[\hat{\mathbf{v}}_{\mathrm{Y}} \hat{\mathbf{v}}_{\mathrm{Y}}^{*}\right] = \frac{1}{\omega^{2}\rho^{2}} \int_S \mu(\mathrm{P}) |A(\mathrm{P})|^{2} |\nabla g(\mathrm{P},\mathrm{Y})|^{2} \mathrm{d}S, \tag{10}$$

and

$$\mathrm{E}\left[\hat{p}_{\mathrm{Y}} \hat{\mathbf{v}}_{\mathrm{Y}}^{*}\right] = \frac{1}{\mathrm{i}\omega\rho} \int_S \mu(\mathrm{P}) |A(\mathrm{P})|^{2} g(\mathrm{P},\mathrm{Y}) \nabla g^{*}(\mathrm{P},\mathrm{Y}) \mathrm{d}S. \tag{11}$$

Substituting equations (9) and (10) in equation (3) and Using relationship $c = \omega/k$, results in

$$\widetilde{e}_{\mathrm{Y}} = \int_S \sigma(\mathrm{P}) \left(\frac{\rho}{64\pi^{2} r^{4}} + \frac{k^{2}\rho}{32\pi^{2} r^{2}} + \frac{\rho}{64\pi^{2} r_1^{4}} + \frac{k^{2}\rho}{32\pi^{2} r_1^{2}}\right) \mathrm{d}S \tag{12}$$

Similarly, substituting equation (11) into equation (4), results in

$$\widetilde{\mathbf{I}}_{\mathrm{Y}} = \int_S \sigma(\mathrm{P})\left(\frac{k^2\rho c}{32\pi^2 r^2}\mathbf{E}_r + \frac{k^2\rho c}{32\pi^2 r_1^2}\mathbf{E}_{r_1}\right)\mathrm{d}S . \tag{13}$$

where $\sigma(\mathrm{P}) = \frac{\mu(\mathrm{P})}{\rho^2\omega^2}|A(\mathrm{P})|^2$ is defined as the strength density of the energy source, which is a frequency-dependent term. Equations (12) and (13) constitute the basic integral equations of the half-space EBEA formulation. The acoustic energy density and intensity at any field point in the half space exterior to the structure can be determined by Equations (12) and (13), respectively. Comparing to the conventional indirect boundary integral method for half-space problems [3, 15], the half-space Green's functions for the acoustic energy density and intensity can be obtained from above two integrals, respectively

$$G(\mathrm{P},\mathrm{Y}) = \frac{\rho}{64\pi^2 r^4} + \frac{k^2\rho}{32\pi^2 r^2} + \frac{\rho}{64\pi^2 r_1^4} + \frac{k^2\rho}{32\pi^2 r_1^2} \tag{14}$$

$$\mathbf{H}(\mathrm{P},\mathrm{Y}) = \frac{k^2\rho c}{32\pi^2 r^2}\mathbf{E}_r + \frac{k^2\rho c}{32\pi^2 r_1^2}\mathbf{E}_{r_1} . \tag{15}$$

NUMERICAL IMPLEMENTATION

In order to develop a numerical solution, the surface S of the radiator (Figure 1) is divided into n quadrilateral or triangular elements. Incoherent infinitesimal energy sources or sinks are distributed on every element. In this work, the source strength density $\sigma_j (j = 1,2,...,n)$ on each element is considered to be constant. Equations (12) and (13) can be rewritten in the discrete form as:

$$\widetilde{e}_{\mathrm{Y}} = \sum_{j=1}^{n}\left[\sigma_j \int_{S_j} G(\xi,\mathbf{Y})\mathrm{d}S\right], \tag{16}$$

$$\widetilde{\mathbf{I}}_{\mathrm{Y}} = \sum_{j=1}^{n}\left[\sigma_j \int_{S_j} \mathbf{H}(\xi,\mathbf{Y})\mathrm{d}S\right], \tag{17}$$

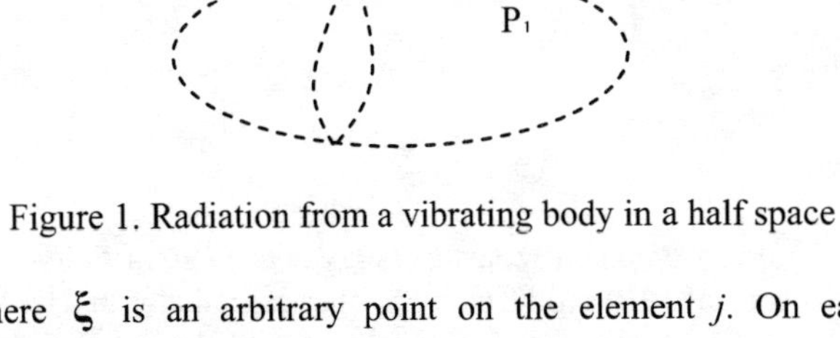

Figure 1. Radiation from a vibrating body in a half space

where ξ is an arbitrary point on the element j. On each element, the source strength density σ_j is constant and can be factored outside of the integral. In order to develop the numerical system of equations that will allow to compute the values of the sources and sinks σ_j over the model, the field point Y is placed on an element q on the outer surface of the radiator and Equation (17) provides the integral equation for the intensity $\widetilde{\mathbf{I}}_Y$ on the element q. In order to avoid the singular integration within the element q in Equation (17), the definition of the averaged acoustic power radiated by element q itself is employed [1]. Since the radiated acoustic power from each element comprises the prescribed boundary condition, the time and frequency averaged intensity on an element integrated over the area of the element must result into the prescribed radiated acoustic power:

$$\int_{S_i} \widetilde{\mathbf{I}}_{\mathrm{Y}} \cdot \mathbf{n}_i \mathrm{d}S = \overline{P}_i , \qquad i=1,2,\ldots,n, \tag{18}$$

where $\overline{P}_i$ is the prescribed boundary condition on element i, $\mathbf{n}_i$ is the normal (outward of the structure) vector of the element i. Substituting equation (17) in above equation, results in the governing equation of half-space EBEA

$$K_{ij}\sigma_j = \overline{P}_i, \qquad i=1,2,\ldots,n, \qquad j=1,2,\ldots,n. \tag{19}$$

For any set of prescribed boundary conditions, the values of the acoustic energy source strength density on each element can be obtained by solving Equation (19). Then by employing Equations (16) and (17), the acoustic energy density and intensity can be determined at any field point in the half space exterior to the structure.

VALIDATION / APPLICATION

In order to demonstrate the validity of the half-space EBEA formulation and its implementation numerical results are compared successfully to analytical solutions and test results for several analyses.

COMPARISON OF THE EBEA WITH ANALYTICAL SOLUTIONS FOR RADIATION IN A HALF SPACE

In this example, acoustic radiation from a spherical source in a half space at high frequency is considered. In order to develop the analytical solution, one point source of unit strength is placed at the center of the sphere. The surface of the radiator is considered originally as transparent. The acoustic energy density and intensity at any field point in the half space produced by this unit strength point source can be calculated using the image method

$$\widetilde{e}_{\mathrm{Y}} = \frac{\rho}{64\pi^2 r^4} + \frac{k^2\rho}{32\pi^2 r^2} + \frac{\rho}{64\pi^2 r_1^4} + \frac{k^2\rho}{32\pi^2 r_1^2}, \tag{20}$$

$$\widetilde{\mathbf{I}}_{\mathrm{Y}} = \frac{k^2\rho c}{32\pi^2 r^2}\mathbf{E}_r + \frac{k^2\rho c}{32\pi^2 r_1^2}\mathbf{E}_{r_1}, \tag{21}$$

where r is the distance between the point source and the field point in the half space, r_1 is distance between the mirror image of the point source and the field point in the half space, $\mathbf{E}_r$ is the unit vector from the point source to the field point, $\mathbf{E}_{r_1}$ is the unit vector from the mirror image of the point source to the field point. The integral of the normal acoustic intensity on the each element of the transparent surface can also be calculated by the analytical formula using the image method

$$\overline{P}_i = \int_{S_i}\left(\frac{k^2\rho c}{32\pi^2 r^2}\mathbf{E}_r\cdot\mathbf{n}_i + \frac{k^2\rho c}{32\pi^2 r_1^2}\mathbf{E}_{r_1}\cdot\mathbf{n}_i\right)\mathrm{d}S \qquad i=1,2,\ldots,n \quad (22)$$

where $\mathbf{n}_i$ is the unit normal (outward of the structure) vector of element *i*. These values for the acoustic intensity comprise the boundary conditions for the half-space EBEA computations. The distribution of energy sources is evaluated on the surface of the EBEA model from the boundary conditions. The computed surface results and Equations (16) and (17) are employed for calculating the acoustic energy density and intensity in the field. The acoustic medium used in the example is air, with the density of 1.21Kg/m^3, and the speed of sound equal to 343m/s. Computations are performed at the center frequency for the 3,000Hz 1/3 octave band. A spherical radiator with the radius of R = 1.0m is analyzed. The EBEA model is presented in Figure 2. It contains 130 elements and 132 nodes. The sphere is located at a distance $b = 2R$ from the infinite rigid plane, where b is the distance from the center of the sphere to the rigid plane as shown in Figure 2.

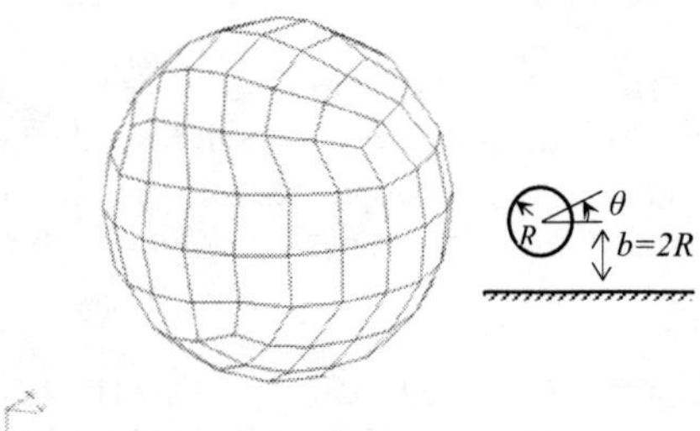

Figure 2. EBEA model for the sphere in the half space

Figures 3 and 4 show the acoustic intensity and the acoustic energy density, respectively, at the distance of 2*R* from the center of the sphere, as a function of polar angle θ (Figure 2) on a plane passing through the center of the sphere and perpendicular to the infinite plane. The half-space EBEA results are compared to the analytical solutions. The polar plots in Figures 3 and 4 differ significantly from the circular polar plot that is characteristic of the pulsating sphere in a full space. The differences are due to the interference between the direct field produced by the sphere and the waves reflected from the rigid plane. The far field points are selected at the distance of 10*R* from the center of the sphere, and the acoustic energy density results are presented in Figure 5 as a function of the polar angle θ. In Figure 5 only the field points above the infinite rigid plane are considered. Figure 6 shows the time and frequency averaged acoustic energy density along the *y* axis on the surface of the infinite rigid plane. The acoustic energy

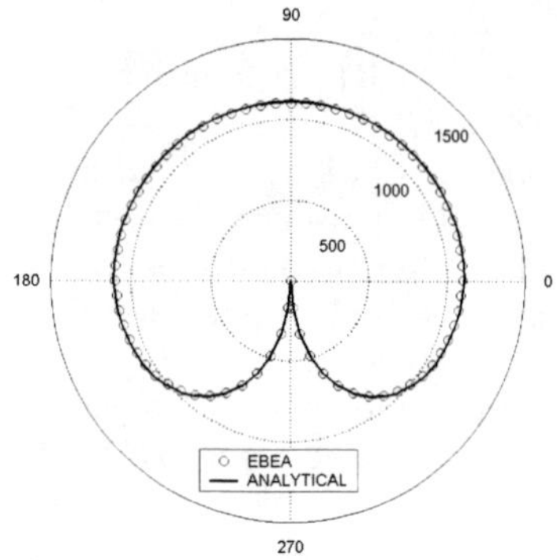

Figure 3. Averaged acoustic intensity at a distance 2*R* produced by a spherical radiator in a half space at the frequency of 3000Hz

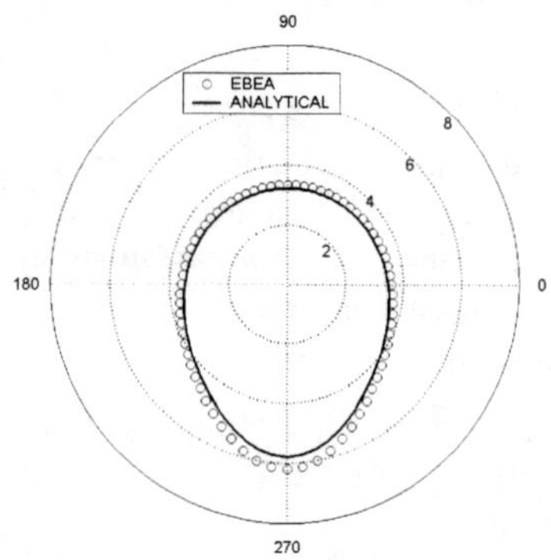

Figure 4. Averaged acoustic energy density at a distance 2*R* produced by a spherical radiator in a half space at the frequency of 3000Hz

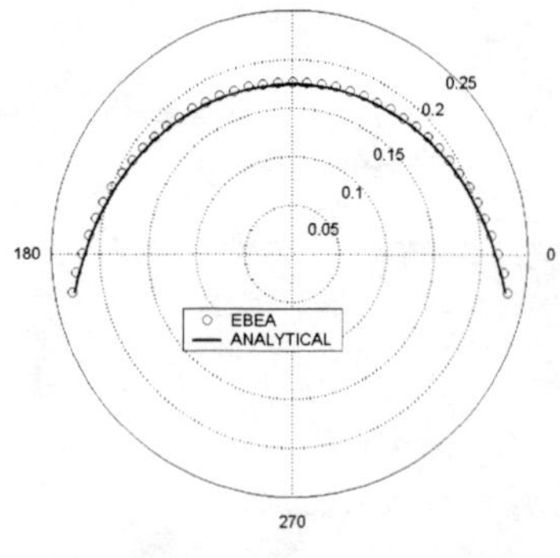

Figure 5. Averaged acoustic energy density at far field (a distance of 10*R*) produced by a spherical radiator in a half space at the frequency of 3000Hz

density at point $y = 0$ (the project of the center of the sphere on the rigid plane) corresponds to the point at 270 deg on the polar plot in Figure 4. Very good correlation is observed between the EBEA results and the analytical solution.

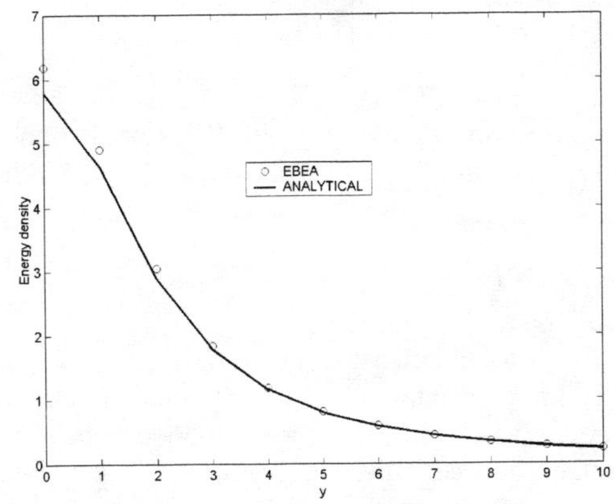

Figure 6. Averaged acoustic energy density along the y axis on the surface of the infinite rigid plane

COMPARISON OF THE EBEA WITH TEST RESULTS FOR AN AUTOMOTIVE APPLICATION

In order to demonstrate and validate the capabilities of the EBEA for automotive applications the acoustic field around a vehicle created from an airborne noise source is simulated. The EBEA mesh of the vehicle is comprised by 6086 elements. The model and the location of the airborne noise source are depicted in Figure 7. Sound pressure levels (SPL) are measured at several field points around the vehicle and the results are summarized in 1/3 octave frequency intervals. The ground is also considered in the computations. Since the acoustic field is generated from an airborne noise source, the surface of the vehicle can also be assumed to be rigid. Thus, the total acoustic field is obtained by enforcing zero normal intensity boundary conditions on the surface of the vehicle. A distribution of energy sources is evaluated on the surface of the EBEA vehicle model. The acoustic energy density and intensity at any field point around the vehicle is determined by equations (20) and (21), respectively.

A dimensionless energy density ratio is used to compare the EBEA results and the test data. At the field points around the vehicle the sound pressure is measured.

The dimensionless energy density ratio is defined as the ratio (dB, ref. 1.0E-12) of the pressure square at a field point over the pressure square measured at the vicinity of the noise source.

$$r_{\text{TEST}} = 10 \times \log\left(\frac{p^2/p_S^2}{10^{-12}}\right), \qquad (23)$$

where p_S^2 indicates the pressure square at the vicinity of the airborne noise source. The same non-dimensional ratio is computed by the EBEA as

$$r_{\text{EBEA}} = 10 \times \log\left(\frac{\tilde{e}/\tilde{e}_S}{10^{-12}}\right), \qquad (24)$$

where $\tilde{e}_S$ indicates the average acoustic energy density in the vicinity of the airborne noise source. The energy densities have replaced the pressure square quantities presented in equation (23). Equations (23) and (24) are equivalent assuming that the averaged energy density can be expressed as

$$\tilde{e} = \frac{p^2}{\rho c^2}. \qquad (25)$$

Comparisons are made at several field points in the vicinity of the vehicle. Figures 8 and 9 present the comparison at 500Hz, from the side view and bottom view, respectively, while results for 4000Hz and 8000Hz are depicted in figures 10 and 11 and figures 12 and 13, respectively. Good correlation is observed for all frequencies and for all the field points.

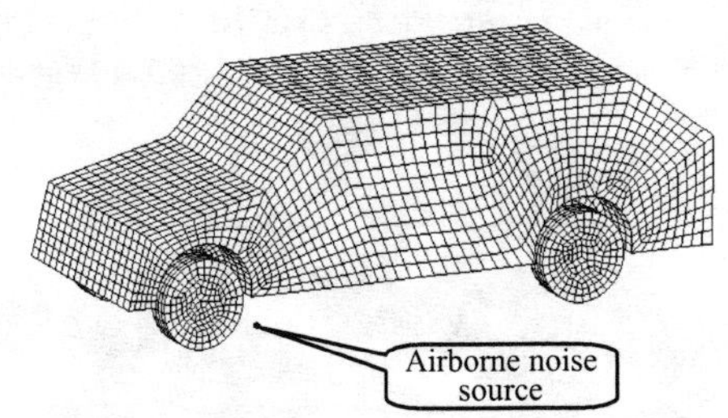

Figure 7. Mesh of the vehicle with an airborne noise source

The EBEA captures correctly both the magnitude and also the distribution of the noise around the vehicle. The EBEA can be used in automotive airborne noise applications for determining the acoustic energy around the vehicle due to airborne noise sources. The acoustic energy on the exterior of the vehicle can be utilized as the input to a SEA vehicle model in order to compute the interior noise due to the airborne sources.

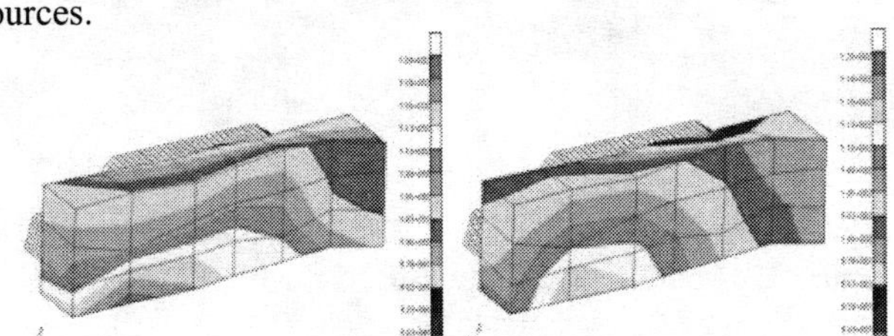

(a) Experimental results (b) EBEA prediction

Figure 8. Contour of the ratio (dB, ref. 1.0E-12) at 500Hz (side view)

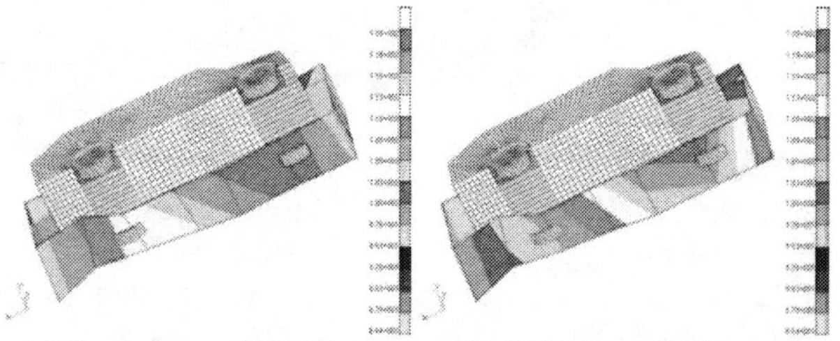

(a) Experimental results (b) EBEA prediction

Figure 9. Contour of the ratio (dB, ref. 1.0E-12) at 500Hz (bottom view)

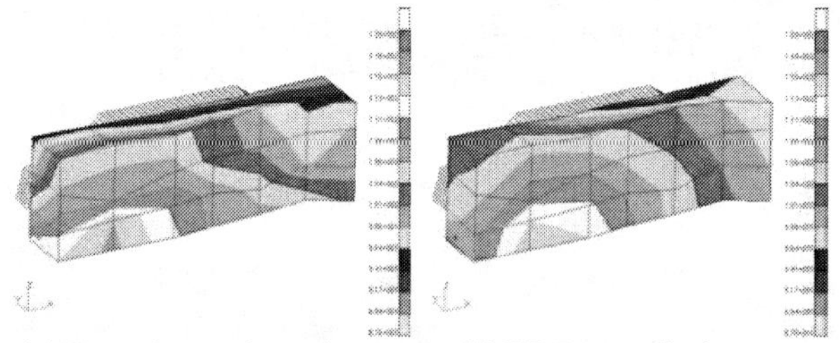

(a) Experimental results (b) EBEA prediction

Figure 10. Contour of the ratio (dB, ref. 1.0E-12) at 4000Hz (side view)

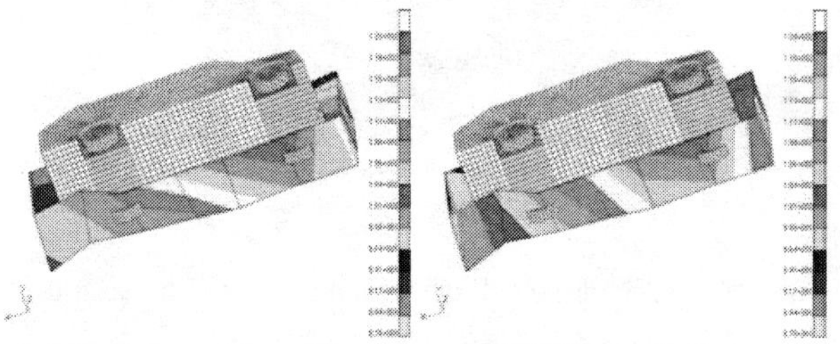

(a) Experimental results (b) EBEA prediction

Figure 11. Contour of the ratio (dB, ref. 1.0E-12) at 4000Hz (bottom view)

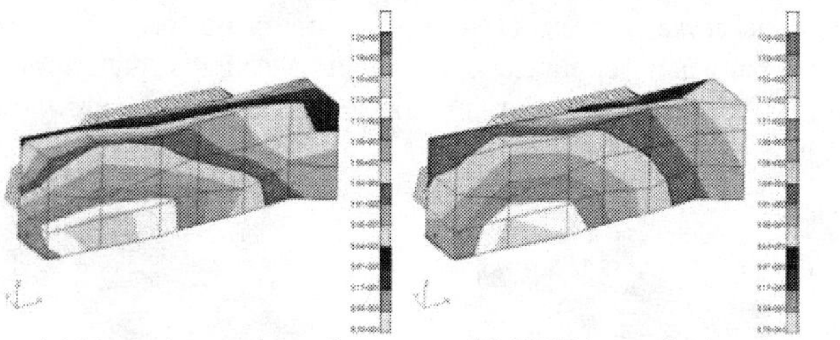

(a) Experimental results (b) EBEA prediction

Figure 12. Contour of the ratio (dB, ref. 1.0E-12) at 8000Hz (side view)

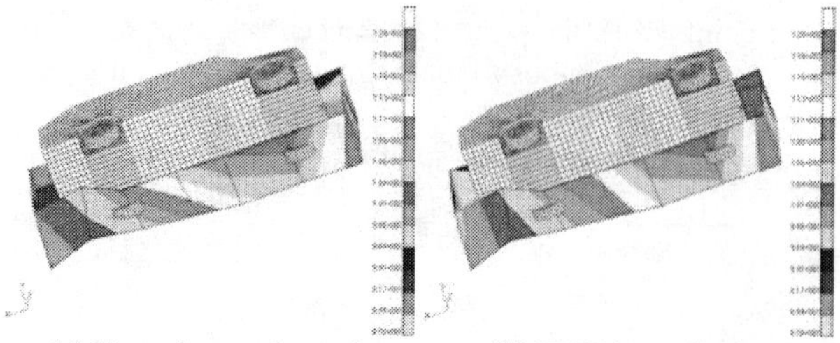

(a) Experimental results (b) EBEA prediction

Figure 13. Contour of the ratio (dB, ref. 1.0E-12) at 8000Hz (bottom view)

CONCLUSION

An EBEA formulation that accounts for a half-space and its numerical implementation are presented for exterior acoustic computations at high frequencies. Good correlation is demonstrated between the EBEA results and analytical solutions. The EBEA is utilized successfully for predicting the exterior acoustic field in the vicinity of a vehicle due to an airborne noise source.

ACKNOWLEDGMENTS

The authors would like to acknowledge the technical and financial support of the Automotive Research Center (ARC) by the National Automotive Center (NAC), and the U.S. Army Tank-Automotive Research, Development and Engineering Center (TARDEC). The ARC is a U.S. Army Center of Excellence for Automotive Research at the University of Michigan, currently in partnership with the University of Alaska-Fairbanks, Clemson University, University of Iowa, Oakland University, University of Tennessee, Wayne State University, and University of Wisconsin-Madison.

REFERENCES

[1]. A. Wang, N. Vlahopoulos and K. Wu 2002 The International Congress and Exposition on Noise Control Engineering, ASME Paper No. 2002-405. An energy boundary element formulation for sound radiation at high frequency.

[2]. C. A. Brebbia, J. C. Telles and L. C. Wrobel 1984 Boundary Element Techniques. Berlin: Springer-Verlag.

[3]. L. H. Chen and D. G. Schweikert 1963 Journal of the Acoustical Society of America 35, 1626-1632. Sound radiation from an arbitrary body.

[4]. R. J. Berhard and J. E. Huff 1999 Journal of Vibration and Acoustics 121, 295-301. Structure-acoustic design at high frequency using the energy finite element method.

[5]. O. M. Bouthier and R. J. Bernhard 1992 AIAA Journal 30, 34-44. Models of space-averaged energetics of plates.

[6]. P. Cho 1993 Energy Flow Analysis of Coupled Structures (Ph. D. Dissertation, Mechanical Engineering Department, Purdue University, Lafayette).

[7]. N. Vlahopoulos, O. G. Luis, and M. Christopher 1999 Journal of Ship Research 43, 143-156. Numerical

implementation, validation, and marine applications of an energy finite element formulation.
[8]. W. Zhang, A. Wang, and N. Vlahopoulos 2002 Finite Elements in Analysis and Design 38, 1095-1113. An alternative Energy Finite Element formulation based on incoherent orthogonal waves and its validation for marine structures.
[9]. W. Zhang, A. Wang, N. Vlahopoulos and K. Wu 2003 Journal of Sound and Vibration 263, 21-46. High frequency vibration analysis of thin elastic plates under heavy fluid loading by an Energy Finite Element Method.
[10]. W. Zhang, A. Wang, and N. Vlahopoulos 2001 Proceedings of the 2001 SAE Noise and Vibration Conference, SAE paper 2001-01-1618. Validation of the EFEA method through correlation with conventional FEA and SEA results.
[11]. N. Vlahopoulos and X. Zhao 2001 Journal of Sound and Vibration 242, 445-473. An investigation of power flow in the mid-frequency range for systems of co-linear beams based on a hybrid finite element formulation.
[12]. R. Lyon 1975 Statistical Energy Analysis of Dynamical Systems: Theory and Application. Cambridge, MA: MIT Press.
[13]. G. W. Benthien and H. A. Schenck 1997 Computers & Structures 65, 295-305. Nonexistence and nonuniqueness problems associated with integral equation method in acoustic.
[14]. A. D. Pierce 1981 Acoustics: An Introduction to its Physical Principles and Applications. New York: McGraw Hill Press.
[15]. A. F. Seybert and B. Soenarko 1988 Journal of Vibration, Acoustics, Stress, and Reliability in Design 110, 112-117. Radiation and scattering of acoustic waves from bodies of arbitrary shape in a three-dimensional half space.
[16]. R. K. Andrew and R. L. Kirlin 2000 IEEE Transactions on Signal Processing 48, 1287-1294. A broadband maximum likelihood imager for a class of extended space-time separable sources.
[17]. J. Li and J. Pascal 1998 Journal of the Acoustical Society of America 103, 962-972. Energy fields of partially coherent sources.
[18]. A. J. Keane 1993 Journal of Sound and Vibration 164, 143-156. A note on modal summations and averaging methods as applied to statistical energy analysis (S.E.A.).
[19]. E. C. N. Wester and B. R. Mace 1996 Journal of Sound and Vibration 193, 793-822. Statistical energy analysis of two edge-coupled rectangular plates: ensemble averages.

Proceedings of IMECE'03
2003 ASME International Mechanical Engineering Congress
Washington, D.C., November 15–21, 2003

IMECE2003-55346

OVERVIEW OF SONIC BOOM NOISE

Victor W. Sparrow
Graduate Program in Acoustics
The Pennsylvania State University
316B Leonhard Bldg.
University Park, PA 16802
Email: vws1@psu.edu

ABSTRACT

This paper provides an overview of sonic boom noise for nonspecialists. A brief description is provided and a source/path/receiver model is emphasized. Significant research has been undertaken in the last few years to develop commercial aircraft (sources) that have sonic booms that are less loud than previous designs, opening the possibility for overland supersonic flight. This work is continuing. However, there are several challenges remaining regarding the propagation (path) of sonic boom noise through the real atmosphere and in the human perception (receiver) of minimized sonic boom waveforms.

NOMENCLATURE

c speed of sound in air
M Mach number $= V/c$
V speed of aircraft

INTRODUCTION

Most transportation noise is easy to conceptualize. Automobiles, trains, and subsonic aircraft all make noise, and the public can understand this noise since they have familiarity with these noise sources. Sonic boom noise, however, is not routinely experienced, except by a few. Hence, there are many misconceptions about where sonic boom noise comes from and what its effects are. Considering the genuine increasing prospects for regular overland flight of low-boom supersonic aircraft, to be described later, it is important that non-specialists have a basic understanding of sonic booms.

Although sonic booms are regularly made by military aircraft during training and operations, that noise is never heard by most Americans since those aircraft usually operate in remote, designated areas. This paper focuses on the prospects for supersonic commercial aircraft.

The purpose of this paper is to provide an up-to-date overview of what is known about sonic booms, appropriate for non-specialists both in and out of acoustics and noise control. The basic sound generation mechanism is first described, leaving detailed technical descriptions to the references.

Gulfstream [1] is one compaony that has advocated the development of a supersonic business jet [2]. The primary benefit is obvious: point to point travel at speeds on the order of twice that of conventional subsonic jets. Certainly some individuals can afford this and would pay a premium price for the resulting enormous time savings.

But along with this time savings comes sonic boom noise. It makes sense that if sonic boom noise is no more annoying than that of conventional aircraft noise, then the public will be accepting of the new generation of supersonic business aircraft. In the past, however, the sonic booms from previous generations of supersonic aircraft have been unacceptable. At this writing, it is against federal law for civil aircraft to fly supersonically over land or to cause a sonic boom over land in the United States.

Since the mid 1960's engineers have been working on methods to mimimize sonic boom noise by very careful shaping of the aircraft. The basic idea will be explained later in this paper. The theory for this sonic boom minimization is very solid, and flight test experiments have been performed to validate the theory.

The new low-boom aircraft will be much more economically viable if they are allowed to fly over land, which will require a change in federal law. The British-French Concorde supersonic passenger airliner, just recently retired from regular service, was only allowed to fly supersonically over water. This was one of many reasons why the Concorde did not see more use. But if overland flight is made legal under certain circumstances, we are probably only a few short years away from low-boom supersonic business jets being rolled out.

It should be mentioned that there are certainly other environmental issues such as engine exhaust emissions and landing and takeoff noise that have to be resolved for a viable supersonic commercial jet. However, those other issues are being dealt with and will not be addressed here.

The efforts of man to finally break the "sound barrier" are well documented, and those historical efforts won't be reviewed here, either. We will assume we can have the technology to build supersonic aircraft on demand.

After the description of sonic booms in the next section, this paper describes some of the efforts to develop low-boom supersonic aircraft. Some of the challenges in flying through the real atmosphere, instead of an idealized atmosphere will be explained, along with the challenges of human perception of sonic boom noise. After a summary conclusion, this paper ends with a brief guide to the open sonic boom literature for readers interested in learning more about the subject.

FIRST PASS LOOK

What is a sonic boom? It is a pressure disturbance created by the passage of an aircraft, or any other object, traveling faster than the speed of sound. In air the speed of sound at 20 °C (68 °F) is approximately 343 m/s (1125 feet per second or 767 miles per hour) and varies with temperature and pressure. A typical non-minimized sonic boom time trace measured at the ground is given in Figure 1. The vertical axis is pressure in pascals and the horizontal axis is time in seconds. This particular graph is an example of a rounded sonic boom waveform produced by an F-15 aircraft taken from the BoomFile database [3]. Notice that the waveform roughly looks like the letter "N." Most conventional sonic boom waveforms look similar to this, although the atmosphere and aircraft maneuvers can alter the shape. Notice there is a distinct beginning and ending to sonic boom waveforms. This is because the aircraft has a finite length.

The peak pressure of this sonic boom is about 85 pascals. 1 poundforce per square inch (psf) is about 48 pascals, giving this boom a peak pressure of 1.77 psf. This is a very small fraction of 1 atmosphere (about 101,325 Pa), but it is still very audible.

The most important thing to take away from anywhere in this paper is that the **sonic boom is continually made the entire time the aircraft is moving supersonically**. This means that along

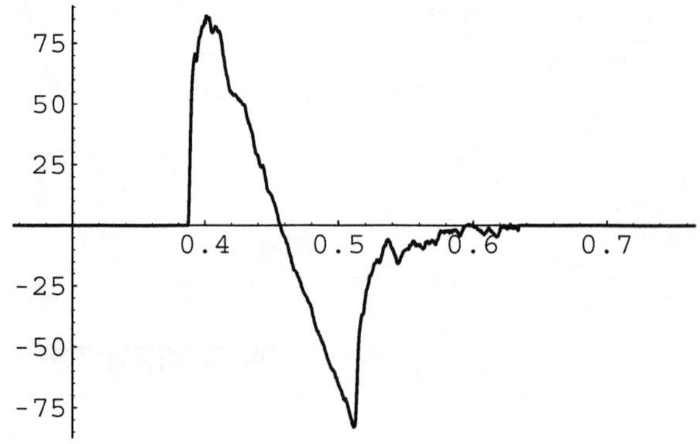

Figure 1. F-15 MEASURED SONIC BOOM WAVEFORM (NON-MINIMIZED).

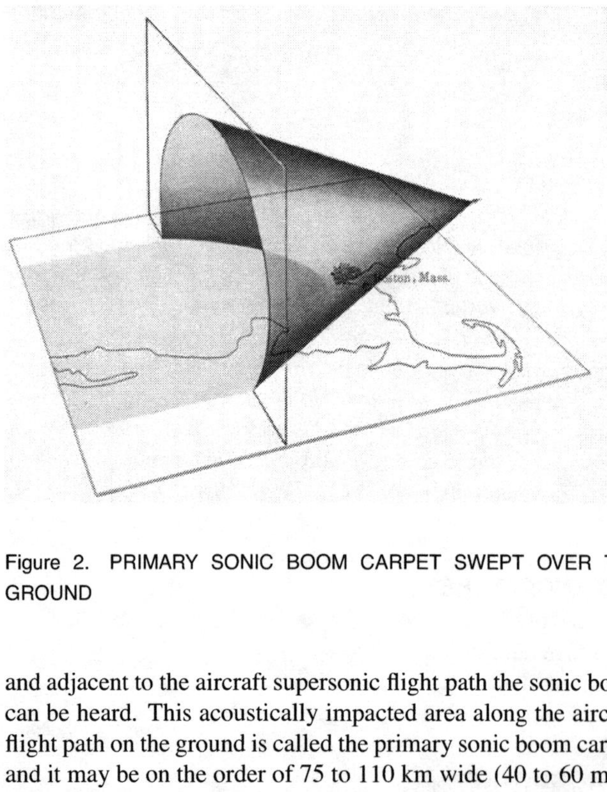

Figure 2. PRIMARY SONIC BOOM CARPET SWEPT OVER THE GROUND

and adjacent to the aircraft supersonic flight path the sonic boom can be heard. This acoustically impacted area along the aircraft flight path on the ground is called the primary sonic boom carpet, and it may be on the order of 75 to 110 km wide (40 to 60 miles wide) depending on the aircraft alititude and the meteorological conditions. It will turn out that the sonic boom wave sweeps out a conical shape as it propagates, and this will be explained in a moment. The ground is impacted by the boom in the shape of a truncated hyperbola.

A good picture from Ref. [4] reproduced in Fig. 2 shows this cone behind the aircraft being swept west to east over southern Massachusetts. Boston is about to be boomed, while Rhode Island already has. The figure gives a reaasonable approximation of the width of the primary boom carpet.

The reader can also see the analogy between a supersonic airplane and a fast moviing speedboat in the water. The speedboat creates a V-shaped wake that spreads out behind the boat and splashes into the nearby shoreline after traveling some distance. The sonic boom is an anaologous conically-shaped wake created in three dimensions by the supersonic aircraft.

One can easily see that because the sonic boom noise is made continuously along the flight path, millions of U.S. residents will regularly hear sonic boom noise if supersonic commercial aircraft become popular and fly overland. The good thing to know about this picture is that the great majority of the sonic boom noise may not be annoying to the public, and that will be described later in this paper.

SOURCE / PATH / RECEIVER MODEL

Noise control engineers often break acoustics problems into three parts, that of the noise source, the noise transmission path, and the receiver. All three portions of the system must be addressed before an optimal solution is achieved to any noise control problem. The remainder of this section will be divided into these three sections.

Source

Sonic booms, like all other sounds, travel through the air at a finite speed, and that speed is slower than the aircraft is traveling. Hence, there is a particular angle at which the sonic boom noise leaves the aircraft, and that angle, called the Mach angle, can be determined by a simple geometrical construction. See Figure 3 inspired by Pierce's book [5]. The Mach angle as meaured as an apex angle from the front of the boom to the aircraft, here shown as a Concorde, is determined from geometry to be

$$\theta_M = \sin^{-1}(1/M) \tag{1}$$

where $M = V/c$ is the Mach number, the relative speed of the aircraft to the speed of sound.

The aircraft is moving forward at a speed V, and the speed of sound c is smaller. Thus in a time Δt the aircraft moves a distance $V\Delta t$, but the sound can only move a distance $c\Delta t$. Correspondingly at a time Δt in the future, the sound will have moved to the dotted line in Figure 3, and the Mach cone will have moved forward with the aircraft.

The angle at which the sound impacts the ground is not necessarily θ_M because sound bends due to winds and the tempararture distribution of the atmosphere, but this gives a good approximation most of the time.

In addition to this simple picture, there are substantial supersonic nonlinear fluid dynamic effects which determine what

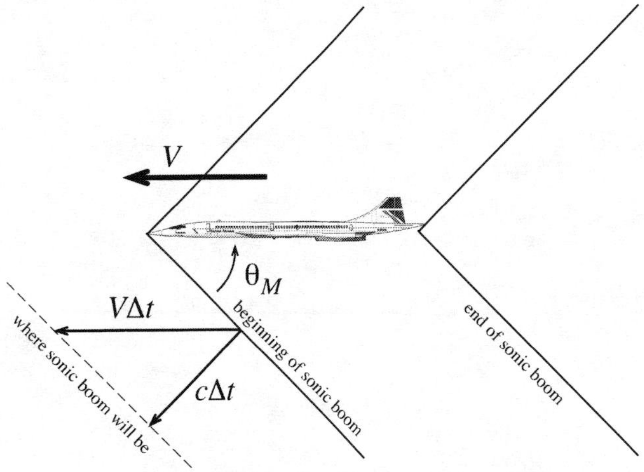

Figure 3. GEOMETRY OF MACH CONE

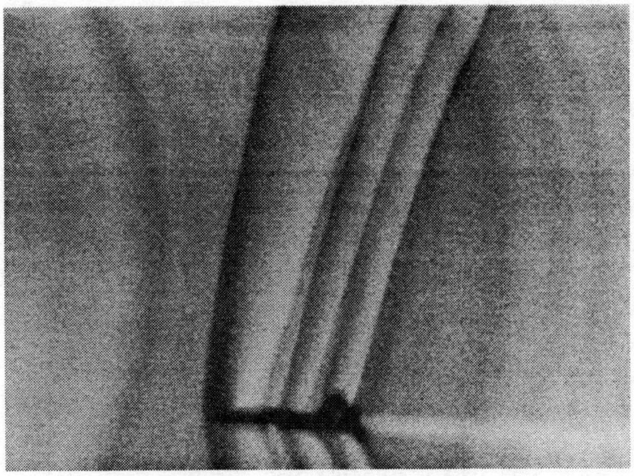

Figure 4. IN-FLIGHT SCHLIEREN IMAGE OF NEAR FIELD SHOCK WAVES

the sonic boom pressure versus time waveform looks like close to the aircraft. There is a series of shock waves which move out from all the geometrical features of the aircraft. An impressive schlieren photographic image captured by L. Weinstein [6] of NASA Langley Research Center of a T-38 aircraft flying at $M = 1.1$ is shown in Figure 4.

No doubt, this *is* rocket science!

Path

How the complicated shock wave flow field develops into a sonic boom waveform like that shown in Fig. 1 is the next link in the source, path, receiver chain.

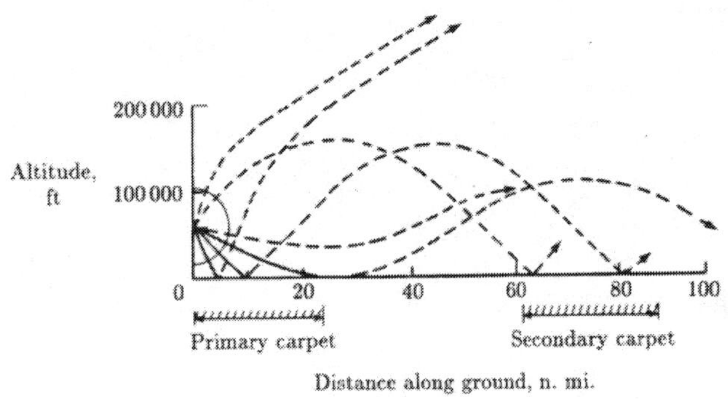

Figure 5. SOUND RAYS TO ONE SIDE OF SUPERSONIC FLIGHT PATH (FRONT VIEW)

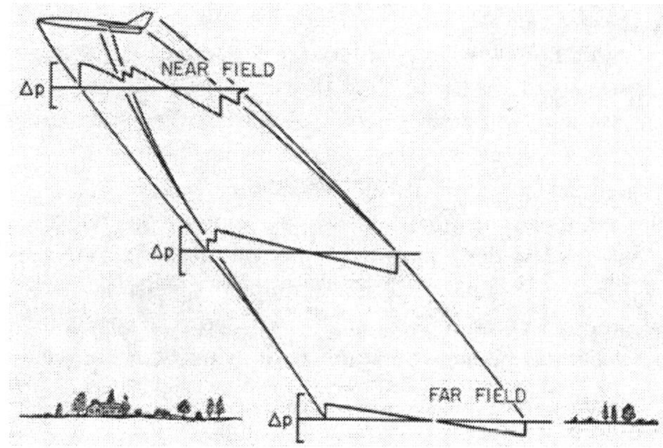

Figure 6. NONLINEAR STEEPENING CHANGES WAVEFORM DURING PROPAGATION FROM AIRCRAFT TO GROUND

Primary path As mentioned earlier and seen in Fig. 2, the sound actually comes off the aircraft cylindrically in 3-dimensions forming a cone emanating from the aircraft. Figures 3 and 4 show vertical slices. One can think of the sound as coming off the aircraft as sound rays, similar to light rays. It is then straightforward to follow those rays, see Fig. 5.

The loudest sounds under the flight path come from rays, shown with solid lines in Fig. 5, which propagate directly from the aircraft to the ground. This makes up the primary boom carpet. We will discuss the other rays in the next section.

What the listener actually hears in the primary carpet is considerably different from the waveform one measures close to the aircraft. That is because the sound propagation is nonlinear, and the waveform's shape can also be distorted by atmospheric effects. Lets deal with these two effects separately.

Firstly, it turns out that for sound of sufficient amplitude (including sonic booms) that higher acoustic pressures travel ever so slightly faster than lower acoustic pressures. If given enough distance to propagate, the conventional, complicated nearfield waveform distorts over the propagation distance into a wave shaped like the letter N, a process called nonlinear steepening. See Fig. 6 [7].

These N shaped waves are in fact observed during atmospheric conditions when there is very little turbulence, like very early in the morning. Atmospheric effects very often distort the N shaped waves. The resulting waveforms can be either spiked, multiply-spiked, or rounded (as seen in Fig. 1). But sometimes the N shaped waves get through undistorted.

There is one instance where an aircraft produces an extra large sonic boom, that when it is accelerating to supersonic speed. The sound produced is called a focused boom, since the sound ray paths cross at the ground and the sound energy is concentrated there. Focused booms occur at the beginning of the supersonic flight path, and can occur in turns or pushovers.

Secondary path During supersonic flight the sound spreads roughly cylindrically from the flight path. The primary sonic boom wave results from sound rays descending dirctly from the aircraft to the ground. But the sonic boom energy also travels along other ray paths. The "over-the-top" sonic boom, also called secondary sonic boom, is the portion of the sonic boom energy that initially travels upward away from the aircraft, but because of atmospheric refraction, travels back downward to impact the ground at substantial distances from the aircraft ground track. This is seen as one of the dotted lines in Fig. 5. The "over the top" sonic boom noise contributes to a secondary sonic boom carpet which can be some 60 nautical miles or more away from the aircraft flight path. Seasonal winds can increase or decrease the impact of "over-the-top" sonic booms.

In addition, sometimes the primary sonic boom wave reflects off the ground, travels back into the atmosphere, and refracts back down into the secondary boom carpet. Because of the increased distance between source and receiver for these secondary booms, most of the high frequency energy has been attenuated in the secondary boom carpet, and all that is left of the waveforms is a distorted, low-frequency rumble.

Receiver

For the receiver part of sonic boom noise people and animals are our primary concerns, but buildings and other structures are also involved.

Human listeners The acceptability of sonic booms is determined solely by man. The reason that supersonic overland flight is currently illegal is that conventional (non-minimized) sonic booms are perceived by the public as annoying. People will hear sonic booms indoors as well as outdoors, and there are some significant differences between these cases. Outdoors an individual will hear the sonic boom at its peak amplitude. Indoors the same person will hear a boom of a lower magnitude because the building acts as a frequency dependent filter. However, the human response to the boom indoors also will be affected by any possible "rattle" of objects or, in extreme cases, structural damage such as the breaking of windows or cracking of plaster. There is a large body of existing work on the psychoacoustics of impulsive noise, see [8, 9] for example, and it is generally accepted that there should be some additional penalty assigned because of the "startle effect." For example for subsonic aircraft, you usually hear the airplane coming initially as a weak signal getting louder. Sonic booms have no weak precursor, so most people are surprised by them.

Wildlife Every study made to date exploring the effect of sonic booms on wildlife in their natural setting have shown no adverse effects on the animals [9]. For example, during regular Concorde flights grey and harbor seals were regularly boomed on Sable Island off Nova Scotia [10]. They showed no substantial effects due to these regular exposures. The thought is that the animals habituated to the sonic boom noise once they understood it was not harmful to them. Sonic boom noise from flights over the ocean does penetrate the water surface [11]. But both experiments [12] and computer simulations [13, 14] clearly show that the sound below the surface never exceeds the sound levels above the surface by more than a few percent at most. In an overwhelmingly large number of studies it has been shown that animals react to sonic booms pretty much the same way people do

Structures The one difference between the reactions of humans and animals to sonic boom noise is that humans sometimes complain and/or bring lawsuits if sufficiently compelled. Usually these lawsuits have to do with structural damage such as chipped plaster or cracked windows in their homes, which fortunately are rare events. There is a large body of research that has gone into determining the extent of sonic boom induced structural damage [15–17].

The structural effects in buildings are usually only on brittle surfaces and secondary structural members [18]. It makes sense that minimized sonic booms with a decreased peak pressure will cause less damage than non-minimized sonic booms. However, one can give no assurances that the triggering of damage can be completely avoided even if the average peak overpressure is reduced [18].

A SHORT HISTORY OF AIRCRAFT PROGRAMS AND BOOM MINIMIZATION

Because N waves are unacceptable to the public, scientists and engineers have been working very hard since the 1960's to develop low-boom aircraft configurations. During the late 1960's and early 1970s there were substantial ideas put forth on how to vary the cross sectional area of an aircraft to minimize the sonic boom waveform on the ground. Unfortunately none of these ideas were incorporated into the U.S. Supersonic Transport (SST) program, since the technology was in its infancy and the program was on such a fast track.

One can make a sonic boom less loud by making it longer, but to be effective this would require an overly long aircraft. One idea from the 1960's which gained popularity in the 1980's and early 1990's was that sonic booms with particular shapes would be perceived as being less annoying than N-shaped booms. A successful method for shaping the boom was developed by George and Seebass [19]. They obtained an optimum solution for the shaping and related that shaping directly to the configuration of the aircraft. These initial ideas were taken up and improved by Darden and others at NASA Langley in the late 1970's and 1980s [20].

With the interest in pursuing a High Speed Civil Transport (HSCT) by NASA, Douglas Aircraft, and Boeing in the late 1980's, substantial psychoacoustic testing took place at NASA Langley Research Center [21]. Hence, the goal of sonic boom minimization became one of obtaining a sonic boom that is considered less obtrusive. Due to a shifting of priorities, the low-boom design principles were abandoned for the HSCT, and no aircraft was ever developed.

In the 1970's and 1980's the designs were made with analytical mathematics using well understood, but overly simplified assumptions. In the 1990's and today computational fluid dynamics (CFD) has become the method of choice for developing a near field sonic boom waveform that will nonlinearly propagate to the ground into a minimized, or low-boom, sonic boom. The coupling of the near field CFD solution to the sonic boom signatures received on the ground has been sucessfully addressed [22].

Concorde, as a benchmark, had a sonic boom wave shaped like an N wave with a peak pressure of about 2 pounds per square foot, 2 psf. The objective of recent low-boom sonic boom researchers in the Quiet Supsersonic Platform (QSP) program, of initially Defense Advanced Research Projects Agency DARPA) and then DARPA and NASA, has been to develop supersonic aircraft that have a peak overpressure of no more than 0.3 psf while having a sonic boom waveform which does not completely steepen into an N wave. Figure 7 shows two prototypical minimized sonic boom signatures compared with the standard N wave [23].

Note that Fig. 7 only shows theoretical general shapes of minimized booms. Recent shaping efforts are summarized by Coen [24].

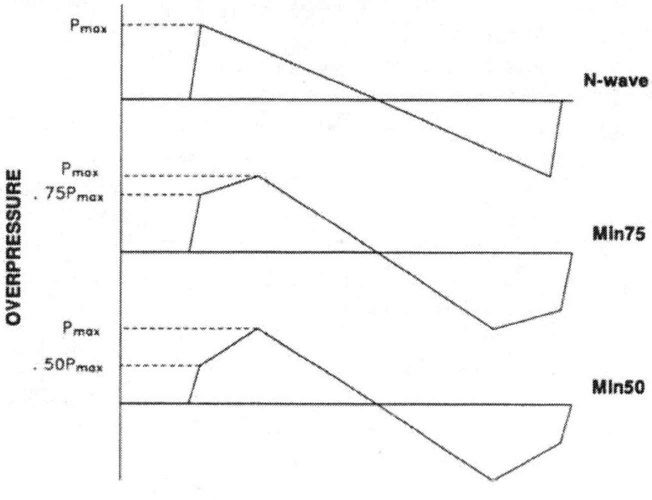

Figure 7. N WAVE AND TWO MINIMIZED SIGNATURES

Figure 8. SHAPED SONIC BOOM DEMONSTRATOR IN FLIGHT

To prove the validity of the sonic boom minimization procedure, a test program has been underway as part of the QSP series of studies. A U.S. Navy F-5E fighter jet aircraft has been outfitted with a modified nose to produce a flat-top sonic boom [25, 26]. The aircraft was called the Shaped Sonic Boom Demonstrator (SSBD), and the program was jointly carried out by Northrup Grumman Corporation, NASA Langley Research Center, and NASA Dryden Flight Research Center. See Figure 8 which is taken from [27]. At the time of this writing we have just learned that the first successful flights of the SSBD have taken place, and that minimized sonic boom waveforms were measured on the ground in late August and early September 2003. This flight test program confirms the substantial previous work on low-boom design involving wind tunnel experiments and numerical simulations. Aircraft shaping works.

CHALLENGES OF WEATHER AND THE ATMOSPHERE

Although most of the research on low-boom aircraft has centered on the source and receiver portions of the problem, less attention has been given to the propagation portion. The recent QSP program, for example, exclusively focused on the source problem. Indeed the major assumption is that through careful area shaping of the aircraft the sonic boom received on the ground will be less annoying than conventional supersonic aircraft without shaping. However, this method assumes that the atmosphere between the aircraft and the ground is a model standard atmosphere. However, we have seen that the real atmosphere can cause peaking and rounding of N wave sonic boom waveforms, and there is no reason why similar distortions won't occur for low-boom sonic booms.

To the author's knowledge there has been no research on how atmospheric turbulence will affect the propagation of low-boom aircraft noise. This seems like a gap that needs to be filled.

Further it makes sense that weather will be another factor affecting the propagation of sonic boom waves to the ground. Our experience with weather and sonic booms is quite limited. Most experiments on sonic boom have been performed in the desert southwest of the United States, and the remainder were performed in good weather conditions elsewhere. Certainly we need to know more about how sonic booms propagate through various weather conditions if we desire unrestricted overland operations.

One scenario which has yet to be addressed is the propagation of sonic boom waves from aircraft to the ground in and around large weather systems, including those with thunderstorms. It is easy to visualize that strong downdrafts due to weather can cause the formation of focus booms far from the area of initial aircraft acceleration. This special focusing should occur, for example, when a supersonic aircraft flies directly toward a large downdraft at an altitude of about half to one-quarter of the aircraft altitude. The converging of the acoustic rays near the ground should cause a sonic boom with a larger than usual amplitude, even if the aircraft is designed for a low-boom configuration. Clearly, detailed theoretical testing using 3-D sound ray tracing software needs to be performed to determine the actual extent of this weather-induced focusing.

Kaouri *et al.* [28] has recently reminded us that "over-the-top" sonic booms have hardly been investigated. This is a very ripe area for research. Plotkin [29] describes two computer programs that have been used to predict such booms, but with mixed results. More work really needs to be done, if one is to predict "over-the-top" booms accurately.

CONCLUSIONS

In summary, this paper has provided an overview of sonic booms necessary for understanding their generation, propagation, and reception. The recent works of many scientists and engineers to develop low-boom configuration aircraft have been highlighted. There is reasonable confidence that low-boom aircraft can be constructed today. Additional work needs to be performed in understanding how the minimized sonic booms propagate through the real atmosphere, including weather, and whether the propagated, minimized booms will be perceived by the public as being annoying.

Engineers will certainly do the best they can in developing low-boom aircraft, but mother nature ultimately will determine what people on the ground will hear.

LEARN MORE ABOUT IT

For individuals interested in learning more about sonic booms, there is plenty of available literature from reliable sources. The best first pass introduction is probably by Maglieri and Plotkin [18]. Two recent reviews include that of Darden [30] and Plotkin and Maglieri [31], the latter describing many recent developments from the QSP program.

The best references for work in the 1960s on sonic boom include the three sonic boom research conferences [32–34]. In the 1970's there were also two early sonic boom symposia sponsored by the Acoustical Society of America [35,36]. An excellent history [must read] of the rise and fall of the U.S. SST program is available online within a history of the U.S. Space Shuttle program [37]. Two easily readable and technicaly correct tutorials on sonic boom from the 1960's include the work by Carlson and McLean [4] and Green [38].

During the late 1980's and early 1990's there was significant NASA interest in sonic booms for the now cancelled HSCT project. There was an initial workshop to kick off the efforts [39] which was later followed by a number of annual High Speed Research sonic boom workshops during 1991–1995 [40–44]. Another Acoustical Society of America sonic boom symposium took place in 1998, and the proceedings were published in 2002 [45]. This was the most recent public forum on sonic booms with a proceedings.

Reports by two national Research Council committees were given in 1997 and 2001 and are available online [46,47]. The first assesses NASA's High Speed Research Program, and the second focuses on future prospects for supersonic commercial flight.

Much of the work in the QSP program during the last few years is proprietary and has not been published. However, many reports stemming from that program have been presented as individual papers at American Institute of Aeronautics and Astronautics (AIAA) meetings, and some magazine articles describing the program have appeared [25].

ACKNOWLEDGMENT

The author thanks Dr. Ken Plotkin of Wyle Laboratories, Arlington, VA for encouragement and for providing useful feedback on an earlier draft of this paper.

REFERENCES

[1] Gulfstream Aerospace Corporation, 2003. See URL `http://www.gulfstream.com`.

[2] Wolz, R., 2003. "A summary of recent supersonic vehicle studies at gulfstream aerospace". In Proceedings of 41st Aerospace Sciences Meeting and Exhibit. AIAA Paper 2003–558.

[3] Lee, R., and Downing, J. M., 1996. "Comparision of measured and predicted lateral distribution of sonic boom overpressures from the United States Air Force sonic boom database". J. Acoust. Soc. Am., **99** (2) [February], pp. 768–776.

[4] Carlson, H., and McLean, F., 1966. "The sonic boom". Intl. Science and Tech. (55) [July], pp. 70–80.

[5] Pierce, A., 1981. *Acoustics: An introduction to its physical principles and applications*. McGraw-Hill, New York. Republished in 1989 by the Acoustical Soc. of America, Melville, NY.

[6] Weinstein, L., 1994. "An optical technique for examining aircraft shock wave structures in flight". In High-speed research – Atmospheric propagation and acceptability studies, D. McCurdy, Ed., NASA Conf. Publ. 3279, NASA Langley Research Center, pp. 1–17.

[7] Carlson, H., 1967. "Experimental and analytic research on sonic boom generation at nasa". In Sonic boom research, A. Seebass, Ed., NASA Special Publ., SP-147, NASA Office of Technology Utilization, p. 10.

[8] Kryter, K., 1985. *The effects of noise on man*, 2nd ed. Academic Press, Orlando, FL.

[9] Kryter, K., 1994. *The handbook of hearing and the effects of noise: physiology, psychology, and public health*. Academic Press, San Diego, CA.

[10] Perry, E., Boness, D., and Insley, S., 2002. "Effects of sonic booms on breeding gray seals and harbor seals on sable island, canada". J. Acoust. Soc. Am., **111** (1, Pt. 2) [January], pp. 599–609.

[11] Sparrow, V., 2002. "Review and status of sonic boom noise penetration into the ocean". J. Acoust. Soc. Am., **111** (1, Pt. 2) [January], pp. 537–543.

[12] Sohn, R., Vernon, F., Hildebrand, J., and Webb, S., 2000. "Field measurements of sonic boom penetration into the

ocean". J. Acoust. Soc. Am., **107** (6) [June], pp. 3073–3083.

[13] Rochat, J., and Sparrow, V., 1997. "Two-dimensional focusing of sonic boom noise penetrating an air-water interface". AIAA J., **35** (1) [January], pp. 35–39.

[14] Rochat, J., and Sparrow, V., 2001. "A computational analysis of sonic booms penetrating a realistic ocean surface". J. Acoust. Soc. Am., **109** (3) [March], pp. 899–908.

[15] Wiggins, J., 1969. *Effects of sonic boom*. J. Wiggins Co., Palos Verdes Estates, CA.

[16] Sutherland, L., and Czech, J., 1992. Evaluation of human response to structural vibrations induced by sonic booms. NASA contractor report 189584, Hampton, VA, May.

[17] Clarkson, B., and Mayes, W., 1972. "Sonic-boom-induced building structure responses including damage". J. Acoust. Soc. Am., **51** (2, Pt. 3) [February], pp. 742–757.

[18] Maglieri, D., and Plotkin, K., 1991. "Sonic boom". In *Aeroacoustics of Flight Vehicles: Theory and Practice*, H. Hubbard, Ed., vol. 1. NASA Langley Research Center, Hampton, VA, Chapter 10, pp. 519–561. NASA Reference Pub. 1258, WRDC Technical Reort 90–3052, reprinted by the Acoustical Society of America, Melville, NY.

[19] Seebass, A., and George, A., 1974. "Design and operation of aircraft to minimize their sonic boom". J. Aircraft, **11** (9) [September], pp. 509–517.

[20] Mack, R., and Darden, C., 1980. "Some effects of applying sonic boom minimization to supersonic cruise aircraft design". J. Aircraft, **17** (3) [March], pp. 182–186.

[21] Leatherwood, J., Sullivan, B., Shepherd, K., McCurdy, D., and Brown, S., 2002. "Summary of recent NASA studies of human response to sonic booms". J. Acoust. Soc. Am., **111** (1, Pt. 2) [January], pp. 586–598.

[22] Plotkin, K., and Page, J., 2002. "Extrapolation of sonic boom signatures from CFD solutions". In 40th AIAA Aerospace Science Meeting and Exhibit. AIAA Paper 2002—0922.

[23] Sullivan, B., and Leatherwood, J., 1994. "Experimental studies of loudness and annoyance response to sonic booms". In High Speed Research: Sonic Boom, vol. 1, NASA Ames Research Center, pp. 153–175. NASA Conference Publication 10132.

[24] Coen, P., 2003. "Configuration shaping for low sonic boom: retrospective and future directions". In 16th AIAA Fluid Dynamics Conference. Slides from AIAA Paper 2003–3573.

[25] Wilson, J., 2002. "The new shape of supersonics". *Aerospace America* [June], pp. 26–33.

[26] Pawlowski, J., 2003. "DARPA QSP supersonic shaped boom flight experiment". In 16th AIAA Fluid Dynamics Conference. AIAA Paper 2003–3577.

[27] Dryden Research Center Photo Gallery, N., 2003. See URL `http://www.dfrc.nasa.gov/ Gallery/ index.html`.

[28] Kaouri, K., Allwright, D., and Dellar, P., 2003. "Aspects of secondary sonic boom propagation". In Proceedings of Euronoise 2003, 5th European Conference on Noise Control, vol. 89, DETEC University of Naples, Italy, p. S70. Abstract in Acta Acustica United with Acoustica, Supplement 1, May/June 2003.

[29] Plotkin, K., 2002. "State of the art of sonic boom modeling". J. Acoust. Soc. Am., **111** (1, Pt. 2) [January], pp. 530–543.

[30] Darden, C., 2002. "Affordable/acceptable supersonic flight: Is it near?". In 16th Intl. Session in 40th Aircraft Symposium, Japan Society for Aeronautical and Space Sciences (JSASS). Yokohama, Japan.

[31] Plotkin, K., and Maglieri, D., 2003. "Sonic boom research: history and future". In Proceedings of 33rd AIAA Fluid Dynamics Conference. AIAA Paper 2003–3575.

[32] Seebass, A., Ed., 1967. Sonic boom research, NASA Special Publ., SP-147, NASA Office of Technology Utilization.

[33] Schwartz, I., Ed., 1968. Second conference on sonic boom research, NASA Special Publ., SP-180, NASA Office of Technology Utilization.

[34] Schwartz, I., Ed., 1970. Third conference on sonic boom research, NASA Special Publ., SP-255, NASA Office of Advanced Research and Technology.

[35] Hubbard, H., 1966. "Sonic-boom symposium: Forword". J. Acoust. Soc. Am., **39** (5, Pt. 2) [May], p. iv.

[36] Ribner, H., and Hubbard, H., 1972. "Sonic-boom symposium: Forword". J. Acoust. Soc. Am., **51** (2, Pt. 3) [February], p. 671.

[37] Heppenheimer, T., Ed., 1999. *The Space Shuttle Decision: NASA's search for a reusable space vehicle.* NASA History Series. NASA History Office, Washington, D.C., Chap. 7, Aerospace recession. Online at `http://history.nasa.gov/SP-4221/sp4221.htm`.

[38] Green, D. M., 1968. "Sonic booms: Jangled jet nerves and the thundering future". Psychology Today, **2** (6) [November], pp. 38–44.

[39] Darden, C., Powell, C., Hayes, W., George, A., and Pierce, A., Eds., 1989. Status of sonic boom methodology and understanding, NASA Conf. Pub. 3027, NASA Langley Research Center.

[40] Whitehead, A., Ed., 1991. First annual high-speed research workshop, vol. 1 of *NASA Conf. Publ. 10087*, NASA Langley Research Center.

[41] Darden, C., Ed., 1992. High-speed research: sonic boom volume 1, NASA Conf. Publ. 3172, NASA Langley Research Center.

[42] Edwards, T., Ed., 1994. High-speed research: sonic boom volume 1, NASA Conf. Publ. 10132, NASA Langley Research Center.

[43] McCurdy, D., Ed., 1994. High-speed research – Atmospheric propagation and acceptability studies, NASA Conf. Publ. 3279, NASA Langley Research Center.

[44] Baize, D., Ed., 1996. 1995 High-speed research program sonic boom workshop Volume 1, NASA Conf. Publ. 3335, NASA Langley Research Center.

[45] Sparrow, V., 2002. "Sonic boom symposium: Foreword". J. Acoust. Soc. Am., **111** (1, Pt. 2) [January], p. 497.

[46] National Research Council, 1997. U.S. supersonic commercial aircraft: Assessing NASA's high speed reseach program. Online book, `http:// www.nap.edu/catalog/5848.html`.

[47] National Research Council, 2001. Commercial supersonic technology: The way ahead. Online book, `http:// www.nap.edu/catalog/10283.html`.

Proceedings of IMECE'03
2003 ASME International Mechanical Engineering Congress
Washington, D.C., November 15–21, 2003

IMECE2003-55391

ON THE CHALLENGES OF QUIETING A FREIGHT LOCOMOTIVE CAB INTERIOR

Mehdi Ahmadian
Professor and Director
Advanced Vehicle Dynamics Laboratory
Department of Mechanical Engineering
Virginia Tech, MC-0238
Blacksburg, VA 24061

ABSTRACT

The primary purpose of this paper is to discuss some of the challenges—both from an engineering standpoint and economically—facing the railway industry in improving the noise and vibration characteristics in locomotive cabs. To this end, we will first establish the vibration characteristics of a typical locomotive cab used in freight locomotives in North America. Next, we will evaluate the effect of separating the cab structure from the reminder of the locomotive structure by elastomeric mounts. This cab configuration is commonly referred to as "soft-mounted cab," in contrast to a "hard-mounted cab," which is rigidly attached to the remainder of the locomotive structure. The structural dynamics of a production locomotive cab is studied in a laboratory environment. The cab is excited by a hydraulic actuator, in a manner closely resembling field inputs. After establishing the baseline vibrations of the cab, the study presents the results of vibration tests for the soft mounted cab. A comparison between the hard-mounted and soft-mounted cabs indicate that soft mounting the locomotive cab can result in substantial noise and vibration reduction at all locations in the cab, and therefore provide

- more crew comfort
- less equipment damage
- manufacturing cost savings

The paper will also discuss the effect of the measures that are needed for quieting the cab, such as soft mounting it, on the locomotive manufacturers from an engineering and business perspective.

INTRODUCTION

Noise, vibration, and harshness (NVH) in vehicles have received increased attention in recent years. In lighter vehicles, such as cars, pickup trucks, and the common utility vehicles, the emphasis has been on improving customer satisfaction and brand image. As vehicle quality has improved across the industry, vehicle manufacturers are searching for other ways of differentiating themselves from their competitors. The advertisements for luxury vehicles have raised customers' awareness and sensitivity to noise and vibrations inside the vehicle. Therefore, the industry has identified NVH as one of the areas that can be used to improve or maintain market share.

The advancements in noise and vibration reduction have further raised customer expectations, forcing the transportation industry to look for yet more improvements.

The heavy vehicle industry is lagging behind the light vehicle industry in this respect by several years; it is only recently that the noise and vibration issues have come to the forefront of engineering development in some sectors of the heavy vehicle industry, such as heavy trucks. This emphasis is partially driven by

- increased customer expectations
- a push for further labor productivity
- litigations and labor negotiations

As their personal automobiles have become quieter, the operators of heavy vehicles have started noticing the noise and vibrations in their heavy vehicles. The difference in noise and vibration between a heavy vehicle and an automobile has increased from a few dBA to several (sometimes many) dBA. This factor has started to become reflected in the buying decisions of customers, and in labor negotiations. Some studies have indicated that in certain sectors of the heavy transportation industry (such as heavy trucks) comfort issues, such as lower noise and vibrations, play an important role in customers' buying decisions, along with more traditional factors such as price, quality, and brand loyalty. Similarly, in negotiations, the labor unions have been demanding a quieter and more comfortable environment inside the vehicles they spend a large portion of their working days.

All of these factors have contributed to the increased sensitivity of the transportation industry to comfort issues, such as reducing interior noise and vibration in vehicles. In this paper, after reviewing some of the past studies related to the topic at hand, we will evaluate the amount of noise and vibration reductions that can be achieved in a locomotive cab through simple engineering means, such as soft mounting the cab. At the end, we will provide an opinion on why some of these seemingly simple solutions are not widespread among the freight locomotive manufacturers, in spite of their vast benefits for lowering interior cab noise and vibrations.

PAST STUDIES

The majority of the studies in this area address the application of various types of suspension systems for reducing vehicle vibrations [1-11]. The majority of these references reflect the vast number of studies that have been conducted in the past two decades in the areas of passive, semi-active, and active primary and secondary suspensions. Primary suspension refers to the suspension system at the vehicle axle, which serves to connect the axle to the vehicle body and control road-induced vibrations to the body. Secondary suspensions are the systems used for connecting other major components, such as the engine, cab, and seat, to the vehicle body. Secondary suspensions are commonly used to isolate the vibration sources from the operator compartment.

Among the articles that we found on cab vibrations, the majority addressed analytical studies relating to designing and optimizing cab suspensions on heavy trucks and farm equipment [12-16]. In references [12] and [16], Hansson and Sankar, respectively, studied cab suspensions for farm equipment. Hansson studied the effectiveness of time invariant state feedback and different observers in terms of vibration isolation and power consumption. The results show that it is possible to design an effective active suspension. Sankar et al., [16], use a finite element model to analyze a farm tractor cab suspended on three mounts. They show that the flexibility of the cab structure is important to evaluating vibration reductions in the cab due to the soft suspensions.

In [13-15], different actively controlled mounts are studied for semi-truck cab suspensions. All of these studies involve the application of active control methods that are based on linear and nonlinear optimal control concepts. Although the results of these analytical studies are promising, their applicability and actual performance benefits are questionable since none of the studies include any experimental or field data. Our experience has shown that the applicability of active control schemes is limited for most vehicle suspension applications due to their complexity and cost burden. Semi-active control schemes, however, have often been applied successfully.

We were only able to find one article, Reference [17], which was related to locomotive cab noise and vibrations. Reference [17] is a study by the Association of American Railroads (AAR) that addresses in-cab noise measurement standards, not a method for reducing noise and vibrations in a locomotive.

VIBRATION TEST SETUP

Cab Configuration: The plan and side views of the locomotive cab that was tested throughout this research are shown in Figures 1 and 2, respectively. The entire structure weighs approximately 22,000 lb., with dimensions 16 ft long, 10 ft wide, and 10 ft high. The cab front section, known as the "nose cab," is constructed predominantly from 0.375 in steel. Inside the front door there is a small area containing the crash posts, which are large steel post structures used to support a head-on collision with another locomotive or a very large object. An internal door is used to separate the immediate entrance area from the remainder of the cab, which provides better sound and temperature isolation. Beyond the internal door, a lavatory is located to the right and a rack for mounting electronic equipment (called an LSI rack) is located to the left. The test cab did not include any of the electronic equipment that is normally installed in the LSI racks or elsewhere in the cab. Most of the internal cabinetry is made of 0.0625 in sheet metal or molded plastic.

The area between the lavatory and LSI racks is called the "alcove." This area leads to a set of three steps that connect the alcove to the operator cab, which is located approximately two feet above the alcove floor. The conductor and engineer's consoles are located immediately to the right and left of the steps, respectively. Each console consists of a plastic counter top, communication equipment, and LED displays for running the locomotive. Several large side windows and a windshield surround the operator cab. The back wall of the cab consists of a small door that leads to a walkway (not included in our test cab), an electronic cabinet (called CA1) that is separated from the cab by a tri-fold door, and a small window. The cab floor is made of 0.125 in sheet steel that is plug-welded to a tubular substructure. This structure is the skeleton of the area that is located directly underneath the operator cab. This area (called the "air brake cab") houses the air brake and HVAC equipment. A foam rubber mat of approximately 0.25 in thick is glued to the cab floor to provide cushion (for operator comfort) and damping effect for vibrations. The engineer and conductor's seats are located on a raised structure to the right and left of the operator cab. The roof of the cab is made of thin-gauge steel with supports placed approximately every 16 inches apart. A thin damping material is installed between the supports to reduce vibrations of the roof plate.

As previously mentioned, the area beneath the cab is the air-brake compartment. Plumbing, air conditioner, ducts, and various other components are located in this compartment. This equipment was not included in our test cab, but weights were included to emulate the affects of the actual equipment. The air-brake compartment is constructed with 4 in x 4 in box tubing in a rectangular configuration. The air brake compartment and the front cab structure are welded to the sill plate, which is constructed of a 1 in thick plate spanning the length and width of the cab. The sill plate is in turn welded to I-Beams (sills), which are the main support.

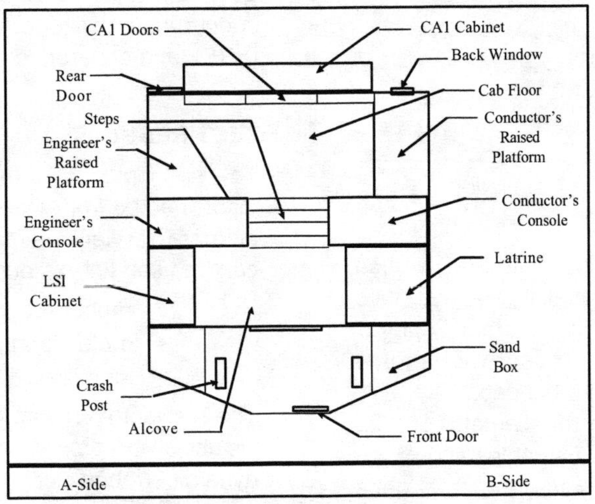

Figure 1. Plan View of Locomotive Cab

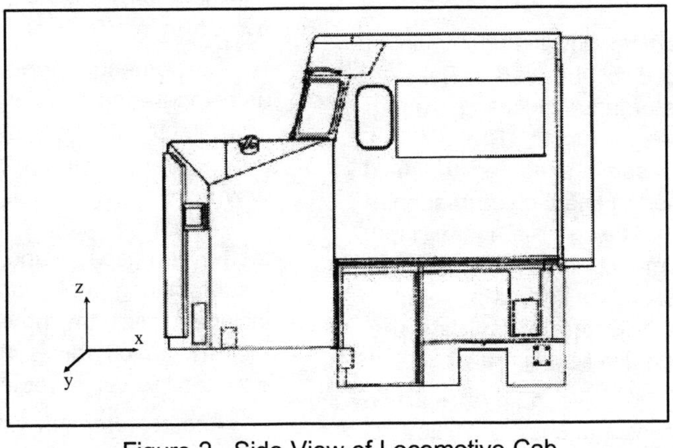

Figure 2. Side View of Locomotive Cab

Cab Installation: The cab was placed on four Goodyear airsprings, model # 1B14-350, in the configuration shown in Figure 3. Each airspring is rated for a 10,500 lb static load. Airsprings are used to isolate the cab from the floor and dynamically simulate actual installation of the locomotive cab. The natural frequency for the cab suspended on the airsprings is in the range of 1-2 Hz. Airsprings are bolted to the sill and the lab floor to ensure that the cab does not slide across the floor, when it is shaken.

Figure 3. Goodyear Airspring Installation

ACTUATION SYSTEM: In order to simulate the vibration input in the field, the cab was excited by a hydraulic actuation system consisting of a hydraulic pump, manifold, and actuator. A hydraulic actuator with a force capacity of 2000 lb was used to shake the cab. The MTS 249 actuator used in our tests has a rod diameter of 1.25 in , and two swivel ends to eliminate lateral loading of the actuator. The actuator is equipped with an internal delta pressure cell and linear voltage differential transformer (LVDT) for measuring force and displacement, respectively.

EXCITATION INPUT: The next step involved determining what type of signal was needed for the tests. The cab was excited with various types of input, including random signal, pink signal, and pure-tone input. The random excitation provides nearly the same effect as the field input, while the pure tone is used to find the cab response at individual frequencies. The sine-random input combines the effect of random and pure tone input to emulate the peaks that are observed in field data. In each case, the intensity is properly adjusted to duplicate the noise and vibration levels observed in field operation.

DATA ACQUISITION SYSTEM: The data acquisition system used for vibration measurements consisted of accelerometers, a data acquisition system, and an HP workstation. The accelerometers were provided by PCB Piezotronics, Inc., [18]. The accelerometer, model number 333A, has a voltage sensitivity of 100 mV/g, along with a measurement range of ±50 g peak. The accelerometers are capable of taking measurements in the frequency range of 1-1000 Hz. The accelerometers were mounted to the cab using either single axial or tri-axial mount blocks with hot glue.

An HP Series 300 workstation and a Zonic 7000 data acquisition system were used to acquire, filter, and store the data. The Zonic 7000 data acquisition system is suited for applications in multi-channel real-time signal processing, multi-shaker modal analysis, acoustic analysis, and structural analysis. The signal processing capabilities provide 4-1000 parallel data acquisition channels, built-in 120 dB/octave anti-aliasing filters, variable FFT block size up to 8K, swept sine, and order tracking. All of these capabilities allow for frequency analysis ranges up to 150 kHz, real-time analysis rates to 20 kHz, and linear spectrum, power spectrum, auto spectrum, cross spectrum, frequency response function (FRF), and coherence measurements.

TEST RESULTS

To soft mount the cab, it must be completely disconnected from the sill structure and mounted on three or more resilient elements. The resilient elements that can be used for this purpose include

- rubber mounts
- metal springs
- airsprings

The resilient elements must be capable of supporting the weight of the cab structure and providing isolation in the dynamic range of isolation. After considering all possibilities, elastomeric mounts were selected for this study. This selection was based on size, installation configuration, shear stability of the cab, and parts availability.

Elastomeric mounts proved to be the most compact, as compared to airsprings and coil springs. Although it is possible to design airsprings and coil springs that can provide the proper static support and dynamics isolation, they have a larger configuration than elastomeric mounts. This larger size could present problems in installation for test purposes and, more importantly, actual manufacturing implementation. The installation problems include finding the physical space to accommodate the mounts, which raises the cab too high and, therefore, exceeds the height requirements of the locomotive.

Another problem with airsprings and coil springs is shear stability. The resilient element is required to provide isolation in the vertical direction and yet have sufficient stability in the longitudinal and lateral directions (i.e., shear directions of the mount) so that the cab does not experience large displacement when it is subjected to lateral and longitudinal forces. Such forces can occur in coupling and uncoupling locomotives to the train, or in curving. Airsprings and coil springs have a minimal amount of shear stiffness and, therefore, are far less stable than elastomeric mounts in these directions. For elastomeric mounts, it is possible to adjust the shear stiffness properties of the mount through proper design of the mount dimensions, elastomeric compound, and shims.

In terms of availability of parts for our testing, the elastomeric mounts were found to be the most favorable. Although a series of mounts were found in a Lord Corporation catalog to satisfy our needs, special coil springs and airsprings were custom ordered at substantial cost. Therefore, two different series of mounts were chosen from Lord Corporation, Erie, PA, [19]. After working with Lord to determine the dynamic properties of the mounts in all 6 directions of motion, we selected the mounts in Table 1 along with the mount requirements in Table 2. Figure 4 illustrates the mount placement in the locomotive cab, while the base mounts and crash-post mounts are shown in Figs. 5 and 6, respectively.

Table 1. Mount Part Numbers and Locations

Part Number	Location
CB - 2205-3	Base Mounts
J-14056-4	Crash-Post Mounts

Table 2. Requirements for Soft Mounts

Requirements	X	Y	Z
Base Mount Stiffness (lb/in)	0.9600E4	0.9600 E4	0.1670 E5
Crash-Post Mount Stiffness (lb/in)	0.8500 E3	0.8500 E3	0.5100 E4

The crash-post mounts were installed to provide a higher stiffness in the longitudinal direction. The shear stiffness of the base-mount parts proved insufficient to prevent large displacement of the cab during coupling and uncoupling of the locomotive, when the longitudinal forces are relatively large.

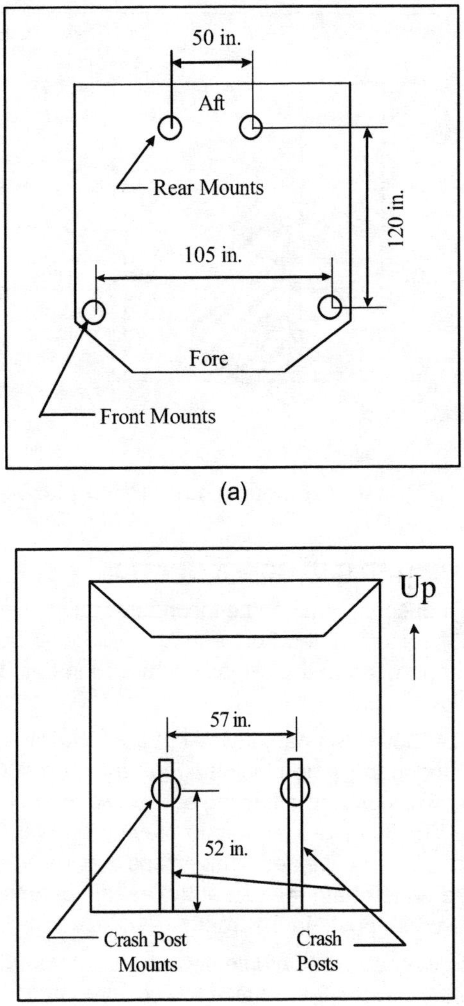

Figure 4. Mount Configuration (a) Plan View (b) Front View

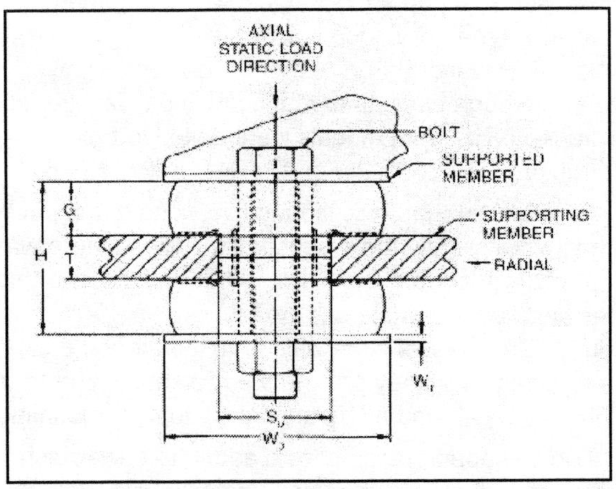

Figure 5. Base Mounts, Lord P/N CB-2205-3 [19]

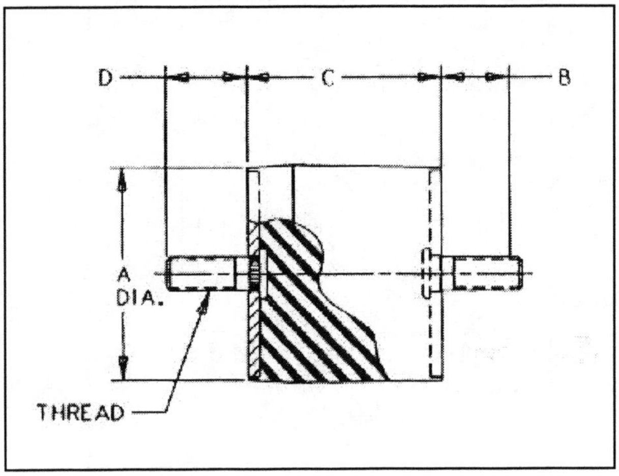

Figure 6. Crash-Post Mounts, Lord P/N J-14056-4 [19]

SOFT-MOUNTED CAB CONFIGURATION

In its original configuration, the locomotive cab is welded to the sill structure at the front and at four needle beams that are connected to the sill, as discussed in detail in this section.

To soft-mount the cab, it was cut from the sill structure along the welded joints. Figure 7 shows the joints that were burned off, using an acetylene torch process. The joints are marked in gray color. Not shown in the figure are the welded connections that were eliminated underneath the cab at the needle beams and along the crash posts in the front.

Once the cab was completely separated from the sill, proper modifications were made to accommodate installing the elastomeric mounts at the four base locations and the two crash posts. As shown in Figure 8, for the front base mounts, an L-shaped bracket was bolted to the cab on each side. The plate thickness for the bracket was selected such that it fit the requirement for the two-piece center-bonded mount that were selected for the base installation. Furthermore, the bracket was stiffened through gussets to reduce residual flexures at the bracket. The front mounts were bolted to the cab and sill structure, as shown in Fig 8.

As shown in Figure 9, the modifications to the back of the cab were slightly more substantial. Since there was no structure adequate for installing the mounts, two thick plates were added to the cab structure. The 10.75 in x 8.0 in x 0.75 in plates were welded to the air brake cab structure such that they provide a sufficient support for the weight of the cab and adequate rigidity for soft mounting.

Two L-shaped brackets were added to each crash post for mounting the tube-form elastomeric mounts that were selected for these locations. As shown in Figure 10, each bracket has a gusset to provide sufficient rigidity, as well as a slot that can easily accommodate the installation of the mounts. The brackets on the cab side were welded after the base mounts had been installed on the cab in order to properly align them to the brackets on the crash post. In a production environment, however, this can be done by properly incorporating the height adjustment into the installation at the crash-post mounts.

Figure 7. Welded Joints Eliminated for Soft-Mounting the Cab

Figure 8. Front-Mounting Bracket Installation

Figure 9. Modifications to the Cab for Installing Aft Base Mounts

Figure 10. Crash Post Mount Brackets

SOFT-MOUNTED CAB TEST RESULTS

A series of tests were run on the soft-mounted cab to establish the benefits of soft mounting in terms of reducing interior noise and vibrations. The tests for the soft-mounted cab are shown in Table 3, along with the tests conducted on the original cab.

<u>Cab Floor Measurements:</u> Thirteen accelerometers are placed on the cab floor, in the arrangement shown in Figure 11. for the purpose of this paper, we will analyze the accelerations at selected locations—specifically at locations 1, 4, 6, 10, and 11—for the soft- and hard-mounted cab. The accelerations at cab floor location 1, Figure 12, show the following characteristics. The soft-mounted case shows that the vibration levels are minimized. The resonance peaks are eliminated across the spectrum, especially at 120 and 160 Hz.

The accelerations at cab floor location 4, Figure 13, show that the vibration levels in the hard-mounted case are higher than in location 1. Soft-mounting the cab virtually eliminates the vibrations and resonance peaks. The large peak at 50 Hz, which was one of the major problem areas, is nearly eliminated.

The accelerations at cab floor location 11, shown in Figure 14, indicate the same trends as those for location 4. Soft-mounting the cab is extremely successful at this point, little additional improvement could be accomplished here.

The accelerations at engineer's console location 10, Figure 15, show the following characteristics. The vibration levels are reduced to relatively zero. The high frequency vibrations are affected the greatest. The two resonance peaks at 12 and 50 Hz are eliminated, which eliminated the resonance at the engineer's console.

Table 3. Soft-Mounted Cab Tests

Vibration Testing	Baseline	Soft-Mounted
Floor	X	X
Roof	X	X
Nose Cab Floor	X	X
CA1 Doors	X	X
CA1 Outside		X
Consoles	X	X
LSI Cabinet		X
Sills		X
Nose Cab	X	
B-Side Outside		X

The accelerations at conductor's console location 6, Figure 16, show that the peaks at 12 and 50 Hz are eliminated at the conductor's console. The consoles originally resonated at 12 Hz. This is eliminated with the soft-mounted structure. Further, a major resonance peak at 220 Hz is also significantly reduced. Although a few of the peaks—for instance, those at 75, 95, and 140 Hz—are shifted, our subjective evaluation did not show any degradation of the noise characteristics inside the cab.

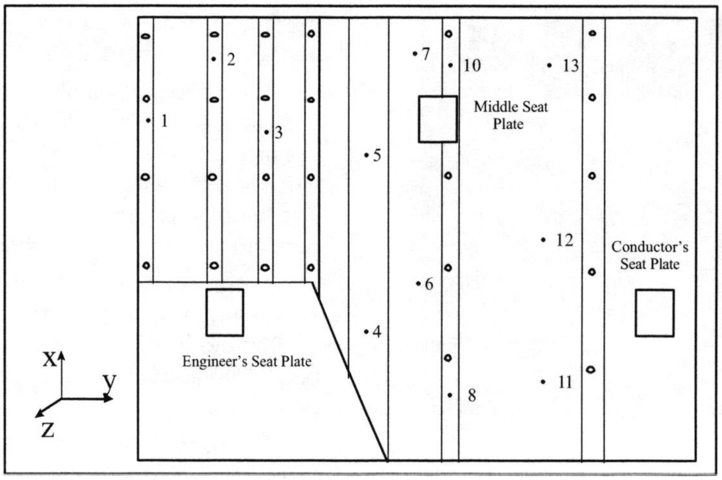

Figure 11. Floor Accelerometer Locations

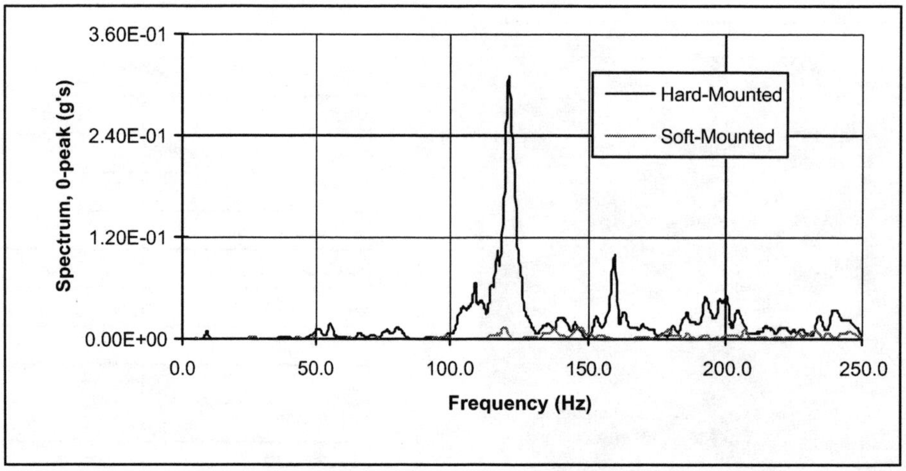

Figure 12. Accelerometer Cab Floor Location 1

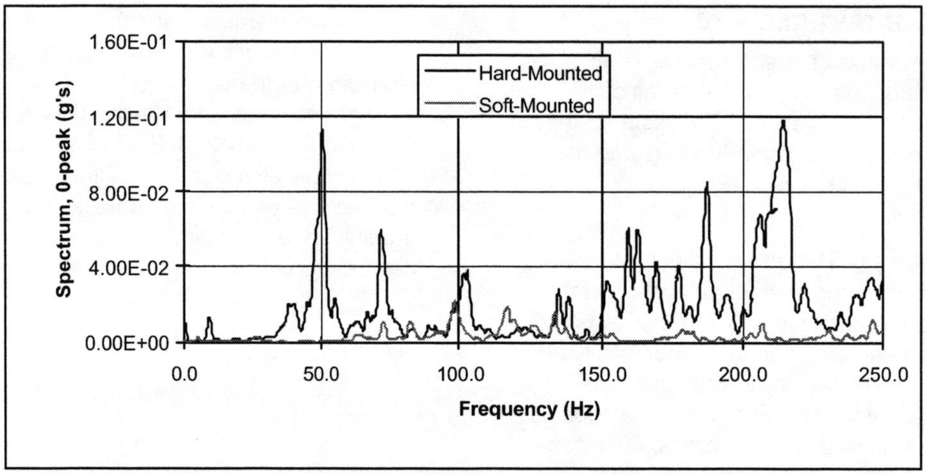

Figure 13. Accelerometer Cab Floor Location 4

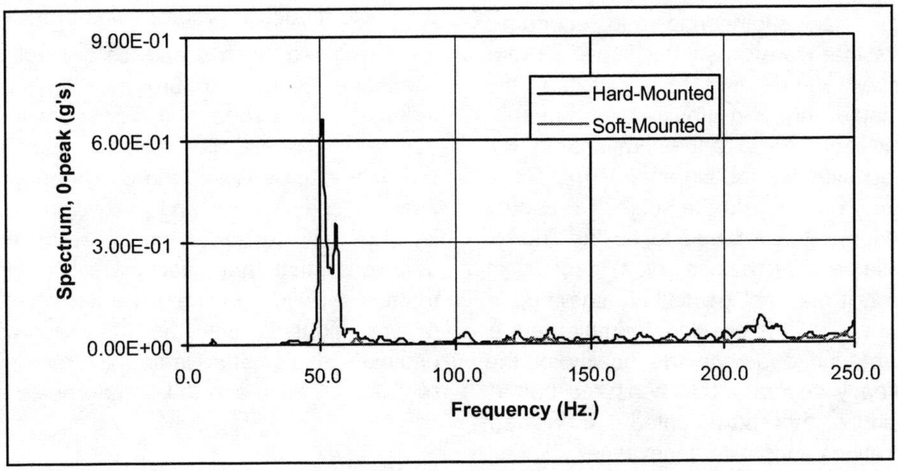

Figure 14. Accelerometer Cab Floor Location 11

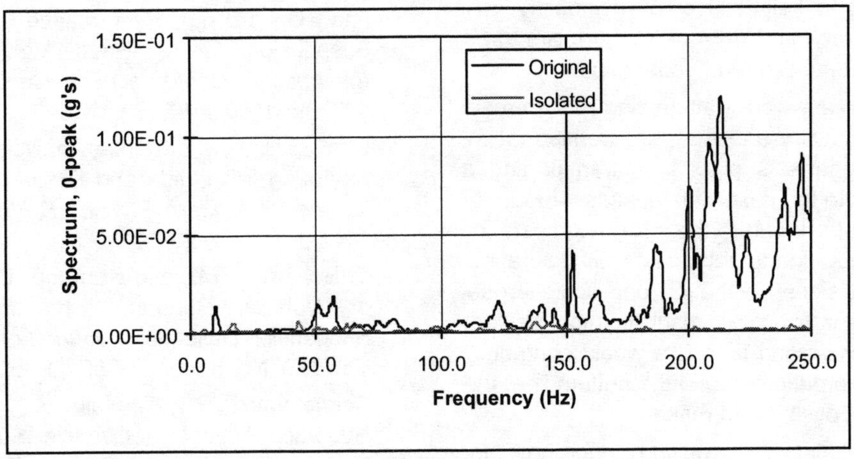

Figure 15. Accelerometer Engineer's Console Location 10

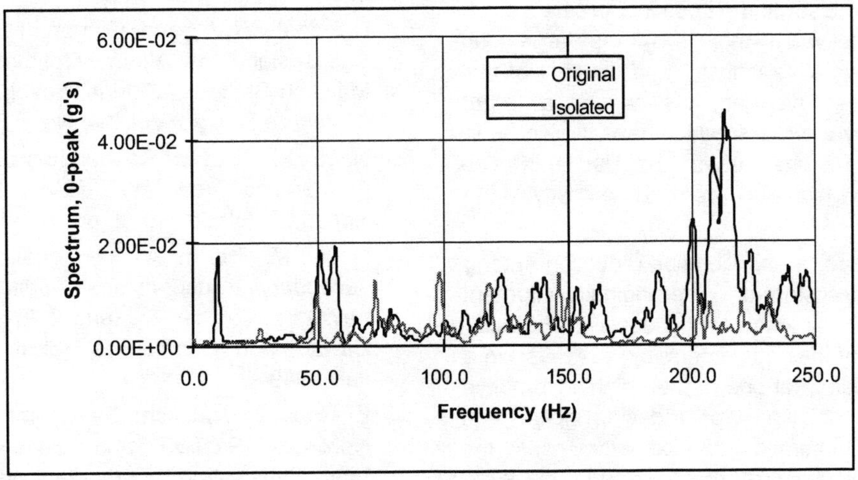

Figure 16. Accelerometer Conductor's Console Location 6

Discussion of Results

We will next undertake an engineering and economic evaluation of the results that were presented above. From an engineering standpoint, there is little doubt that the solution that was evaluated here can provide a remarkable amount of reduction in both noise and vibrations in the locomotive cab. The acceleration measurements that were presented in this study do not capture the subjective aspect of the noise and vibration, often referred to as the "noise quality." A low noise quality has a high degree of annoyance, even though it may not particularly show up in the noise spectra. In practice, it is possible to implement a solution that has a profound effect on the quality of the noise, but does not greatly decrease the measured noise. Our subjective evaluation of the soft-mounted cab showed that it lowered both vibrations and noise annoyance.

It is, however, worth noting that in spite of the vast improvements that are documented here, soft mounted cab are not widely adopted in freight locomotive industry. A reasonable question to ask is: Why not? There are three reasons: economics, economics, and economics!

As the U.S. railways are attempting to remain competitive with their industry and the road transportation industry, the main emphasis is placed on transporting a cargo from one location to the next, with the least amount of expenses. Anything that will add to the operating costs, or does not directly contribute to the bottom line will take a low priority in operating decisions. Although one could argue that lower noise and vibrations will result in more rested, satisfied, and happier crew; and therefore would contribute to the railway industry's productivity and the bottom line, the link is difficult—if not impossible—to make.

At the railways level, the buying decisions for locomotives are not made by the crew, but by executives who squarely have their eyes on the bottom line. Considering improved noise and vibrations in the recent decade have been motivated mostly by the large gap between the freight locomotive industry and the rest of the transportation industry. This gap has caused a larger number of complaints by the ranks, which have shown up in labor negotiations. In some cases, the gap has also brought about lawsuits, which of course can be costly to the industry.

The low priority for noise and vibration reduction among the railways is directly transferred to the original equipment (OE) manufacturers. Their engineering decisions directly reflect the preferences of their customers: the railways. In a recent development program at one of the freight locomotive OEs for incorporating solutions for reducing noise and vibrations in the next generation of locomotive cabs, the question was asked by one of the executives as to "Have we lost any orders as a result of not having this option in our locomotives?" To which the answer was "No." Shortly after, the noise reduction program was promptly cancelled.

Of course, such scenarios are not limited to noise and vibrations issues, nor are they limited to the locomotive industry. As anyone who has spent some time in industry can attest, there is often a disconnect between the business decisions and the engineering facts. The issue of interior cab noise and vibration reduction in freight locomotive industry is no different than other business decisions that are disconnected from the engineering matters. Until more improvements are demanded by the railway industry, there is no reason by the locomotive original equipment manufacturers to change their engineering approach to lowering noise and vibrations in the cab.

REFERENCES

1. Gao, J., and Leighton, N.J., and Morgan, C., "Modeling and Open-Loop Performance of a Novel Active Suspension for a Road Vehicle," *Proc. UKACC International Conference on Control 1996*, Vol. 2, No. 427, pp. 1160-1165, Sept. 1996.
2. Choi, S.H., "Performance Analysis of an Engine Mount Featuring ER Fluids and Piezoactuators," *International Journal of Modern Physics B*, Vol. 10, No. 23-24, pp. 3143-3157, July 1995.
3. El Beheiry, E.M., and Karnopp, D.C., "Optimal Control of Vehicle Random Vibration with Constrained Suspension Deflection," *Journal of Sound and Vibration*, Vol. 189, No. 5, pp. 547-564, Feb. 1996.
4. Shimamune, R., Tanifuji, K., "Application of Oil-Hydraulic Actuator for Active Suspension of Railway Vehicle: Experimental Study," *Proceedings of the 34th SICE Annual Conference*, pp. 519, 1335-1340, July 1995.
5. Dukkipati, R.V., and Osman, M.O.M., "Adaptive and Suspension to Attain Optimal Performance and Maintain Static Equilibrium Level," *International Journal of Vehicle Design*, Vol. 14, No. 5-6, pp. 471-496, 1993.
6. Nagai, M., "Recent Researches on Active Suspensions for Ground Vehicles," *JSME International Journal*, Series C, Vol. 36, No. 2, pp. 161-170, 1993.
7. Nagai, M., Moran, A, and Tanaka, S., "Optimal Active secondary Suspension Applied to Magnetically Levitated Vehicle Systems," *Proceedings of the 28th SICE Annual Conference*, Vol. II, pp. 1396, 1167-1170, July 1989.
8. El Madany, M.M., and El-Tamimi, A., "On a subclass of Nonlinear Passive and Semi-Active Damping for Vibration Isolation," *Computers and Structures*, Vol. 36, No. 5, pp. 921-931, 1990.

9. von Flotow, A. H., "An Expository Overview of Active Control of Machinery Mounts," *Proceedings of the 27th IEEE Conference on Decision and Control*, Vol. 3, pp. 2029-2032, 2508, Dec. 1988.
10. Grass, D. E., "Flexible Air Springs do Multiple Duty," *Machine Design*, Vol. 60, No. 7, pp. 107-110, April 1988.
11. Alexandridis, A. A., and Weaver, T. R., "Active Vibration Isolation of Truck Cabs," *Proceedings of the 1984 American Control Conference*, Vol. 3 pp. 1199-1208, June 1984.
12. Hansson, P. A., "Control Algorithms for Active Cab Suspensions on Agricultural Tractors," *Vehicle System Dynamics*, Vol. 25, No. 6, pp. 431-461, June 1996.
13. Hiromatsu, T., Inaba, T., and Matsuo, Y., "Development of an Electromechanical Active-Cab-Suspension," *Proceedings of the IECON 1993*, Vol. 3, pp. 2132-2137, Nov. 1993.
14. Matsuo, Y., Inaba, T., and Hiromatsu, T., "Control of an Active Suspension for a Truck Cab," *Transactions of the Society of Instrument and Control Engineers*, Vol. 30, No. 1, pp. 54-63, Jan 1994.
15. El Madany, M. M., and El Razaz, Z. S., "Performance of Actively Suspended Cabs in Highway Trucks-Evaluation and Optimization," *Journal of Sound and Vibration*, Vol. 126, No. 3, pp. 423-435, Nov. 1988.
16. Sankar, S., Rakheja, S. and Venkataraman, K., "Computer-Aided Design Tool for Analysis of Off-Road Vehicle Cab Suspension," *Proceedings of the 7th World Congress*, Vol. 2, pp. 949-953, Sept. 1987.
17. Clark, R. M. " Stationary Locomotive In-Cab Noise Measurements," *Sound and Vibration*, Vol. 12, No. 5, pp. 18-19, 22-23, May 1978.
18. PCB Piezotronics, Inc., Depew, NY 14043-2495, (716) 684-0001.
19. *Engineering Guide and Catalog for Vibration, Shock and Noise Control Products*, Lord Corporation, 2000 West Grandview Blvd., P.O. Box 10038, Erie, PA, 16514-0038, (814) 868-3109.

Proceedings of IMECE'03
2003 ASME International Mechanical Engineering Congress
Washington, D.C., November 15–21, 2003

IMECE2003-41812

Event Data Recorder Pre-Crash Data Sources for General Motors Vehicles

John C. Steiner; KEVA Engineering

ABSTRACT

The Event Data Recorder (EDR) found in some 1994 model year and newer General Motors (GM) passenger vehicles has the ability to record up to five seconds of pre-crash data. Such as vehicle speed, engine speed, percent throttle application and brake application before a predetermined deceleration event as well as crash data such as delta-v's. The pre-crash and crash data can be downloaded by properly equipped and trained personnel using the Vetronix Crash Data Retrieval (CDR) System. However, this data must not simply be taken at face value; the Accident Analyst must be aware of the nature of the differing types and sources of the data, and must ensure that the systems supplying the data was in a normal operating mode and not in a default mode or in a "limp home" mode due to pre existing problems.

This paper discusses how different environments and scenarios that the vehicle can be operated in changes how the vehicle will respond to driver inputs thus effecting pre-crash data recorded by the Event Data Recorder.

EDR HISTORY

For many years the National Highway Traffic Safety Administration (NHTSA) has actively collected data on collision dynamics and pursued research of occupant injuries resulting from vehicle collisions. According to the National Center for Statistics and Analysis, there were 37,338 fatal crashes and 6,266,000 nonfatal crashes in the year 2000 [5]. If useful and reliable data could be obtained from these real-life accidents, a significant amount of knowledge could be gained in the study of vehicle dynamics and occupant injuries during an accident.

For this reason, the National Transportation Safety Board (NTSB), and the National Aeronautics and Space Administration's Jet Propulsion Laboratory (JPL) in 1997 urged NHTSA to look into the feasibility of equipping passenger cars with black box type devices to help understand the dynamics of a vehicle collision and occupant injuries.

General Motors (GM) has taken a leading role in the Event Data Recorder (EDR) field and has equipped a wide number of makes and models with these devices. Additionally, the hardware and software package used to download this data is readily available to the general public for purchase and use with proper training.

SYSTEMS OVERVIEW

As vehicle manufacturers worked to meet stringent State and Federal regulations on efficiency and pollution, vehicles have become more and more sophisticated with numerous on-board sensors that constantly supply the on-board processors with data that is used to run the vehicle in a clean and efficient manner. A large amount of data is flowing back and forth throughout the vehicle's "local area network" or Serial Data Bus any time the vehicle is running. The existence of these sensors and processors made it easy for General Motors to add an Event Data Recorder function to its airbag control modules known as Sensing and Diagnostics Modules (SDM).

This data, which is the vehicle's operating parameters, can have a "snapshot' taken that represents up to five seconds before a deceleration event. The data is collected and recorded from the vehicle's operating systems such as the Powertrain Control Module and Brake Control Module and is stored in the memory of the EDR. It is also important to understand that these systems and the technology they are based on are constantly changing.

When a late-model General Motors vehicle is in operation, each of the vehicle's systems are in constant communication with each other: to run the engine and transmission in the most optimal configuration possible for the current load and environmental conditions and to keep the driver and passengers comfortable.

For example, the Vehicle Speed sensor (VSS) is reading the number of revolutions of the transaxle output shaft to calculate the vehicle's current speed and display it for the driver on the speedometer. That same data is being sampled and compared to other parameters such as throttle position, torque requests from the Electronic

Brake and Traction Control Module (EBTCM) and torque outputs. Even the radio is sampling the vehicles velocity sensor to adjust the radio's volume as the speed increases. The use of the speedometer cable to provide useful information for the driver has been replaced by the VSS, which not only provides the same information for the driver but supplies the vehicle and all of its systems with the same data as well. The VSS is just one of the sensors from which the vehicle's Sensing and Diagnostics Module (SDM) samples and records pre-crash data.

All of this is not done with just one on-board computer. A late-model GM vehicle can have two or more computers working in unison that allow the vehicle to operate in a variety of operational states.

THE VEHICLE'S "LOCAL AREA NETWORK"

There are two different formats of serial data that are broadcast over the vehicle's serial data bus. Universal Asynchronous Receiver/Transmitters (UART) devices were widely used on General Motors vehicles and have mostly been phased out by 1996. Starting with most 1996 and newer GM vehicles, Class 2 serial data is widely used. However, there are still some vehicles today that utilize both UART serial buses and Class 2 serial buses.

The nature of UART serial data is that it travels along wires in "single file line". So, one command at a time is the only way that data can travel through a UART serial data bus. Very similar to a party line, if a UART device needs to send information to another UART device, the sending device must wait for the line to clear before it can send its data.

UART data is an organization of bits that are composed of a header bit, message length, mode bit, the actual data and a check sum bit.

Class 2 serial data is still limited in speed by the limitations of "one bit at a time". However the Class 2 serial bus is more advanced than the UART serial bus in that instead of operating similar to a "Party Line", there is a "Switchboard Operator" that is in between each Class 2 device and the serial bus. The "Operator" controls when each Class 2 device can communicate over the bus by monitoring not only when the bus is in use but also what devices with different assigned priorities are attempting to communicate over the bus.

The pre-crash data is constantly being sampled by the SDM until the SDM senses a 1 to 2 g deceleration and gives an "Algorithm Enable" (AE) command. This triggers the SDM to write it to the SDM's electrically erasable programmable read-only memory (EEPROM). In recent crash tests, it has been found that the SDM can continue to capture data after the AE command and if the crash pulse is long enough, actual crash data can be recorded in the pre-crash variable data tables [1]. Because of the nature of the serial bus and the data traveling "one bit at-a-time", each channel of the pre-crash data can be skewed from each other by up to one second.

ON BOARD DIAGNOSTICS (OBD) SYSTEMS OVERVIEW

There are two or more control modules that supply pre-crash data to the SDM. Those systems are the Powertrain Control Module (PCM) and the Electronic Brake and Traction Control Module (EBTCM).

Every time the vehicle is started by the operator, each control module performs its own diagnostics test to ensure that all systems are functioning normally. If a fault is found anywhere in the system, the control module in question will store a Diagnostic Trouble Code (DTC) into memory and depending on the type of DTC stored, the "Malfunction Indicator Lamp" (MIL) could be turned on. There are some DTC's that can be set by any of the modules that will not trigger the malfunction indicator lamp to come on. The Electronic Brake and Traction Control Module (EBTCM) will illuminate the "ABS" MIL if a problem is found. The Sensing and Diagnostics Module will illuminate the "AIRBAG" MIL if a problem with the airbag system is found.

POWERTRAIN CONTROL MODULE (PCM)

The vehicle's drivetrain (engine and transmission) can be segregated more or less from the rest of the vehicle. The operation of the powertrain is monitored and controlled by the Powertrain Control Module (PCM). Powertrain control is accomplished by monitoring input parameters from a long list of sensors mounted throughout the vehicle such as intake air temperature, barometric pressure, manifold absolute pressure, throttle position, engine speed, engine timing, engine coolant temperature, etc. All of these parameters and many more are monitored by the PCM so that it can accomplish its four main tasks, which are:

- Fuel Metering System
- Transaxle/Transmission shifting
- Ignition Timing
- On-Board Diagnostics of powertrain functions.

The PCM is in constant communications with a number of sensors to control all of the systems that have a direct affect on vehicle performance. General Motor's refers to this model of PCM as a PCM 32U with an integral Knock Sensor (KS) System.

The PCM uses Electrically Erasable Programmable Read-Only Memory (EEPROM) to store programs, program information and data calibrations that are required for powertrain operations and diagnostics. The EEPROM is not removable from the PCM therefore if an internal problem arises within the data stored in the EEPROM, the vehicle's PCM must be re-programmed.

If a fault is detected somewhere in the PCM, a Diagnostics Trouble Code (DTC) is written to the PCM's memory and the PCM will trigger the Malfunction Indicator Lamp (MIL) on the instrument panel to illuminate which reads "SERVICE ENGINE SOON" or "CHECK ENGINE" notifying the driver/owner of the vehicle that service is required.

The PCM is located inside the intake air cleaner housing assembly at the left/front of the engine compartment in some GM vehicles.

ELECTRONIC BRAKE AND TRACTION CONTROL MODULE (EBTCM)

A majority of General Motors vehicles come equipped with Antilock Braking (ABS) systems and some vehicles are also equipped with Traction Control (TC). The processor that controls antilock braking functions and traction control functions is known as the Electronic Brake Traction Control Module (EBTCM). If the vehicle is not equipped with Traction Control, the control module is then referred to as the Electronic Brake Control Module (EBCM). The EBTCM has its own system of sensors mounted at each wheel of the vehicle to monitor for wheel lockup and wheel spin. The primary functions of the EBTCM can be summarized in the following list:

- To detect any wheel slip
- To detect wheel speed differences
- Controls the brake system while in ABS mode or TC mode
- Controls the Variable Effort Steering System (MAGNASTEER) (if equipped)
- Constantly monitors system for proper function

For most vehicles, the Antilock Brake Function of the EBTCM is accomplished with the Delphi 6 Antilock Brake System. The ABS system utilizes the wheel speed sensors to detect excessive reduction of wheel speeds, and modulate the application of brakes to prevent wheels from locking up during hard brake applications or braking on low mu (friction) surfaces.

There are two traction control systems that a GM vehicle can be equipped with. Standard system is simply known as the Enhanced (Electronic) Traction System (ETS) and can be found on lower trim level cars. A more advanced traction control function found on higher trim GM vehicles is the Traction Control System (TCS).

The ETS monitors the ABS wheel speed sensors for wheel spin. If one or both of the drive wheels begin to spin faster than the non-driven wheels, the EBTCM can send a requested wheel torque value to the PCM via a Pulsewidth Modulated (PWM) signal. The PCM can generate a lower torque output from the drivetrain by implementing one or all four of the methods below:

- Ignition timing is retarded
- Up to three fuel injectors can be shut off
- Up-shifting the transmission
- Shut off supercharger (if equipped)

These four methods will be repeated by the PCM until the wheel spin can be controlled.

The TCS also monitors the ABS wheel sensors for excessive speed at one or both of the drive wheels. When the TCS detects that one or both of the drive wheels are spinning faster than the non-driven wheels, the TCS can request lower torque outputs from the PCM just as the ETS does. The TCS can also apply brakes to either one or both of the drive wheels to control wheel spin by manipulating the inlet and outlet valves of the Brake Pressure Modulator Valve (BPMV).

If any problems are found with the TCS or ETS systems, both systems will be disabled by the EBTCM, a DTC is stored in the memory of the EBTCM and the ABS MIL is illuminated as well as the "TRAC OFF" warning lamp. The TCS/ETS system can also be disabled by the PCM if any problems within the PCM are detected.

In most vehicles, the EBTCM is mounted to the firewall at the left/rear of the engine compartment in the vicinity of the brake master cylinder.

SENSING AND DIAGNOSTICS MODULE (SDM)

Perhaps the module of greatest interest for crash analysis is the airbag module or the Sensing and Diagnostics Control Module (SDM). This processor has the function of determining the readiness of the airbag system, sensing decelerations for airbag deployment and writing to memory DTC's and illuminating the "AIRBAG" warning light if any problems are detected.

The SDM is only one component of the entire Supplemental Inflatable Restraint (SIR) System. The SIR system also consists of the inflatable restraint steering wheel module, inflatable restraint steering wheel module coil, inflatable restraint IP module and the AIRBAG warning lamp located in the instrument cluster.

The SDM has many critical tasks to accomplish when the vehicle is in operation. During normal vehicle operation and during vehicle starting, the SDM performs self-diagnostics tests to check for any possible problems with the SIR system. If any problems are found, the SDM will set a Diagnostic Trouble Code (DTC) and notify the driver of a malfunction in the SIR system by illuminating the AIRBAG Malfunction Indicator Lamp.

The SDM is also responsible for maintaining an approximate 23-volt power reserve to fire the airbags in the event that the vehicle's power is compromised during the collision. This backup power circuit is referred to as the 23-VLR energy supply and its status can be checked with a scan tool.

Frontal crash detection is no longer accomplished with two or three deceleration sensors mounted at the front of the vehicle. The SDM has a built in accelerometer that senses forward decelerations. These decelerations are converted into an electrical signal and compared to values stored in memory. If the input signal is greater than the values programmed into memory, the SDM can issue an "Algorithm Enable" command as well as a fire command to deploy the airbags.

The SDM also has a built in "Event Data Recorder" to record pre-crash and crash data in the event of a significant deceleration event.

Typically, the SDM can be found mounted either below the driver's seat or below the right/front passenger seat. In some vehicles such as the Chevrolet Corvette, the SDM is mounted at the center, base of the instrument panel below the radio and HVAC controls. Some disassembly of the vehicle is required to reach the SDM if data cannot be retrieved through the Diagnostics Link Connector (DLC).

THE EVENT DATA RECORDER

The Event Data Recorder has been designed to record three different types of events:

- Non Deployment Event
- Deployment Event
- Deployment Level Event

A "Non Deployment Event" is a deceleration event that the vehicle in question experiences which is strong enough to "wake up" the SDM or place the SDM in "Algorithm Enable (AE)" mode however the deceleration is NOT strong enough to deploy the airbags. Typically, a "Non Deployment" can be triggered by a deceleration of the vehicle of 1.5g's for 2 milliseconds.

"Non Deployment" Events are not permanently stored by the SDM. They can be erased by two methods. A Non Deployment Event can be erased by another Non Deployment event of greater severity than the currently stored Non Deployment or it will be erased after 250 ignition cycles. One ignition cycle counts as the ignition being turned from the "OFF/LOCKED" position to the "RUN" position and allowing the SDM to go through its "boot-up" phase that takes approximately 7 to 10 seconds.

It is very important to keep in mind that when the SDM has been removed from the vehicle and has been disconnected from a 12-Volt power source, it is still active for up to two minute after being disconnected. The SDM's power reserve module will have power to keep the accelerometer functioning and the EDR portion of the SDM. So if the SDM is mishandled, the accelerometer could register the mishandling as a vehicle deceleration and record a Non Deployment or Deployment Event and overwrite any previous near deployments event.

The second type of event that will trigger the SDM to record data is a true airbag deployment event. A deployment event is recorded if the vehicle has decelerated enough to deploy the vehicle's airbags. A deployment event cannot be erased from an SDM unlike a near deployment event. Once an SDM fires the airbags, the Event Data Recorder records pre-crash and crash data and a "latched DTC" is written, thus making the SDM unusable again in a vehicle.

A Deployment Level event is a second Deployment Level event that occurs after the airbags have already been fired but within 5 seconds of the original Deployment Event.

A significant quantity of data is being recorded to the SDM in the event of a non deployment or a deployment event so where exactly in all of the above-mentioned systems is this data coming from? How reliable is this data and how can these systems be checked?

SDM DATA SOURCES

VEHICLE SPEED

In the analysis of a collision event, having insight into the involved vehicles' pre-impact speeds can be crucial in the reconstruction of the accident. A significant number of airbag equipped GM vehicles have the ability to store pre-impact velocity on the SDM when a near deployment or deployment event is logged onto the memory of the SDM.

The velocity data is generated by the Vehicle Speed Sensor (VSS) that is generally mounted in the transaxle case. The VSS is basically a pulse generator that consists of a speed sensor mounted in the transaxle case, which reads a toothed rotor that is pressed into either the left or right output shaft of the transaxle.

As the vehicle rolls forward, the final drive assembly and the output shafts begin to rotate as well. This produces a voltage signal in the speed sensor, which is proportional to the vehicle's velocity. The speed sensor takes the voltage signal and feeds the Powertrain Control Module (PCM) AC voltage from 0.5 volts at 100 RPM (at the transaxle output shaft) to 100 volts or more at 6000 RPM. With this AC voltage, the PCM can calculate the vehicle's speed (in either MPH or KPH) and control the powertrain with the data and display the vehicle's speed on the speedometer for the driver.

It is important to remember that this system operates under the assumption that stock tire and rim sizes are mounted on the vehicle. Any time the tire sizes have been modified, the output of the VSS will be affected. Consumers who own light trucks and sports cars such as the Camaro, Firebird or Corvette might change rear axle ratios for performance changes. If this occurs, the PCM must be updated with the new ratio information just as if the tire sizes have been changed. The EBTCM cannot compensate for drastic changes in the size of the tires and rims mounted on the vehicle; however the EBTCM can compensate for the spare tire. Some vehicles can be reprogrammed as to what size tire is mounted on the vehicle.

The full scale of the vehicle speed that can be recorded by the SDM is from 0 MPH to 158.4 MPH with a resolution of 0.6 MPH. The accuracy of the vehicle speed data is +/- 4% and the vehicle speed data is updated if the vehicle speed changes by 0.1 or more MPH. [2]

When the SDM's crash data is downloaded and the pre-crash vehicle speed data is analyzed, it is important to correlate the recorded pre-impact speed with the outcome of the accident. Does the damage for that velocity make sense? Do the post crash trajectories of the vehicles make sense for the closing speeds? It is important to understand how the physical placement of the sensor can cause excessively high or low readings to be recorded by the SDM due to the conditions the vehicle is operating in and due to inputs requested by the driver during evasive maneuvers and how the vehicle responded. If the pre-crash speed data does not make sense, the environment that the vehicle was operating in at the time of the accident could explain it.

In many GM vehicles, the driver has the option of shutting off the TCS by depressing the TRAC button on the instrument panel. If the driver deactivates the TCS, the "TRAC OFF" warning lamp will be illuminated on the instrument cluster until the driver reactivates the system. This is an important factor to consider when analyzing the SDM's Pre Crash Data as the traction control system can have a significant effect on the vehicles response to the driver's inputs thus affecting the pre-crash data that could be recorded by the SDM.

For example, if a driver of a 2000 Pontiac Grand Prix equipped with the 3.8L V-6 (capable of 200+ ft-lb. of torque) stops on a set of railroad tracks and realizes that a train is just seconds from broad siding his or her vehicle. The driver's response could be to demand wide-open throttle (WOT) to get out of the way of the train. When the driver inputs wide open throttle, the High Output (HO) V6 equipped Pontiac will possibly break the drive wheels loose due to poor traction because of loose gravel or wet roads thus giving a false vehicle speed reading. This occurs because the driver has input WOT and the drive wheels lost traction in response to the high torque request of the driver. The wheels and the drivetrain are spinning very fast (the VSS is generating a high number of pulses) and generating a speed signal higher than the vehicle's actual velocity. So when the pre-crash data is reviewed, the speed data could possibly jump from 0 MPH to 15 or 20 MPH in 1 second.

Another scenario to discuss is what types of deceleration will be noted in the pre-crash vehicle speed data for ABS equipped and non-ABS equipped vehicles. If a 4-wheel ABS equipped, front wheel drive vehicle is approaching an intersection on wet roads and the driver brakes heavily before some type of collision. An investigating engineer will download the EDR data and most likely find a steeply sloped plot of the vehicle's velocity as well as the brake switch changing from 0 (OFF) to 100 (ON). However, the recorded pre-crash speed will be underreported because of the difference in wheel speed (due to the ABS function) from the actual vehicle speed

If the same scenario is used as discussed above but with a non-ABS equipped front wheel drive vehicle the recorded pre-crash vehicle speed data can be entirely different. Therefore, if the driver attempts to make a panic stop, let's assume that the front wheels (drive wheels) lock up two to three seconds prior to a collision. When the investigating engineer downloads the EDR data and looks at the pre-crash vehicle speed what will be discovered? It will look like the vehicle came to a stop two to three seconds *prior* to the approximate time of the collision at the T=-1 seconds point on the pre crash data plot. This fact can be correlated to other reviewed evidence such as photographs of skid marks from the accident scene or documentation of evidence in the Traffic Collision Report.

Some other scenarios to consider and that will be considered in future papers is what kind of pre-crash speed data would be expected from a vehicle in a heavy yaw just prior to an impact? What if that same yawing vehicle was also under the influence of the driver having the throttle wide open?

If the data cannot be explained by the circumstances of the accident, then the Powertrain Control Module should be analyzed for any pre-existing problems. If no speed data is recorded it is most likely that because of damage from the accident, a power loss occurred and/or the data bus that passes the VSS data to the SDM could have been compromised in the accident. If pre-crash speed data is present but it does not make sense, the PCM can be checked to see if any Diagnostic Trouble Codes (DTCs) have been stored in the PCM indicating any possible problems with the VSS itself.

If a DTC is "set" in the memory of the PCM due to a problem with the VSS, the PCM will illuminate the Malfunction Indicator Lamp (MIL), the transmission will default to operation in 3rd and the PCM will inhibit the Torque Converter Clutch (TCC) from operating.

ENGINE SPEED

The engine speed (in RPM) is another channel of variable data that is recorded by the SDM as pre-crash data. The primary use of the engine speed data is for the Electronic Ignition Control (IC) and the fuel metering components of the PCM to accurately control engine timing and fuel injection. The RPM signal is generated from one of two Hall Effect switches that are mounted adjacent to the crankshaft balancer. The two components of the Hall Effect switch are the Crankshaft Position Sensor (CKP) mounted in an aluminum bracket generally on the front/left of the engine timing cover and partially behind the crankshaft balancer. The crankshaft balancer is basically integral to the serpentine belt pulley assembly that is attached to the engine crankshaft and powers all other engine accessories that are powered via the serpentine belt.

The second portions of the Hall Effect switch are the crankshaft balancer interrupter rings that spin with the crankshaft balancer. There are two different interrupter rings on the crankshaft balancer that mate with the two crankshaft position sensors. However both interrupter rings are quite different. One interrupter ring has 18 evenly spaced blades and windows and the other Interrupter ring has 3 unevenly spaced blades and windows.

The Hall Effect switch works with a magnet to generate a magnetic field near the switch and grounds the switch out placing the switch in an 'ON' position. When an object such as one of the blades on the crankshaft balancer interrupter ring comes between the magnet and the switch, the magnetic field is interrupted and the switch is in an OFF position. So as the crankshaft and the 18-window interrupter ring spins through one revolution, 18 equal length ON – OFF pulses are generated. The same goes for the 3-window interrupter ring, which generates 3 ON – OFF pulses of unequal length for every revolution of the crankshaft sensor.

The Crankshaft Position Sensor (CKP) is made up of both of these described Hall Effect switches with the switch and 18-window Interrupter Ring being called the CKP 18X portion of the CKP and it generates the 18X reference PCM input signal. The Hall Effect Switch and 3-window interrupter Ring is called the CKP 3X portion of the CKP and it generates the 3X reference PCM input signal.

The 3X reference PCM input is what is actually used by the PCM to calculate the engines rotational speed in revolutions per minute (RPM). This data is primarily used to control ignition and fuel delivery to the engine as well as supply data to the tachometer mounted in the instrument panel (if the vehicle is equipped with a tachometer).

The engine speed data included in the SDM pre-crash data might not seem to be very important however it is a rather interesting "double check" on two other sets of variable data that are recorded as pre-crash data. Those two sets of data would be the vehicle speed and the percent throttle applied by the driver. If in the pre-crash data the engine speed is climbing quickly, the vehicle speed *should* theoretically be climbing as well. The engine speed data should also correlate with the percent throttle data. If the driver has wide-open throttle (WOT) applied, the engine speed should be increasing significantly with the vehicle speed. However, it is important to recognize what systems the vehicle in question is equipped with. If the vehicle is equipped with traction control, and the downloaded EDR data shows WOT but the engine speed is not continuously increasing, this could be explained by the EBTCM's inhibiting the engine output to control wheel spin. If the vehicle is not equipped with traction control, the PCM should be analyzed with Scan Tool to see if any Diagnostics Trouble Codes relating to misfires or some other problem with the vehicle is present in the PCM's memory.

A vehicle system that is widely used by American drivers is the vehicle's cruise control system. The fact that the cruise system could be in use just prior to a collision could potentially have an interesting effect in the vehicle's pre-crash data. For example, a vehicle traveling on a slightly elevated roadway with the cruise on and the driver swerves and exits the roadway becoming airborne just prior to a collision what would you expect to see in the pre-crash data? Keep in mind what the function of the cruise control system is; to maintain a driver's defined set speed. If the vehicle becomes airborne, the load on the vehicle's wheels becomes essentially zero. The response that the cruise control will give is closed throttle or very little throttle to maintain the vehicle speed. When the investigating engineer downloads the EDR data, he or she will find that just prior to the collision, the vehicle's speed remained constant up to the point of impact however the throttle position drops to 0%.

The full scale of the engine speed that can be recorded by the SDM is from 0 RPM to 16383 RPM with a resolution of 0.25 RPM. The accuracy of the engine speed data is +/- 1 RPM and the engine speed data is updated if the engine speed changes by 32 or more RPM. [2]

PERCENT THROTTLE

Percent throttle is a set of variable data recorded by the SDM in the pre-crash data section. The Throttle Position (TP) sensor mounted in the engine compartment on the end of the throttle butterfly between the engine air cleaner and the intake manifold generates the "Percent Throttle Data". The TP Sensor is probably one of the most important sensors monitored by the PCM to run the engine's fuel delivery system. Based on throttle position and the throttle valve angle, the PCM calculates how much fuel to deliver to the engine.

The Throttle Position Sensor is simply a potentiometer mounted to the end of the throttle butterfly located at the inlet of the intake manifold. The accelerator pedal is controlling the butterfly angle via a cable. On some vehicles, the cable is being replaced by an electric motor that is controlled via "fly-by-wire" style controls from the accelerator pedal.

The TP Sensor provides a voltage signal to the PCM that increases as the throttle butterfly angle increases to WOT. At 0% throttle (idle) the TP sends a 1 volt or less signal to the PCM and approximately 4 volts signal at WOT. The PCM is constantly monitoring the TP and comparing the TP signal with what the PCM is calculating as a predicted throttle position based on engine speed.

The full scale of the throttle position data that can be recorded by the SDM is from 0% to 100% WOT with a resolution of 0.4%. The accuracy of the throttle position data is +/- 5% and the throttle position data is updated if the throttle position changes by 5.0 or more percent. [2]

The percent throttle data is very interesting to know during the 5 seconds prior to a near deployment or a deployment event. If a driver claims that he or she was attempting to brake hard to avoid a collision and the vehicle kept accelerating, that situation could be explained with this pre-crash data.

Another scenario that the percent throttle data could be useful in analyzing are unintended acceleration claims. What if the driver becomes disoriented in a panic situation and accidentally applies full throttle when he or she meant to apply the brakes causing the vehicle to lurch forward and collide with another object. Pedal application will be recorded in the EDR and the most likely sequence of driver inputs prior to the collision will be known.

If any pre-existing problems occurred with the TP sensor, there would most likely be complaints of poor idling and poor performance by the driver since the TP sensor is crucial to the operation of the engine.

BRAKE DATA

The brake switch variable data is probably one of the simplest data sources of the four pre-crash data sources. However, different control modules can monitor the brake switch data depending on the vehicle; the Powertrain Control Module (PCM), the Electronic Brake Traction Control Module (EBTCM) or even the Body Control Module (BCM) could supply the monitored brake switch data. The physical switch that is monitored for data is simply the vehicle's brake light switch. The switch is a simple plunger type switch that is either ON or OFF. For this reason, the brake switch pre-crash variable data will only record at 0 for OFF or 100 for ON.

Since the switch is not a potentiometer, percent-applied brake data is *not* available.

ON-BOARD DIAGNOSTICS SCANNERS

An OBD-II scanner is necessary in order to scan the vehicle systems supplying pre-crash data. These tools can be invaluable to see how key variables of the vehicle are configured such as what tire size is selected. This is important to check because this affects wheel speed data that is crucial to ABS operation and recorded pre-crash data.

Vehicle systems and their sensors can also be probed for proper operation and any pre-existing problems.

Several scanner tools are available on the market but only two are discussed. The first is the Vetronix Tech-2 General Motors OBD-II scanner, and is what all General Motors dealer service centers use to diagnose vehicles.

The second scanner is the Snap-On MT2500 OBD-II scanner. The MT2500 is a generic OBD-II scanner that can interface with many different makes of vehicles and can perform only a few of the functions that the Tech-2 can.

Both scanners are equipped with bi-directional controls. This allows for testing of most individual system components such as valves, relays and switches to be toggled on and off by the scanner for diagnostic purposes.

VETRONIX TECH-2

The Vetronix Tech-2 OBD-II vehicle scanner is the most in-depth scanner for General Motors vehicles. The Vetronix Tech-2 can be used with all General Motors vehicles including SAAB, Opel, Vauxhall and Holden.

Software is also available separately that will allow the Tech-2 to interface with other brands of cars.

When properly equipped, the Tech-2 can be used not only to diagnose a GM vehicle, but also to "reprogram" such vehicle systems as the heating and air conditioning system operational parameters or updating what tire sizes and axle ratios are used to keep vehicle speed accurate when tires or wheels are changed.

The Tech-2 is equipped with bi-directional controls and is able to download current DTC's as well as a DTC history for each vehicle system. Bi-directional controls are necessary for troubleshooting vehicle systems.

SNAP-ON MT2500 OBD-II SCANNER

The Snap-On MT2500 OBD-II Scanner is a generic vehicle scanner with bi-directional controls. The MT2500 scanner will not only interface with GM vehicles but also with Daimler-Chrysler, Ford, Asian and European vehicles. When interfacing with a GM vehicle, the MT2500 can perform most of the functions that the Tech-2 can. However, the MT2500 cannot display nor change any vehicle configuration parameters such as tire size or rear axle ratio.

Most vehicle systems can be checked for any current diagnostic trouble codes as well as the DTC history. Some systems can be activated and de-activated by the scanner to insure proper operation just as the Tech-2 can.

CONCLUSION

The GM Event Data Recorded has the potential to be an invaluable tool for research and analysis. However, this is true only if it is used as a tool in addition to the normal analytical and empirical tools available to the trained engineer.

It is also important to understand how the physical location and operation of the vehicle's sensors supplying pre-crash data can give, what seems to be, abnormal data. If the environmental conditions are known at the time of the accident a better understanding of how the vehicle will respond to driver inputs will be obtained.

ACKNOWLEDGMENTS

Many, many thanks to Roger E. Clark, P.E., Neil Clark, Stein Husher, Michael Varat, Chris Armstrong, Susan Lutzweit and Dr. Robert McElroy for all of their help and support in this endeavor.

Many thanks especially to Mr. Don Floyd of General Motors Corporation for his help and insight into this paper.

REFERENCES

[1] R. Haight, D. Floyd, 2001, "Vetronix CDR Tool Training Seminar"

[2] Augustus "Chip" Chidester, NHTSA; John Hinch, NHTSA; Thomas C. Mercer, GM Corp.; Keith S. Schultz, GM Corp, 1999, "Recording Automotive Crash Event Data," International Symposium on Transportation Recorders, Arlington, VA

[3] National Highway Traffic Safety Administration, 1974, "Automotive Recorder Research – A Summary of Accident Data and Results"

[4] Vehicle EE Systems Diagnostics Standards Committee, "OBD-II Scan Tool"

[5] "National Highway Traffic Safety Administration Website" (www.nhtsa.gov)

[6] "National Highway Traffic Safety Administration Research & Development Website" (www-nrd.nhtsa.dot.gov)

[7] "Event Data Recorder (EDR) Applications for Highway and Traffic Safety," National Highway Traffic Safety Administration (www-nrd.nhtsa.dot.gov/edr-site/)

[8] North American Operations, General Motors Corporation, © 1999, "General Motors W-Platform Service Manual"

[9] North American Operations, General Motors Corporation, © 1999, "General Motors E/Ksp – Platform Service Manual"

Proceedings of IMECE'03
2003 ASME International Mechanical Engineering Congress
Washington, D.C., November 15–21, 2003

IMECE2003-42227

Energy Absorption Properties of Various Square FRP Tubes

Masanori Okano[1,*], Hiroshi Saito[1], Asami Nakai[2], Hiroyuki Hamada[2]
[1]Graduate School of Kyoto Institute of Technology, Matsugasaki, Sakyo-ku, Kyoto 606-8585, JAPAN
[2]Kyoto Institute of Technology, Matsugasaki, Sakyo-ku, Kyoto 606-8585, JAPAN

Abstract

In this study, the static and impact energy absorption characteristics and crushing mechanism based on precise cross-sectional observation of crush zone in two kinds of square FRP tube were investigated. The specific energy absorption value of unidirectional tube showed higher value than that of knitted fabric tube in both static and impact test. Energy absorption mechanism changed from central crack progressive mode with fiber fracture to delamination mode without fiber fracture for unidirectional tube, and from central crack progressive mode to collapse mode with large fragments of inner fronds for knitted fabric tube with the change in testing speed from static to impact

Keywords; Energy absorption, Square tube, FRP, static, impact, unidirectional, knitted fabric

INTRODUCTION

Safety is one of the prime concerns among various requirements in automotives, rail vehicles and aircrafts. In order to protect passengers from accidents or collisions, vehicle bodies should possess high crashworthiness performance. Traditionally, metals can absorb impact energy by deforming plastically and therefore have been used for crashworthy structural applications in automotives. On the other hand, there is a considerable amount of published data on the energy absorption characteristics of polymer composite materials. It is well established that the polymer composite materials with proper design achieve higher specific energy absorption than the conventional metals [1]. The fracture mode is called progressive crushing, which proceeds under constant load. This unique potential of polymer composite materials has generated considerable interest in recent years for crashworthy structural applications in aircraft and automotive vehicles.

In previous study, the effects of both reinforcing fiber and matrix resin, types of reinforcements, testing speed and geometries on the energy absorption performance of polymer composite tubes with circular cross section were clarified by Hamada and Ramakrishna [2-3]. The crushing mechanism of polymer composite tubes was very complicated because it involved the combination of various fracture modes such as fiber fracture, delamination and cracking. Therefore the crushing mechanism has been clarified by precise cross-sectional observation of crush zone.

For vehicle application, square tubes in which flat plate face can be easily assembled with other component parts in the automotive construction, have geometrical advantage than circular tubes. Therefore polymer composite square tubes have been studied. However, in the square tubes, the crushing mechanism based on precise cross-sectional observation of crush zone is not clarified well.

* To whom correspondence should be addressed. E-mail: mokano02@ipc.kit.ac.jp
TEL: +81-75-724-7844, FAX: +81-75-724-7800

In this study, the static and impact energy absorption characteristics and crushing mechanism based on precise cross-sectional observation of crush zone in two kinds of square FRP tube; one was unidirectional tube, and the other was knitted fabric tube were investigated.

MATERIALS

Two kinds of square tube with each different reinforcing configuration were used. Both tubes were fabricated by the pultrusion process. The unidirectional tube consisted of five layers of unidirectional glass roving, glass mat and polyester non-woven mat. A construction of the tube is shown in Figure 1. Polyester non-woven mat was placed at the skin layer. The matrix was unsaturated polyester resin.

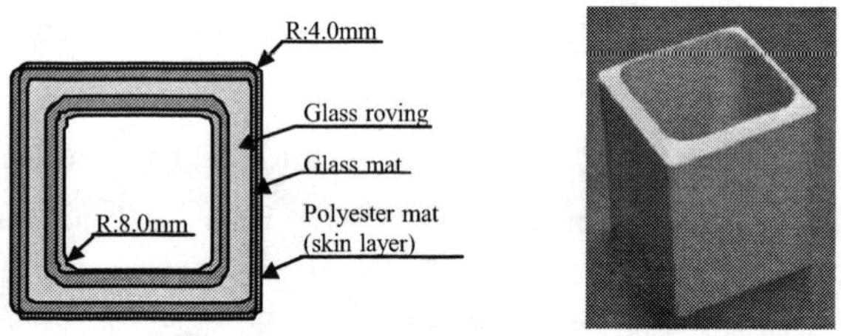

Figure 1. Construction and photograph of the unidirectional tube

The other tube was multi-axial warp knitted fabric tube. Multi-axial warp knitted fabric allows the placement of warp, weft, and off-axis materials directly into fabric structure. Moreover, it can be combined with not only the unidirectional fiber bundles, but also chopped strand mat. The reinforcement was glass fiber and the structure of laminate was [90 degree/0 degree]. A construction of the tube is shown in Figure 2. Also, glass mat was formed outside of the tube. The matrix was unsaturated polyester resin.

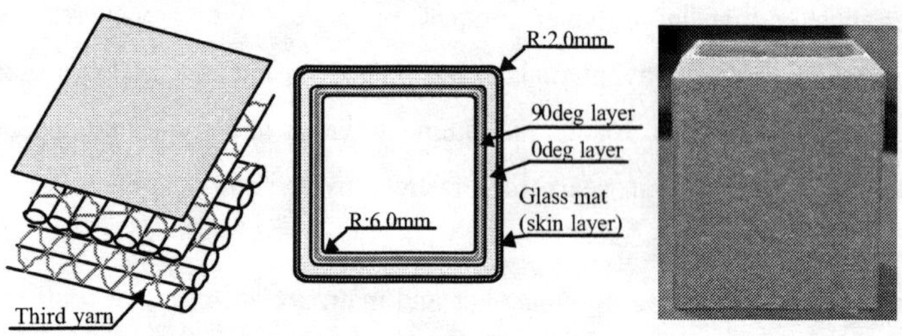

Figure 2. Construction and photograph of the knitted fabric tube

The fiber volume contents of both tubes were about 50%. The geometric information and density of specimen are shown in Table 1. The fabricated tubes were cut into specimens of 50mm height in static test and specimens of 200mm height in impact test. In order to initiate progressive crushing, 45 degree chamfer was made at one end of each tube specimen.

Table 1. Geometric information and density of each specimen

tube type	Side length [mm]	Thickness [mm]	Outer trimmed radius [mm]	Inner trimmed radius [mm]	Cross-sectional area [mm^2]	density [g/cm^3]
Unidirectional	60.0	4.2	4.0	8.0	873.0	1.9
Knitted fabric	50.0	4.0	2.0	6.0	762.6	1.9

EXPERIMENTAL METHODS

Static tests were performed by using INSTRON testing machine at a constant cross-head speed of $0.83*10^{-4}$ m/s (5.0 mm/min). Composite tubes were axially crushed between parallel steel flat platens. Impact tests were performed by using sled test facility of GP Daikyo Corporation., JAPAN. Weight of impactor was approximately 1.5 ton. The impact tests were performed at two impact velocities; the case of unidirectional tube specimen was 3.16 m/s and the case of knitted fabric tube specimen was 2.20 m/s.

RESULTS AND DISCUSSION

STATIC TEST RESULTS

Typical load-displacement curves obtained from static testing of unidirectional tube and knitted fabric tube are shown in Figure 3. Initially the load increased rapidly to a maximum value with respect to both tubes. After the initial rapid increase, the load decreased drastically. The peak load value of unidirectional tube was higher than that of knitted fabric tube. After peak load, the re-increase in load of unidirectional tube was earlier than that of knitted fabric tube. After all, in unidirectional tube, progressive crushing occurred at mean crush load of 61.0kN which was higher than the 40.1kN mean crush load in knitted fabric tube. The energy absorption performances of both tubes are given in Table 2 The unidirectional tube absorbed approximately 75% higher energy than knitted fabric tube.

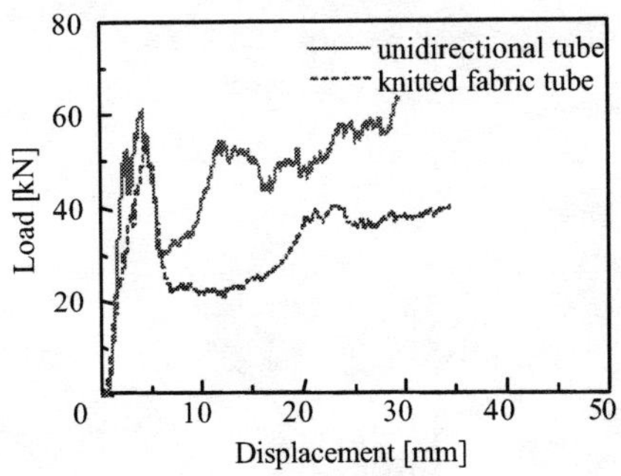

Figure 3. Load-displacement curve

Table 2. Results of mean crush load P and Specific energy absorption Es

Tube type	$\overline{P}$ [kN]	Es [kJ/kg]
Unidirectional	61.0	37.0
Knitted fabric	40.1	21.2

The photographs of static tested unidirectional tube and knitted fabric tube are shown in Figure 4 respectively. In both tubes, crush zone was split into four fronds along circumference of the tube. These fronds spread out into the inside and outside of the tube.

(a) Unidirectional tube (b) Knitted fabric tube

Figure 4. Photographs of static tested tube

Detailed photographs of cross section cut through the crush zones parallel to the axis of the each tube are shown in Figure 5 and Figure 6 respectively. In observation of the cross-sections, two locations were cut. One was flat plate face and the other was corner face. (a) is photograph of flat plate face, and (b) is photograph of corner face. The schematic illustrations of the fracture aspects are also shown.

For the unidirectional tube, in a flat plate face, the fronds spread out towards outer side and inner side of the tube. Especially the fronds at the inner side showed bending and buckling with central crack which progressed longitudinally. On the other hand, the fronds in a corner face were not spread out widely compared with those in flat plate face. A central crack was observed.

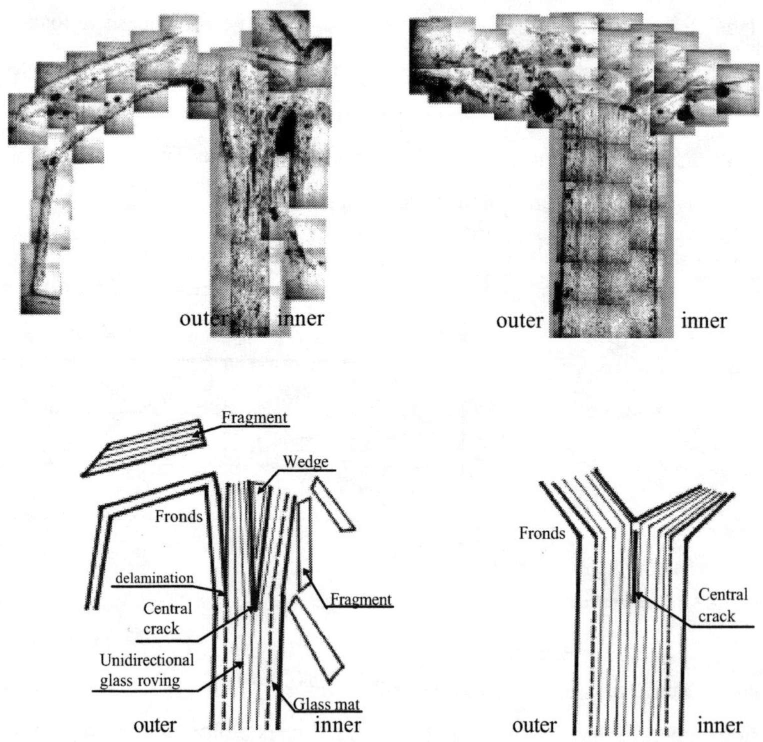

(a) In flat plate face (b) In corner face

Figure 5. Photographs and schematic illustrations of the cross section for unidirectional tube

For the knitted fabric tube, in a flat plate face, the fronds spread out towards outer side and inner side of the tube with central crack as same as unidirectional tube. 90 degree layer at innermost layer was almost exfoliated from a 0 degree layer and was pushed into inner side of the tube. On the other hand, the fronds in a corner face were not spread out widely compared with those in flat plate face as same as unidirectional tube. However, the central crack in corner face was longer than that in flat plate face. Many coarse fragments of a 90 degree layer were observed and pushed into inner side of the tube.

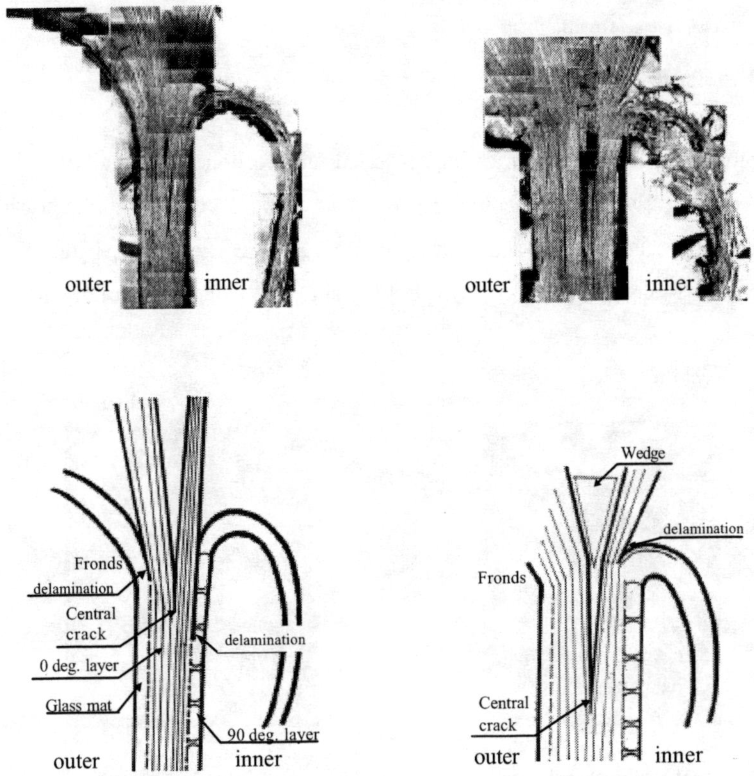

(a) In flat plate face (b) In corner face

Figure 6. Photographs and schematic illustrations of the cross section for knitted fabric tube

IMPACT TEST RESULTS

Typical load-displacement curves obtained form impact testing are shown in Figure 7. The mean crush loads obtained from the L-d curve and the specific energies of impact tested tubes are summarized in Table 3. The unidirectional tube exhibited about 60% higher Es than the knitted fabric tube. In both tubes, the Es values were decreased at 30% for the unidirectional tube and at 25% for knitted fabric tube with the change in testing speed from static to impact

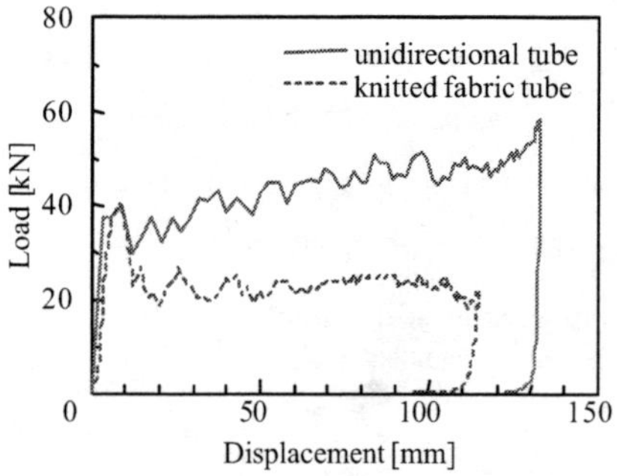

Figure 7. Load-displacement curve

Table 3. Results of mean crush load P and specific energy absorption Es

Tube type	$\overline{P}$ [kN]	Es [kJ/kg]
Unidirectional	47.5	26.1
Knitted fabric	22.5	16.1

The outward photographs of crush zones of impact tested unidirectional tube and knitted fabric tube are shown in Figure 8 respectively. Typical photographs of cross sections cut through the crush zones parallel to the axis of the unidirectional tube and knitted fabric tube are shown in Figure 9 and Figure 10 respectively. Below these photographs of crush zone, the schematic illustrations of the fracture aspects are also shown.

(a) Unidirectional tube

(b) Knitted fabric tube

Figure 8. Photographs of impact tested tube

For the unidirectional tube, in a flat plate face, a large delamination between unidirectional fiber and glass mat at both inner and outer side was observed. The fronds in both sides were not spread out widely compared with those in static. The central crack which was found in static was not observed. On the other hand, in corner face, it was observed that the delamination between unidirectional fiber and glass mat at inner side was smaller and that at outer side was longer than those in flat plate face. A central crack was not observed as same as flat plate face.

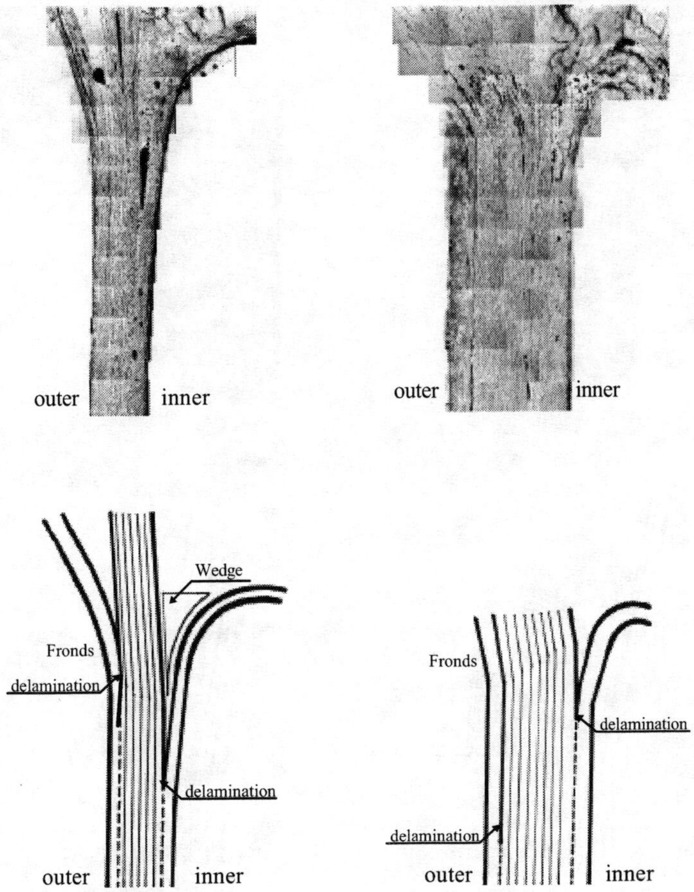

(a) In flat plate face (b) In corner face

Figure 9. Photographs and schematic illustrations of the cross section for knitted fabric tube

For the knitted fabric tube, in a flat plate face, a central crack was observed and the fronds were spread out widely. It was observed that the fronds at inner side failed in shear. On the other hand, in corner face, a central crack as same as flat plate face was observed and many large fragments in inner side were found.

DISCUSSION

A total energy (U_T) absorbed during crushing is given by (1)

$$U_T = U_{lc} + U_{sf} + U_{ff} + U_{fr} + \text{others}, \tag{1}$$

here U_c is the energy required for longitudinal cracking of the tube, U_{sf} is the energy absorbed by the splitting of fronds into beams, U_{ff} is the energy required for fiber fracture, and U_{fr} is the energy associated with frictional heating. In addition to these four main mechanisms, there are other fracture processes, such as splitting of the tube into fronds along the circumference of the tube, which contribute to the overall energy absorption.

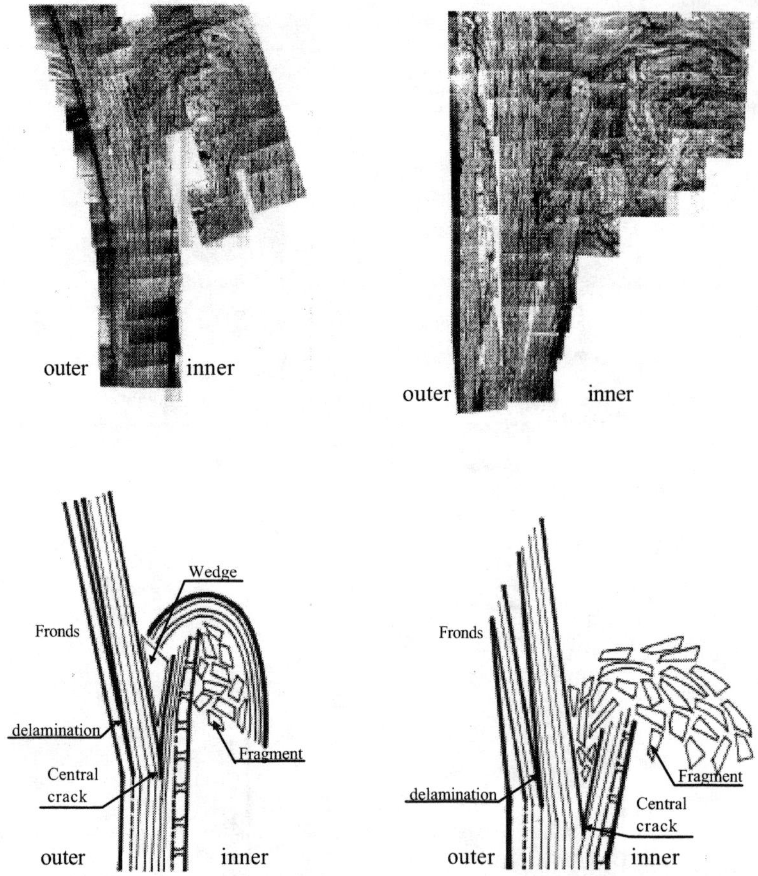

(a) In flat plate face (b) In corner face

Figure 10. Photographs and schematic illustrations of the cross section for knitted fabric tube

The specific energy absorption value, Es, of unidirectional tube was higher than that of knitted fabric tube in both static test and impact test. Es is associated with fiber orientation, the structure of laminate and interlaminar fracture toughness. For knitted fabric tube, inner side of tube is consisted of 90 degree layer, which possesses the lowest strength in the off-axis of unidirectional composites. Therefore, it was expected that knitted fabric tube absorbed lower energy than unidirectional tube. In static test, the re-increase in load of unidirectional tube was earlier than that of knitted fabric tube. This might be related to the length of central crack. For unidirectional tube, a central crack which was generated in initial stage might be smaller and shorter than that of knitted fabric tube. And in impact tests, the inner fronds of knitted fabric tube failed in collapse aspects with large fragments, so that it was possible that the energy U_{fr} required for friction between impactor and fronds was decreased considerably. Consequently, unidirectional tube showed higher energy absorption characteristic than knitted fabric tube.

The Es values were decreased at 30% for the unidirectional tube and at 25% for knitted fabric tube with the change in testing speed from static to impact. For unidirectional tube, the fronds were spread out widely by fiber fracture in case of static. On the other hand, in case of impact, no fiber fracture but a large delamination between unidirectional fiber and glass mat were found. As a result, it was considered that energy U_{ff} required for fiber fracture at impact could be lower than that of static.

For knitted fabric tube, the fronds were spread out widely by central crack and delamination in case of static. However, in case of impact, large fragments of collapse aspects were caused by fracture of inner fronds which was not continuity. After all, as mentioned above, energy U_{fr} required for friction between impactor and fronds was decreased considerably.

Hull [4] reported experimentally that if a fiber rate for axial-to-hoop direction in case of circular woven glass cloth-epoxy tube was changed, it showed various energy absorbing capabilities. This can be one important explanation of obtained results. For knitted fabric tube, inner side of tube is consisted of 90 degree layer. If fiber orientation angle of outer side layer is 90 degree, the fronds spreading out towards outside can be restrained and fine fragments may be generated and energy absorption properties can be higher. And if knitting yarn with higher strength is used, a large delamination between 0 degree layer and 90 degree layer is possible to be restrained.

CONCLUSIONS

In this study, the static and impact energy absorption characteristics and crushing mechanism based on precise cross-sectional observation of crush zone using various square FRP tubes were investigated. The specific energy absorption value of unidirectional tube showed higher value than that of knitted fabric tube in both static test and impact test. For unidirectional tube, energy absorption mechanism changed from central crack progressive mode with fiber fracture to delamination mode without fiber fracture with the change in testing speed from static to impact. For knitted fabric tube, energy absorption mechanism changed from central crack progressive mode to collapse mode with large fragments of inner fronds. Energy U_{ff} required for fiber fracture and energy U_{fr} required for friction were dominant for energy absorption capabilities.

ACKNOWLEDGEMENTS

We are grateful to E. Fukui (FUKUI GYOMO CO.,LTD) and S. Sugihara (NIHHON POLYESTER CO.,LTD) for supply of unidirectional tube and knitted fabric tube.

REFERENCES

[1] H. Hamada, S. Ramakrishna and H. Satoh, 1995, "Crushing mechanism of carbon fibre/PEEK composite tubes" COMPOSITES, Vol.26, Number11, pp.749-755

[2] H. HAMADA, J. C COPPOLA, D. HULL, Z. MAEKAWA and H. SATO, 1992, "Comparison of energy absorption of carbon/epoxy and carbon/PEEK composite tubes" COMPOSITES, Vol.23, Number4, pp.245-252

[3] H. HAMADA, S. RAMAKRISHNA and Z. MAEKAWA, 1995, "Environmental effects on the progressive crushing behavior of glass cloth/epoxy composite tubes" Composite Interface, Vol.3, No.1, pp.23-39

[4] D. Hull, 1991, "A Unified Approach to Progressive Crushing of Fiber-Reinforced Composites Tubes" Composite Science and Technology 40, pp.377-422

AUTHOR INDEX

TRANSPORTATION 2003:
Making Tracks for Tomorrow's ransportation

Book Number: I00699